T0134460

Modelling Human Motion

Nicoletta Noceti · Alessandra Sciutti ·
Francesco Rea
Editors

Modelling Human Motion

From Human Perception to Robot Design

Springer

Editors
Nicoletta Noceti
MaLGa Center - DIBRIS
Università di Genova
Genoa, Italy

Alessandra Sciutti
Contact Unit
Istituto Italiano di Tecnologia
Genoa, Italy

Francesco Rea
RBCS Unit
Istituto Italiano di Tecnologia
Genoa, Italy

ISBN 978-3-030-46734-0 ISBN 978-3-030-46732-6 (eBook)
https://doi.org/10.1007/978-3-030-46732-6

This Springer imprint is published by the registered company Springer Nature Switzerland AG
The registered company address is: Gewerbestrasse 11, 6330 Cham, Switzerland

Contents

Contributors

Muneeb Imtiaz Ahmad Social Robotics Group, Computer Science Department, Heriot-Watt University, Edinburgh Center for Robotics, Edinburgh, Scotland

Yiannis Aloimonos University of Maryland, College Park, MD, USA

Lucia Amoruso Basque Center on Cognition, Brain and Language, San Sebastian, Spain;
IKERBASQUE, Basque Foundation for Science, Bilbao, Spain

Paola Ardón Social Robotics Group, Computer Science Department, Heriot-Watt University, Edinburgh Center for Robotics, Edinburgh, Scotland

Emilia I. Barakova Eindhoven University of Technology, Eindhoven, The Netherlands

Pasquale Cardellicchio IIT@UniFe Center for Translational Neurophysiology, Istituto Italiano di Tecnologia, Ferrara, Italy

Fabio Cuzzolin Oxford Brookes University, Oxford, United Kingdom

Alessandro D'Ausilio Section of Human Physiology, Università di Ferrara, Ferrara, Italy

Giuseppe Di Cesare Istituto Italiano di Tecnologia, Genoa, Italy

Christian Dondrup Social Robotics Group, Computer Science Department, Heriot-Watt University, Edinburgh Center for Robotics, Edinburgh, Scotland

Cornelia Fermüller University of Maryland, College Park, MD, USA

Alessandra Finisguerra Scientific Institute, IRCCS E. Medea, Pasian di Prato, Udine, Italy

Kanishka Ganguly University of Maryland, College Park, MD, USA

Paul Hemeren University of Skövde, Skövde, Sweden

Carlos Herrera Perez Eindhoven University of Technology, Eindhoven, The Netherlands

Pauline M. Hilt IIT@UniFe Center for Translational Neurophysiology, Istituto Italiano di Tecnologia, Ferrara, Italy

Junhwa Hur Department of Computer Science, Technische Universität Darmstadt, Darmstadt, Germany

Allison Langer Recanati School for Community Health Professions, Department of Physical Therapy, Ben-Gurion University of the Negev, Beer-Sheva, Israel

Hagen Lehmann Università degli Studi di Macerata, Macerata, Italy

Shelly Levy-Tzedek Recanati School for Community Health Professions, Department of Physical Therapy, Ben-Gurion University of the Negev, Beer-Sheva, Israel;
Zlotowski Center for Neuroscience, Ben-Gurion University of the Negev, Beer-Sheva, Israel;
Freiburg Institute for Advanced Studies (FRIAS), University of Freiburg, Freiburg, Germany

Katrin Lohan Social Robotics Group, Computer Science Department, Heriot-Watt University, Edinburgh Center for Robotics, Edinburgh, Scotland;
EMS Institute for Development of Mechatronic Systems, NTB University of Applied Sciences in Technology, Buchs, Switzerland

Maurizio Mancini School of Computer Science and Information Technology, University College Cork, Cork, Ireland

Nicoletta Noceti Università di Genova, Genoa, Italy

Èric Pairet Social Robotics Group, Computer Science Department, Heriot-Watt University, Edinburgh Center for Robotics, Edinburgh, Scotland

Ana Paiva INESC-ID & Instituto Superior Técnico, University of Lisbon, Lisbon, Portugal

German I. Parisi University of Hamburg, Hamburg, Germany

Francesco Rea Istituto Italiano di Tecnologia, Genoa, Italy

Tiago Ribeiro INESC-ID & Instituto Superior Técnico, University of Lisbon, Lisbon, Portugal

Pier Giuseppe Rossi Università degli Studi di Macerata, Macerata, Italy

Stefan Roth Department of Computer Science, Technische Universität Darmstadt, Darmstadt, Germany

Yves Rybarczyk Dalarna University, Falun, Sweden

Suman Saha Computer Vision Lab (CVL), ETH Zurich, Zurich, Switzerland

Michael Sapienza Think Tank Team, Samsung Research America, Mountain View, CA, United States

Alessandra Sciutti Istituto Italiano di Tecnologia, Genoa, Italy

Gurkirt Singh Computer Vision Lab (CVL), ETH Zurich, Zurich, Switzerland

Janny Christina Stapel Donders Institute for Brain, Cognition, and Behaviour, Radboud University, Nijmegen, The Netherlands

Philip H. S. Torr Department of Engineering Science, University of Oxford, Oxford, United Kingdom

Cosimo Urgesi Scientific Institute, IRCCS E. Medea, Pasian di Prato, Udine, Italy; Laboratory of Cognitive Neuroscience, Department of Languages, Literatures, Communication, Education and Society, University of Udine, Udine, Italy

Giovanna Varni LTCI, Télécom Paris, Institut polytechnique de Paris, Paris, France

Alessandro Vinciarelli School of Computing Science, University of Glasgow, Glasgow, Scotland

Konstantinos Zampogiannis University of Maryland, College Park, MD, USA

Chapter 1
Modeling Human Motion: A Task at the Crossroads of Neuroscience, Computer Vision and Robotics

Nicoletta Noceti, Alessandra Sciutti, and Francesco Rea

Abstract Human motion understanding has been studied for decades but yet it remains a challenging research field which attracts the interest from different disciplines. This book wants to provide a comprehensive view on this topic, closing the loop between perception and action, starting from humans' action perception skills and then moving to computational models of motion perception and control adopted in robotics. To achieve this aim, the book collects contributions from experts in different fields, spanning neuroscience, computer vision and robotics. The first part focuses on the features of human motion perception and its neural underpinnings. The second part considers motion perception from the computational perspective, providing a view on cutting-edge machine learning solutions. Finally, the third part takes into account the implications for robotics, exploring how motion and gestures should be generated by communicative artificial agents to establish intuitive and effective human–robot interaction.

1.1 Introduction

The new frontiers of robotics research foresee future scenarios where artificial agents will be leaving the laboratory to progressively take part in our daily life activities. This will require robots to have very sophisticated perceptual and action skills in many intelligence-demanding applications, with particular reference to the ability to seamlessly interact with humans. For the next generation robotics, it will be crucial to understand the human partners and at the same time be intuitively understood

N. Noceti (✉)
Università di Genova, Genoa, Italy
e-mail: nicoletta.noceti@unige.it

A. Sciutti · F. Rea
Istituto Italiano di Tecnologia, Genoa, Italy
e-mail: alessandra.sciutti@iit.it

F. Rea
e-mail: francesco.rea@iit.it

N. Noceti et al. (eds.), *Modelling Human Motion*,
https://doi.org/10.1007/978-3-030-46732-6_1

by them [1]. In this context, a deep understanding of human motion is essential for robotics applications, where the ability to detect, represent and recognize human dynamics and the capability of generating appropriate movements in response sets the scene for higher-level tasks. **This book wants to provide a comprehensive view on this historical yet challenging research field, closing the loop between perception and action and between human studies and robotics.**

The book is organized in three main parts. The first part focuses on human motion perception, with contributions analyzing the neural substrates of human action understanding, how perception is influenced by motor control, and how it develops over time and is exploited in social contexts.

The second part considers motion perception from the computational perspective, providing a view on cutting-edge solutions available from the computer vision and machine learning research fields, addressing low to higher-level perceptual tasks.

Finally, the third part takes into account the implications for robotics, with chapters on how motor control is achieved in the last generation artificial agents and how such technologies have been exploited to favor human–robot interaction.

Overall, the book considers the entire ideal human–robot cycle: from humans and their strategies to perceive motion and act in the world, to artificial agents and their models for motion perception and control. In this respect, **the purpose of the book is to offer insights into the perception and action loop in humans and machines, joining together aspects that are often addressed in independent investigations**. As a consequence, this book positions itself in a field at the intersection of different disciplines, as robotics, neuroscience, cognitive science, psychology, computer vision and machine learning. The aim is to provide a reference for different research domains, in the attempt of bridging them, by establishing common reference points among researchers interested in human motion for different applications and from different standpoints.

To achieve this ambitious objective, the book collects contributions from experts in different fields, with a shared interest in understanding human motion for various reasons: to design computational models of perception, from the perspective of traditional human motor control or to improve robots and their ability to interact naturally. Additionally, the book is designed to discuss principles, methods and analytical procedures, which could apply to the whole variety of human (or robot) movements. More precisely, the book complements the models derived from human motor control and perception with parallels in the field of robotics and computational vision, providing a comprehensive view of similarities and dissimilarities in action-perception skills between human and robots.

1.2 Book Structure

A more detailed look at the three parts of the book clearly highlights the three different perspectives of motion modeling (human-centered, computational vision-centered and robot-centered).

1.2.1 Part I: Motion Perception in Humans

Part I explores the neurophysiological bases of movement perception and their implications for the understanding of the actions and the attitudes of other agents, from childhood to adult life.

A fundamental role in supporting action perception is played by parieto-premotor brain network, which is engaged both during action execution and during the perception of the same actions performed by other agents. Such overlap between neural activations in the mirror neuron system has been associated with action anticipation and comprehension and hence to social cognition [2]. Hilt et al. in their chapter introduce the history of the discovery of these neural mechanisms, first in monkeys and then in humans, and then propose a novel route to expand further the current understanding. In particular, the authors suggest to translate the methodological and theoretical principles of motor control literature to the study of action perception. The authors exemplify the validity of this approach by demonstrating the relevance of two core motor control principles also in action observation: the presence of a modular control strategy [3] and a role of motor inhibition in case of unpredictable action outcomes [4]. The main claim of the chapter is that the difficulties still present in understanding the coding of the mirror system could be overcome by a better comprehension of the activity of the motor cortex during action execution.

The second chapter further expands the analysis of the neurophysiological underpinnings of action understanding, but with a specific focus on the role that different action features might play. The authors provide a description of the studies on mirror responses during action observation, considering the different levels at which an action can be represented. In particular, following the hierarchical model of action representation [5], actions can be described at the muscular level—encoding the pattern of muscular activity; at the kinematic level—encoding the spatio-temporal properties of the motion of the effector; at the level of the goal—encoding the short-term aim of the action; or at the level of the intention—encoding the long-term purpose of the action. Although traditionally mirror activation has been conceived as an inner replica of *low-level* motor aspects [6], more recent studies verify that this low-level mirroring could be affected by other *high-level* factors, such as expectations or even top-down contextual effects [7, 8]. In this perspective, context guides action comprehension under situations of perceptual ambiguity [9] and aids action recognition by signaling which intentions are more likely to drive upcoming actions. The take home message is that human actions are embedded in internal contexts (such as personality traits or previous experience) and external contexts (e.g., objects, affordances, environmental constraints and opportunities). As a consequence, also action understanding and its neural correlates can be modulated by a wide range of high-level contextual factors, as suggested by the mounting neurophysiological evidence reviewed in the chapter.

Beyond the neural activity supporting action understanding, it is important to investigate the perceptual component of the process, i.e., how different levels of visual processing are involved when we are faced with biological motion. This point is addressed in the third chapter. Hemeren and Rybarczyk review a series of studies showing the strong human sensitivity to motion patterns created by other individuals, even when they are presented only in the forms of groups of dots moving coherently, the so-called point light displays [10]. They then present two categories of critical visual features for biological motion perception: the *global* features, which refer to action concepts and categories, and the *local* features, referring to specific movement kinematics parameters. These different features contribute to visual and cognitive processing in relation to action category prototypes and action segmentation [11]. The authors then focus on the interaction between our own movement patterns and their impact on our visual perception of biological motion, presenting evidence of the fact that features proper of human motor control, when present in an observed motion, may facilitate its discrimination by a human observer. Among them, the authors cite the Fitts' law [12], the minimum jerk [13] and the power law [14] and present evidence that the (lack of) their implementation on robots can have detrimental effect on robot action perception from a human partner [15, 16].

The ability to understand others' actions is acquired swiftly by humans during the first years of life. Stapel guides us through this astonishing process in the third chapter of this part. She describes how initially action perception "gets a head start" through early preferences for faces [17], manipulable objects, visible mouth movements [18] and biological movement in general [19]. Such initial tendencies to look at relevant visual aspects provide infants with important input, which is then transformed into useful building blocks through learning and interaction with the environment. The development of action perception is then allowed by a number of capacities, among which categorization [20], (potentially) mental rotation [21], statistical learning [22], sensorimotor development [23] and imitation [24]. Stapel then presents how both sensorimotor experience [25] and purely perceptual, observational experience [26] are utilized as building blocks for action perception development, clarifying the advantages and disadvantages of these two different types of learning. As a result of this process, infants acquire the ability to dissociate observed actions from each other, segment action streams into smaller parts, form expectations and predict others' actions, and ultimately form an understanding of others' actions and become a proficient social partner.

The last chapter of this part delves into another important aspect of action understanding. In fact, *how* an action is performed, beyond its goal (the *what*) and the intention driving it (the *why*), plays a crucial role in interaction. The very same gesture, as a hand shake, acquires a very different meaning if it is performed gently or vigorously, as it communicates the positive or negative attitude of the agent. These different action forms have been named "vitality forms" by Stern [27]. Di Cesare in this chapter provides an overview of action and speech vitality forms highlighting the neural substrates underpinning their encoding. He presents a series of studies conducted with functional magnetic resonance imaging (fMRI) demonstrating the role of the dorso-central insula in the processing of vitality forms. In particular, he provides evidence that this area is active both during the observation and the execution of action vitality forms [28] and when listening to action verbs pronounced with gentle and rude vitality forms [29]. Moreover, he shows with a behavioral study that during social interactions, the vitality form expressed by one agent influences the motor response of the partner [30]. In summary, *how* an action is performed allows people to understand and communicate their attitudes, enabling mutual adaptation. The author concludes suggesting that vitality forms could become a future fundamental source of social communication to promote also human–robot interaction.

All chapters, beyond providing a summary of the most important findings on human action understanding, introduce the reader to the vast range of different methodologies adopted in the investigation of this topic, ranging from fMRI to Transcranial Magnetic Stimulation, from psychophysical studies with point light displays to methodologies specific to investigate infant action understanding. This helps to grasp the complexity of the question addressed and the limits that these methodologies of investigation might have. In this respect, the contribution of other fields to this topic—as detailed in the two next parts—can prove to be fundamental in addressing the questions still open in the field.

1.2.2 Part II: Computational Models of Motion Understanding

In the last decades, modeling and understanding human motion from visual data has gained an increasing importance in several applications, including human–machine interaction, gaming, assisted living and robotics. Although the significant advances of the last years, where as in other domains deep learning techniques has gained momentum [31], the tasks remain among the most challenging, for an intrinsic complexity due to the extreme variability of the dynamic information and its appearance [32], and still a lot of work needs to be done to approach human performance.

The biological perceptual systems remain the gold standard for efficient, flexible and accurate performance across a wide range of complex real-world tasks, relying on an amazing capability of saving and organizing the appropriate amount of information, giving the room to new concepts when needed and exploiting an efficient sharing of information. A natural solution for devising computational perception models is thus to elect as a reference and inspiration the mechanisms underlying human motion perception, and the knowledge derived from the cognitive and neuroscience fields. In robotics, where the ambition is to replicate the overall human model, this inspiration is especially relevant and powerful.

At the earliest processing stage, motion analysis can be cast into a detection problem, where the aim is to identify spatio-temporal image regions where the motion is occurring. A well-established method to obtain salient low-level motion information relies on the computation of the optical flow [33], an estimation of the apparent motion vectors in the image plane. The estimated fields show strong connections with the behavior of human brain areas involved in the perception of motion, as the V1 and the MST, where neurons were found to be directionally selective with dedicated receptive fields [34]. The second part of the book, thus, starts from this task: with their chapter, Junhwa Hur and Stefan Roth address the problem of estimating optical flow. In the contribution, it is discussed how recent advances in deep learning have significantly influenced the literature on the topic. After a historical perspective discussing the transition between the use of classical energy-based models and modern convolutional neural networks (CNNs) (see e.g., [35, 36]), the authors focus on two main families of CNNs approaches for optical flow estimation. The first category uses CNNs as a feature extractor followed by a classical energy minimization problem; the latter adopts CNNs as end-to-end architectures for regression tasks. The chapter reports a comparison between different methods based on deep architectures that have been proposed in the last years—as [37–39] just to name some—with a technical and experimental analysis, that extends also to less conventional approaches, shaped as semi-supervised or unsupervised problems, to cure overfitting issues of a fully supervised framework. The discussion highlights how, although CNNs-based methods provide nowadays state-of-the-art results, they also have a number of limitations, as the poor ability of generalizing to unseen domains, leaving the room for future developments.

A building block of biological motion understanding is the ability to precisely identify spatio-temporal start and end points of actions, i.e., *when* and *where* the motion is occurring. In the domain of motor planning, these concepts are related to the notion of motion primitive [40, 41], a main structural element of actions. In the computational domain, instead, these tasks are often referred to as the problems of detecting and segmenting actions [42, 43]. While most state-of-the-art human action recognition methods work on already trimmed video sequences in which only one action occurs per frame [32], the tasks mentioned above remain a challenge and at the same time an essential element for fully automatic motion recognition systems. The problem is tackled in the chapter by Saha et al., where the authors refer to real-world scenarios in which often times videos contain multiple action instances concurrently, that may belong to different action classes. The contribution introduces a deep learning method in which action detection is casted as an energy maximization problem using RGB images and optical flow maps in input. As other similar approaches [43, 44], the method includes two stages: in the first one, region proposals are spatio-temporally associated with action classes; in the second stage, the detections at frame-level are temporally linked according to their class-specific scores and spatial overlaps, to finally obtain class-specific action tubes. Lastly, each pixel within the detection windows is assigned to a class and instance-aware label to obtain instance segmentation. The experimental analysis on a benchmark dataset shows state-of-the-art results and speaks in favor of the generalization capability of the proposed solution.

The task of motion segmentation is further explored in the chapter by Zampogiannis et al., where specific attention is posed on manipulation actions (meaning object manipulation, also extended to a generic interaction with the environment) in human–robot interaction (HRI) scenarios. Contact and motion are two fundamental aspects of manipulation, and they naturally encode crucial information about the action [45]. Indeed, the contact encodes *where* the object is touched or grasped, and *when* and for how long the interaction takes place; on the other hand, motion conveys information on *what* part of the environment is involved in the interaction and how it moves [46].

The authors refer in particular to the framework of imitation learning [47], in which robots act in less controlled workspaces and thus require an effective and efficient mechanism for learning how to perform manipulation tasks. The solution proposed in the chapter refers to an unsupervised, bottom-up scenario, in which the actor-object contact is used as an attention mechanism for the detection of moved objects and the understanding of their motion in RGBD sequences. The method allows not only to detect actor-environment contact locations and time intervals, but also to segment motion of the manipulated objects and finally obtain an estimate of its 6D pose.

In the experimental analysis, the authors qualitatively demonstrate that the approach performs successfully on a variety of videos, and describe in detail the relevant role this method has for imitation tasks in the robotics domain, where it can bridge the gap between action observation and planning.

Considering again HRI scenarios, the chapter by Parisi discusses efficient human action understanding with neural network models based on depth information and RGB images. Particular emphasis is put on *flexibility*, *robustness* and *scalability*, to act in a scenario where, as in human experience, data are continuously acquired, and real-time capabilities are required for the perceptual tasks.

Continual learning, also referred to as lifelong learning, refers to the ability of a system to continuously acquire and fine-tune knowledge and skills over time while preventing catastrophic forgetting, i.e., the fact the abilities already acquired on previously learned tasks abruptly decrease when novel tasks are learned [48].

Considering the issues above and inspired by biological motivations [49, 50], the models at the basis of the strategies presented in the chapter use hierarchical arrangements of different variants of growing self-organizing networks for learning action sequence [51]. More specifically, two classes of growing network are discussed in the contribution: the first class refers to the Grow When Required (GWR) model that can be updated (i.e., neurons are grown and removed) in response to a time-varying input distribution; the second class concerns the Gamma-GWR, which extends the previous one with temporal context for the efficient learning of visual representations from temporally correlated input. The type of dynamic events considered in the chapter includes short-term behaviors (as daily life activities) but also longer-term activities. The latter are in particular relevant for motion assessment in healthcare, where a more prolonged analysis in time may help identifying cues about the overall wellbeing of individuals.

Going beyond classical action recognition, the Part II closing chapter by Varni and Mancini considers the analysis of movement expressivity as a fundamental element toward machines with human-like social and emotional intelligence [52]. Movement Expressivity is the whole-body motor component of emotional episodes, sometimes described as an unintentional action component of emotion expression [53, 54]. It can also be referred to as the dynamic movement component in affect perception, in contrast to the static form component [55]. Widely studied in the domains of psychology, sociology and neuroscience, in the last decades, it started receiving increasing attention also in the computer science field [56, 57]. In this sense, a main goal is to devise computational solution to replicate on a machine the humans' capacity of being aware during interaction of one's own and others' feelings and integrate that information in communicating with others. After providing their definition from a psychological and sociological point of view, the chapter presents a survey on computational approaches to address movement expressivity analysis, touching different aspects ranging from the ability to "sense" human motion, to an overview on datasets of human expressive movements available to the community, to continue with an approach to address the analysis proposed by the authors. The chapter is closed with a discussion on how such models can be employed in robotics scenarios, providing an ideal opening for the last part of the book.

1.2.3 Part III: The Robotic Point of View

The last part of the book explores how robots can use their movements to interact with their human partners. The interaction is designed in relation to two different aspects: movement as form of communication, and movement as form of synchronization. In the first chapter, by Ribeiro and Paiva, the authors show how to bridge the world of animation artists and the world of robotics. Animators can be part of the process of developing social robots, providing expressive and emotional traits for the next generation of interactive robots. Social robots might indeed need to reuse a collection of principles defined in animation to improve their ability to be recognized as animated communicative agents with character and personality. In this context, it is important to know which tools enable such alternative design. The authors of the first chapter give a detailed description of the most popular tools that promote animacy in robots [58–60].

The character of the social robot can promote its acceptability, but it is also important to establish effective and intentional coordination between the parts. In this context, the movement plays the important role of mean of communication, which helps the human and the robot in the preparation of a coordinated act [61]. In the chapter by Lohan et al., communication is studied in its spatial domain and plays the role of director in the action–reaction paradigm. For this application, the communication strategies depend on action and reaction expressed in the form of movements (gestures, facial expressions, eyes movement and navigation) and in the form of behaviors (conversation and dialog). In particular, the authors express how communication can build on movements of mobile robots [62, 63]. Such robots can communicate through their proximity to the partner and through path planning. The generated complex behavior exhibited in these situations requires equally complex perception and understanding of the human counterpart. In order to explain this, the authors propose strategies that measure the cognitive load on the partner. Perceiving the level of cognitive load allows the robot to plan the opportune rhythm. The robot can use the same perception to improve its own skills, especially in the production of implicit learning behaviors. The authors indicate how movements enable communication that promotes robot learning of manipulation tasks [64] and object grasp affordances [65].

In robotics, movement assumes a special meaning because the robot has a body. In the third chapter of the Part III, Barakova et al. propose an embodied dynamic interaction. The authors explain the important consideration that if the robot has a body, the movement of the entire body has a social meaning for a human observer. In their view, the specialized behaviors involving the entire body not only are expressive but they constitute a fundamental mean of interaction [66, 67]. This embodied dynamic interaction creates a relationship with the other social agents in the world, is intrinsically expressive and more importantly provides contextual cues. Such a claim is further demonstrated in the chapter with the discussion of the communicative role of robot dance. Dance is considered as playground for the deep understanding of expressivity in the movement [68]. The reason is that dance unveils the unconscious structure of movement understanding and of the attribution of meaning even if the performer is a robotic agent. In this context, systems adopted for choreographic design can be reused for robotic communicative behavior design. The Laban system introduced in the chapter is an example on how dance researchers can produce systematic hypotheses that can be exploited by social robotics.

Having discussed about the role of robotic movement in the process of communication and synchronization with the human partner, the fourth chapter exploits the reuse of such insights in a specific application. Lehmann and Rossi explore how to transfer an enactive robot assisted approach to the field of didactics. The authors project the enactivism concept [69], the assumption that cognition arises through a dynamic interaction between an agent and the environment, to the robot as interactive agent in the environment. The natural consequence of such argument is that structural behavioral coupling in interaction is the result of enabling complex self-organizing systems which are dynamically embedded in a complex self-organizing environment. In other words, changes in the dynamics of the environment generate perturbations in the dynamic of the robotic agent. Self-regulative behaviors of the agent aim to compensate the perturbations and generate actions that react to these changes in order to reach a system equilibrium. This is an evidence of how robots should be designed to be accepted into mixed human–robot ecologies. In the chapter, the enactive theory for robots is presented also through one form of its applications: enactive robot assisted didactics [70, 71]. Enactive didactics assumes close interaction between the teacher and the student in the knowledge creation process. In this process, the robot can assume the role of tutor that mediates in the enactive didactics process between the student and the teacher. The application shows the possibilities of social robotic tutors that facilitate the didactic approach thanks to its nonverbal communication competencies mainly based on movement.

In the last chapter, Langer and Levy-Tzedek explore one important aspect of movement: the timing. In general, movement is characterized by different parameters such as the speed, the level of biological plausibility or the vitality. Timing is another key aspect that has to be encoded in the robot behavior. In the last chapter of the Part III, Langer and Levy-Tzedek give an interpretation on why timing plays such important role in human–robot interaction. The timing of the robot movement widely impact on the users' satisfaction. In general, the impact on the human response is so tangible that robotic movements opportunely programmed can elicit a "priming" effect. Priming is explained as the impact that individual's movement has on how the observer moves. Recent studies demonstrated how priming is also present when a robotic agent, with opportune behavior, shows movement to the human observer [72, 73]. Both the role of timing and priming seem to be related to the tendency in human observers to anthropomorphize the partner even if this is a robotic agent [74]. Individuals seem to anthropomorphize robot effortlessly and this is due to their embodiment and their physical presence [75]. This seems an important determinant for the future generation of social robots interacting with humans. Embodiment and physical presence of robots that communicate through movement seem important factors in all the chapter contributions in the Part III of this book.

In summary, similar topics are investigated in the three different book parts, with different methodologies and different experimental constraints, but with a common goal of understanding humans, either to gain a better comprehension of ourselves or to build machines and robots that are proficient at interacting with us [76].

References

1. Sandini, G., & Sciutti, A. (2018). Humane robots—From robots with a humanoid body to robots with an anthropomorphic mind. *ACM Transactions on Human-Robot Interaction, 7*(1), 1–4. https://doi.org/10.1145/3208954.
2. Rizzolatti, G., & Sinigaglia, C. (2010). The functional role of the parieto-frontal mirror circuit: Interpretations and misinterpretations. *Nature Reviews Neuroscience, 11*(March), 264–274. https://doi.org/10.1038/nrn2805.
3. Hilt, P. M., Bartoli, E., Ferrari, E., Jacono, M., Fadiga, L., & D'Ausilio, A. (2017). Action observation effects reflect the modular organization of the human motor system. *Cortex*. https://doi.org/10.1016/j.cortex.2017.07.020.
4. Cardellicchio, P., Hilt, P. M., Olivier, E., Fadiga, L., & D'Ausilio, A. (2018). Early modulation of intra-cortical inhibition during the observation of action mistakes. *Scientific Reports*. https://doi.org/10.1038/s41598-018-20245-z.
5. Hamilton, A. F de C., & Grafton, S. T. (2007). The motor hierarchy: From kinematics to goals and intentions. In *Sensorimotor foundations of higher cognition* (pp. 381–408). Oxford: Oxford University Press. https://doi.org/10.1093/acprof:oso/9780199231447.003.0018.
6. Naish, K. R., Houston-Price, C., Bremner, A. J., & Holmes, N. P. (2014). Effects of action observation on corticospinal excitability: Muscle specificity, direction, and timing of the mirror response. *Neuropsychologia*. https://doi.org/10.1016/j.neuropsychologia.2014.09.034.
7. Amoruso, L., Finisguerra, A., & Urgesi, C. (2016). Tracking the time course of top-down contextual effects on motor responses during action comprehension. *Journal of Neuroscience*. https://doi.org/10.1523/JNEUROSCI.4340-15.2016.
8. Cretu, A. L., Ruddy, K., Germann, M., & Wenderoth, N. (2019). Uncertainty in contextual and kinematic cues jointly modulates motor resonance in primary motor cortex. *Journal of Neurophysiology*. https://doi.org/10.1152/jn.00655.2018.
9. Kilner, J. M. (2011). More than one pathway to action understanding. *Trends in Cognitive Sciences*. https://doi.org/10.1016/j.tics.2011.06.005.

10. Johansson, G. (1973). Visual perception of biological motion and a model for its analysis. *Perception and Psychophysics, 14*(2), 201–211. https://doi.org/10.3758/BF03212378.
11. Hemeren, P. E., & Thill, S. (2011, January). Deriving motor primitives through action segmentation. *Frontiers in Psychology,* 1, 1–11. https://doi.org/10.3389/fpsyg.2010.00243.
12. Fitts, P. M. (1954). The information capacity of the human motor system in controlling the amplitude of movement. *Journal of Experimental Psychology*. https://doi.org/10.1037/h0055392.
13. Viviani, P., & Flash, T. (1995). Minimum-Jerk, two-thirds power law, and isochrony: Converging approaches to movement planning. *Journal of Experimental Psychology. Human Perception and Performance, 21*(1), 32–53. https://doi.org/10.1037/0096-1523.21.1.32.
14. Lacquaniti, F., Terzuolo, C., & Viviani, P. (1983). The law relating the kinematic and figural aspects of drawing movements. *Acta Psychologica, 54*(1–3), 115–130. https://doi.org/10.1016/0001-6918(83)90027-6.
15. Bisio, A., Sciutti, A., Nori, F., Metta, G., Fadiga, L., Sandini, G., et al. (2014). Motor contagion during human-human and human-robot interaction. *PLoS ONE, 9*(8), e106172. https://doi.org/10.1371/journal.pone.0106172.
16. Maurice, P., Huber, M. E., Hogan, N., & Sternad, D. (2018). Velocity-curvature patterns limit human-robot physical interaction. *IEEE Robotics and Automation Letters*. https://doi.org/10.1109/LRA.2017.2737048.
17. Morton, J., & Johnson, M. H. (1991). CONSPEC and CONLERN: A two-process theory of infant face recognition. *Psychological Review*. https://doi.org/10.1037/0033-295X.98.2.164.
18. Lewkowicz, D. J., & Hansen-Tift, A. M. (2012). Infants deploy selective attention to the mouth of a talking face when learning speech. *Proceedings of the National Academy of Sciences of the United States of America*. https://doi.org/10.1073/pnas.1114783109.
19. Simion, F., Regolin, L., & Bulf, H. (2008). A predisposition for biological motion in the newborn baby. *Proceedings of the National Academy of Sciences of the United States of America*. https://doi.org/10.1073/pnas.0707021105.
20. Kloos, H., & Sloutsky, V. M. (2008). What's behind different kinds of kinds: Effects of statistical density on learning and representation of categories. *Journal of Experimental Psychology: General*. https://doi.org/10.1037/0096-3445.137.1.52.
21. Estes, D. (1998). Young children's awareness of their mental activity: The case of mental rotation. *Child Development*. https://doi.org/10.2307/1132270.
22. Haith, M. M., Hazan, C., & Goodman, G. S. (1988). Expectation and anticipation of dynamic visual events by 3.5-month-old babies. *Child Development*. https://doi.org/10.1111/j.1467-8624.1988.tb01481.x.
23. Adolph, K. E., Cole, W. G., & Vereijken, B. (2014). Intraindividual Variability in the development of motor skills in childhood. *Handbook of Intraindividual Variability Across the Life Span*. https://doi.org/10.4324/9780203113066.
24. Meltzoff, A. N., & Keith Moore, M. (1977). Imitation of facial and manual gestures by human neonates. *Science*. https://doi.org/10.1126/science.198.4312.75.
25. Heyes, C. (2001). Causes and consequences of imitation. *Trends in Cognitive Sciences*. https://doi.org/10.1016/S1364-6613(00)01661-2.
26. Hunnius, S., & Bekkering, H. (2010). The early development of object knowledge: A study of infants' visual anticipations during action observation. *Developmental Psychology*. https://doi.org/10.1037/a0016543.
27. Stern, D. N. (2010). *Forms of vitality: Exploring dynamic experience in psychology, the arts, psychotherapy, and development*. Oxford: Oxford University Press. https://doi.org/10.5860/choice.48-4178.
28. Di Cesare, G., Di Dio, C., Rochat, M. J., Sinigaglia, C., Bruschweiler-Stern, N., Stern, D. N., et al. (2014). The neural correlates of 'vitality form' recognition: An FMRI study. *Social Cognitive and Affective Neuroscience, 9*(7), 951–960. https://doi.org/10.1093/scan/nst068.
29. Di Cesare, G., Fasano, F., Errante, A., Marchi, M., & Rizzolatti, G. (2016, August). Understanding the internal states of others by listening to action verbs. *Neuropsychologia, 89,* 172–179. https://doi.org/10.1016/j.neuropsychologia.2016.06.017.

30. Di Cesare, G., De Stefani, E., Gentilucci, M., & De Marco, D. (2017, November). Vitality forms expressed by others modulate our own motor response: A kinematic study. *Frontiers in Human Neuroscience, 11*. https://doi.org/10.3389/fnhum.2017.00565.
31. Asadi-Aghbolaghi, M., Clapes, A., Bellantonio, M., Escalante, H. J., Ponce-Lopez, V., Baro, X. et al. (2017). A survey on deep learning based approaches for action and gesture recognition in image sequences. In *Proceedings—12th IEEE International Conference on Automatic Face and Gesture Recognition, FG 2017*.
32. Zhang, H. B., Zhang, Y. X., Zhong, B., Lei, Q., Yang, L., Xiang, D. J., et al. (2019). A comprehensive survey of vision-based human action recognition methods. *Sensors (Switzerland)*. https://doi.org/10.3390/s19051005.
33. Fortun, D., Bouthemy, P., & Kervrann, C. (2015). Optical flow modeling and computation: A survey. *Computer Vision and Image Understanding*. https://doi.org/10.1016/j.cviu.2015.02.008.
34. Burr, D. C., Morrone, M. C., & Vaina, L. M. (1998). Large receptive fields for optic flow detection in humans. *Vision Research, 38*(12), 1731–1743. https://doi.org/10.1016/S0042-6989(97)00346-5.
35. Bailer, C., Varanasi, K., & Stricker, D. (2017). CNN-based patch matching for optical flow with thresholded hinge embedding loss. In *Proceedings—30th IEEE Conference on Computer Vision and Pattern Recognition, CVPR 2017*. https://doi.org/10.1109/CVPR.2017.290.
36. Weinzaepfel, P., Revaud, J., Harchaoui, Z., & Schmid, C. (2013). DeepFlow: Large displacement optical flow with deep matching. *Proceedings of the IEEE International Conference on Computer Vision*. https://doi.org/10.1109/ICCV.2013.175.
37. Dosovitskiy, A., Fischery, P., Ilg, E., Hausser, P., Hazirbas, Cner, Golkov, V., et al. (2015). FlowNet: Learning optical flow with convolutional networks. *Proceedings of the IEEE International Conference on Computer Vision*. https://doi.org/10.1109/ICCV.2015.316.
38. Ilg, E., NMayer, S., Saikia, T., Keuper, M., Dosovitskiy, A., & Brox, T. (2017). FlowNet 2.0: Evolution of optical flow estimation with deep networks. In *Proceedings—30th IEEE Conference on Computer Vision and Pattern Recognition, CVPR 2017*. https://doi.org/10.1109/CVPR.2017.179.
39. Sun, D., Yang, X., Liu, M. Y., & Kautz, J. (2018). PWC-Net: CNNs for optical flow using pyramid, warping, and cost volume. *Proceedings of the IEEE Computer Society Conference on Computer Vision and Pattern Recognition*. https://doi.org/10.1109/CVPR.2018.00931.
40. Flanagan, J. R., Bowman, M. C., & Johansson, R. S. (2006). Control strategies in object manipulation tasks. *Current Opinion in Neurobiology, 16*(6), 650–659. https://doi.org/10.1016/j.conb.2006.10.005.
41. Stulp, F., Theodorou, E. A., & Schaal, S. (2012). Reinforcement learning with sequences of motion primitives for robust manipulation. *IEEE Transactions on Robotics*. https://doi.org/10.1109/TRO.2012.2210294.
42. Lea, C., Flynn, M. D., Vidal, R., Reiter, A., & Hager, G. D. (2017). Temporal convolutional networks for action segmentation and detection. In *Proceedings—30th IEEE Conference on Computer Vision and Pattern Recognition, CVPR 2017*. https://doi.org/10.1109/CVPR.2017.113.
43. Peng, X., & Schmid, C. (2016). Multi-region two-stream R-CNN for action detection. In *Lecture Notes in Computer Science (Including Subseries Lecture Notes in Artificial Intelligence and Lecture Notes in Bioinformatics)* (pp. 744–59). https://doi.org/10.1007/978-3-319-46493-0_45.
44. Gkioxari, G., & Malik, J. (2015). Finding action tubes. *Proceedings of the IEEE Computer Society Conference on Computer Vision and Pattern Recognition*. https://doi.org/10.1109/CVPR.2015.7298676.
45. Weinland, D., Ronfard, R., & Boyer, E. (2011). A survey of vision-based methods for action representation, segmentation and recognition. *Computer Vision and Image Understanding, 115*(2), 224–241. https://doi.org/10.1016/j.cviu.2010.10.002.
46. Aksoy, E. E., Abramov, A., Dörr, J., Ning, K., Dellen, B., & Wörgötter, F. (2011). Learning the semantics of object-action relations by observation. *International Journal of Robotics Research*. https://doi.org/10.1177/0278364911410459.

47. Hussein, A., Gaber, M. M., Elyan, E., & Jayne, C. (2017). Imitation learning. *ACM Computing Surveys, 50*(2), 1–35. https://doi.org/10.1145/3054912.
48. Chen, Z., & Liu, B. (2016). Lifelong machine learning. *Synthesis Lectures on Artificial Intelligence and Machine Learning*. https://doi.org/10.2200/S00737ED1V01Y201610AIM033.
49. Nelson, C. A. (2000). Neural plasticity and human development: The role of early experience in sculpting memory systems. *Developmental Science*. https://doi.org/10.1111/1467-7687.00104.
50. Willshaw, D. J., & Von Der Malsburg, C. (1976). How patterned neural connections can be set up by self organization. *Proceedings of the Royal Society of London - Biological Sciences*. https://doi.org/10.1098/rspb.1976.0087.
51. Marsland, S., Shapiro, J., & Nehmzow, U. (2002). A self-organising network that grows when required. *Neural Networks*. https://doi.org/10.1016/S0893-6080(02)00078-3.
52. Breazeal, C. (2003). Toward sociable robots. *Robotics and Autonomous Systems, 42*(3–4), 167–175. https://doi.org/10.1016/S0921-8890(02)00373-1.
53. Giraud, T., Focone, F., Isableu, B., Martin, J. C., & Demulier, V. (2016). Impact of elicited mood on movement expressivity during a fitness task. *Human Movement Science*. https://doi.org/10.1016/j.humov.2016.05.009.
54. Sherer, K. R. (1984). On the nature and function of emotion: A component process approach. In *Approaches to Emotion* (p. 31).
55. Kleinsmith, A., & Bianchi-Berthouze, N. (2013). Affective body expression perception and recognition: A survey. *IEEE Transactions on Affective Computing, 4*(1), 15–33. https://doi.org/10.1109/T-AFFC.2012.16.
56. Piana, S., Staglianò, A., Odone, F., & Camurri, A. (2016). Adaptive body gesture representation for automatic emotion recognition. *ACM Transactions on Interactive Intelligent Systems, 6*(1), 1–31. https://doi.org/10.1145/2818740.
57. Varni, G., Volpe, G., & Camurri, A. (2010). A system for real-time multimodal analysis of nonverbal affective social interaction in user-centric media. *IEEE Transactions on Multimedia, 12*(6), 576–590. https://doi.org/10.1109/TMM.2010.2052592.
58. Balit, E., Vaufreydaz, D., & Reignier, P. (2018). PEAR: prototyping expressive animated robots a framework for social robot prototyping. In *VISIGRAPP 2018—Proceedings of the 13th International Joint Conference on Computer Vision, Imaging and Computer Graphics Theory and Applications*. https://doi.org/10.5220/0006622600440054.
59. Bartneck, C., Kanda, T., Mubin, O., & Mahmud, A. A. (2009). Does the design of a robot influence its animacy and perceived intelligence? *International Journal of Social Robotics*. https://doi.org/10.1007/s12369-009-0013-7.
60. Gray, J., Hoffman, G., & Adalgeirsson, S. (2010). Expressive, interactive robots: Tools, techniques, and insights based on collaborations. In *HRI 2010 Workshop: What Do Collaborations with the Arts Have to Say about HRI*.
61. Lorenz, T., Mortl, A., Vlaskamp, B., Schubo, A., & Hirche, S. (2011). Synchronization in a goal-directed task: Human movement coordination with each other and robotic partners. In *2011 RO-MAN* (pp. 198–203). New York: IEEE. https://doi.org/10.1109/ROMAN.2011.6005253.
62. Althaus, P., Ishiguro, H., Kanda, T., Miyashita, T., & Christensen, H. I. (2004). Navigation for human-robot interaction tasks. *Proceedings—IEEE International Conference on Robotics and Automation*. https://doi.org/10.1109/robot.2004.1308100.
63. Ciolek, T. M., & Kendon, A. (1980). Environment and the spatial arrangement of conversational encounters. *Sociological Inquiry, 50*(3–4), 237–271. https://doi.org/10.1111/j.1475-682X.1980.tb00022.x.
64. Pairet, E., Ardon, P., Mistry, M., & Petillot, Y. (2019). Learning generalizable coupling terms for obstacle avoidance via low-dimensional geometric descriptors. *IEEE Robotics and Automation Letters, 4*(4), 3979–3986. https://doi.org/10.1109/LRA.2019.2930431.
65. Ardon, P., Pairet, E., Petrick, R. P. A., Ramamoorthy, S., & Lohan, K. S. (2019). Learning grasp affordance reasoning through semantic relations. *IEEE Robotics and Automation Letters*. https://doi.org/10.1109/LRA.2019.2933815.

66. Brooks, R. A. (1991). Intelligence without representation. *Artificial Intelligence*. https://doi.org/10.1016/0004-3702(91)90053-M.
67. Rizzolatti, G., & Arbib, M. A. (1998). Language within our grasp. *Trends in Neurosciences*. https://doi.org/10.1016/S0166-2236(98)01260-0.
68. Hagendoorn, I. (2010). Dance, language and the brain. *International Journal of Arts and Technology*. https://doi.org/10.1504/IJART.2010.032565.
69. Varela, F. J., Thompson, E., Rosch, E., & Kabat-Zinn, J. (1991). *The embodied mind: Cognitive science and human experience. The embodied mind: Cognitive science and human experience*. Cambridge: MIT Press.
70. Mubin, O., Stevens, C. J., Shahid, S., Al Mahmud, A.., & Dong, J.-J. (2013). A review of the applicability of robots in education. *Technology for Education and Learning,* 1(1). https://doi.org/10.2316/Journal.209.2013.1.209-0015.
71. Tanaka, F., Isshiki, K., Takahashi, F., Uekusa, M., Sei, R., & Hayashi, K. (2015). Pepper learns together with children: Development of an educational application. *IEEE-RAS International Conference on Humanoid Robots*. https://doi.org/10.1109/HUMANOIDS.2015.7363546.
72. Kashi, S., & Levy-Tzedek, S. (2018). Smooth leader or sharp follower? Playing the mirror game with a robot. *Restorative Neurology and Neuroscience, 36*(2), 147–159. https://doi.org/10.3233/RNN-170756.
73. Vannucci, F., Sciutti, A., Lehman, H., Sandini, G., Nagai, Y., & Rea, F. (2019). Cultural differences in speed adaptation in human-robot interaction tasks. *Paladyn, Journal of Behavioral Robotics, 10*(1), 256–266. https://doi.org/10.1515/pjbr-2019-0022.
74. Darling, K. (2017). *Who's Johnny? Anthropomorphic framing in human–robot interaction, integration, and policy. Robot Ethics 2.0: From autonomous cars to artificial intelligence* (Vol. 1). Oxford: Oxford University Press. https://doi.org/10.1093/oso/9780190652951.003.0012.
75. Kidd, C. D., & Breazeal, C. (2005). Human-robot interaction experiments: Lessons learned. In *AISB'05 Convention: Social Intelligence and Interaction in Animals, Robots and Agents—Proceedings of the Symposium on Robot Companions: Hard Problems and Open Challenges in Robot-Human Interaction* (pp. 141–42).
76. Sciutti, A., Mara, M., Tagliasco, V., & Sandini, G. (2018). Humanizing human-robot interaction: On the importance of mutual understanding. *IEEE Technology and Society Magazine,* 37(1). https://doi.org/10.1109/MTS.2018.2795095.

Part I
Motion Perception in Humans

Chapter 2
The Neurophysiology of Action Perception

Pauline M. Hilt, Pasquale Cardellicchio, and Alessandro D'Ausilio

Abstract Action perception relies on a parieto-premotor brain network engaged during both perception of actions performed by conspecifics and actual execution of actions. Despite important overlap between neural activations during action observation and action execution, the functional relevance of these activities remains debated. In this chapter, we will discuss how the study of action perception may effectively be enriched by applying core principles of motor control. By doing so, we present evidences in favour of: (i) the presence of a modular control strategy in action observation; (ii) the role of motor inhibition in coping with unpredictable action outcomes. We conclude that reaffirming the strong parallel with motor control would provide important insight into the investigation of action perception mechanisms.

2.1 Action Perception

Mirror neurons were originally described as visuomotor neurons that are engaged both during visual presentation of actions performed by conspecifics, and during the actual execution of these actions [1]. These neurons were first discovered using single-cell recordings in monkey premotor cortex (area F5; [2]) and later within monkey inferior parietal cortex (PF/PFG; [3, 4]).

Since then, there has been a growing interest in mirror neurons both in the scientific literature and the popular media. The widespread interest was in particular driven by their potential role in imitation and thus in a fundamental aspect of social cognition [5, 6]. In follow-up studies, neurons with mirror properties have been found in different parietal and frontal areas of monkeys and other species, including humans [7].

The mirror neuron system has also been associated to action perception. In fact, others' action anticipation and comprehension might be achieved both by the ventral

P. M. Hilt · P. Cardellicchio
IIT@UniFe Center for Translational Neurophysiology, Istituto Italiano di Tecnologia, Ferrara, Italy

A. D'Ausilio (✉)
Section of Human Physiology, Università di Ferrara, Ferrara, Italy
e-mail: alessandro.dausilio@gmail.com

N. Noceti et al. (eds.), *Modelling Human Motion*,
https://doi.org/10.1007/978-3-030-46732-6_2

route (Middle Temporal Gyrus—MTG—and the anterior Inferior Frontal Gyrus—aIFG), and the dorsal route (Inferior Parietal Lobule—IPL—and the posterior Inferior Frontal Gyrus—pIFG). The dorsal stream may support this process by reactivating the most likely action needed to achieve the predicted goal. In line with this account, action discrimination could rely on internal forward models [8, 9] to anticipate the unfolding of a given action [10].

2.2 Mirror Neuron System in Humans

Immediately following the initial reports of mirror neurons in the macaque brain, the existence of an analogous mechanism in humans was discussed.

While some authors argued that clear evidence of a human mirror neuron system was still lacking (e.g. [11–13]), further and numerous results coming from various techniques such as transcranial magnetic stimulation (TMS; [14, 15]), electroencephalography (EEG; [16]), functional magnetic resonance imaging (fMRI; [17]) and human single-cell recordings [18] revealed the existence of a fronto-parietal network with mirror-like properties in humans [6].

Based on human brain-imaging data [19–21] and cytoarchitecture [22], the ventral premotor cortex and the pars opercularis of the posterior inferior frontal gyrus (Brodmann area 44) were assumed to be the human homologues of macaque mirror area F5. Later, the rostral inferior parietal lobule was identified as equivalent of the monkey mirror area PF/PFG [1, 23].

In parallel, EEG research showed that event-related synchronization and desynchronization of the mu rhythm (rolandic alpha band) were linked to action performance, observation and imagery [16, 24]. These results suggest that Rolandic mu event-related desynchronization [25, 26] during action observation reflects activity of a mirror-like system present in humans [16, 27, 28].

Finally, single-pulse TMS over the primary motor cortex (M1) and motor evoked potentials (MEPs) amplitude were employed as a direct index of corticospinal recruitment (Corticospinal Excitability—CSE). Using this technique, several studies showed a modulation of MEPs amplitude during action observation matching various changes occurring during action execution [29]; for a review please see: [14, 15, 30].

2.3 Mirror Neuron System: Transcranial Magnetic Stimulation Studies

Although all these measures have been instrumental to investigate mechanisms underlying action observation in humans, the most useful to determine the degree of matching between observed and executed action remains TMS. Differently from other techniques that either measure slow metabolic signals (e.g. PET, SPECT, fMRI, fNIRS) or the electrical mass activity of an extended brain networks (EEG-MEG), TMS allows the direct measurement of motor activity. In fact, a single pulse of TMS noninvasively stimulate the human motor cortex to instantaneously assesses the magnitude of the descending motor drive to muscles [31]. MEP size reflects the net facilitatory and inhibitory input to the pyramidal projecting neurons [32], thus providing an instantaneous read-out of the activity of an extended motor network.

This approach has been classically used to evaluate corticospinal excitability during motor imagery [33, 34], action observation [14, 35] as well as planning or execution of an action [36, 37]. The high temporal resolution of the technique allows effective exploration of motor recruitment in these tasks. Indeed, this technique provides direct comparison between the unfolding of motor processes during action observation and the timing of real muscle activation during action execution [38, 39]. Until today, more than a hundred TMS studies have shown that the neural match between action execution and observation [14, 15, 29] is characterized by an important degree of temporal and somatotopic congruency between the motor representations elicited by two conditions. Indeed, MEPs are modulated by observed low-level movement features such as finger aperture in a grasping action [40], the amplitude of muscle activities over time [41, 42] and the forces required to lift objects of different weights [43, 44]. In parallel, modulation of CSE amplitude was also shown for higher level features such as action goals [45, 46]. For example, MEPs amplitude did not seem to depend on the effector used to attain the same action goal [47–49], suggesting independence from low-level movement features.

Recently, other studies investigated the factors that modulate these action observation effects. Data shows that these effects depend on the instructions provided to subjects, attention [50], action context [51], TMS timing [52], recorded muscle [53], as well as motor learning [54, 55]. The data we have so far describe the huge variability of these effects. A variability that question fundamental methodological and theoretical aspects related to the role played by motor activities during action observation.

Indeed, we proposed that these controversies arise from a poorly defined description of what the activity of the motor cortex, as well as the relationship between motion kinematics and muscle recruitment, is during action execution [30]. In other words, the difficulties in understanding the mirror coding may directly stem from how we conceptualize the workings of the motor system during action execution.

Here we will present two consolidated research areas in motor control that have been substantially neglected in action observation studies. The two following sections will discuss recent investigations aiming at the translation of methodological and theoretical principles from motor control literature to the study of action perception.

2.4 Modularity in the Motor System Could Be Exploited in Action Observation

One of the most fundamental questions in motor control concerns the mechanisms that underlie muscle recruitment during the execution of movements. The complexity of the musculoskeletal apparatus as well as its dynamical properties is huge. For example, synergic muscles contribute to the mobilization of the same joint. Each one is characterized by different point of insertion and thus slightly different action. Furthermore, muscle viscoelasticity and torque is both length- and velocity-dependent as well as characterized by different properties during passive stretching or active concentric and eccentric contraction. More importantly, actions often involve the coordinated control of multiple interdependent links. For instance, to extend fingers we need to stabilize the wrist, otherwise finger extensor muscles would exert no effect. To extend the arm and reach for an object, we need to anticipatorily activate the gastrocnemius to keep the center of gravity within the base of support. All in all, the combination of muscle activity to produce a specific postural configuration has to deal with a complexity that is mastered by the nervous system via efficient control strategies [56].

On the one hand, motor redundancy suggests that different joint configurations can equally be used to reach the same behavioral goal [57, 58]. At the same time, due to the aspects briefly outlined earlier, the same kinematic configuration can be achieved via largely different underlying muscle activation patterns [59, 60]. In fact, in a realistic scenario (e.g., movement execution to reach an object), small changes such as those caused by a change in height of the table may have a dramatic influence on the temporal evolution and recruitment of the same muscles in the same action towards the same goal. It follows that the same amount of EMG activity in one muscle is present in many different actions and it is not necessarily predictive of the action goal.

Therefore, due to motor control redundancy, there is no simple mapping between muscle activity and visual appearance of action (e.g. kinematics) [61]. In this regard, it is not clear how an observer could reconstruct the fine motor details of the action executed by someone else, solely based on a partial and noisy description of kinematics (e.g. including occlusions and hidden body parts). Notably, in simple motor tasks (e.g. finger's abduction/adduction), the presence of a unique motor mapping directly translate into meaningful action observation effects (e.g. [62]). Instead, in multi-joint actions (e.g. upper-limb reaching to-grasp movement), the muscle-to-movement mapping has many solutions, and this translate into greater amount of noise in the data. These facts may explain why robust group-level CSE modulations are harder to find in the observation of complex multi-joint actions [63–65].

In action execution, humans and animals do not control each muscle independently and rather rely on a modular control architecture [66–69]. In fact, the nervous system flexibly activates a combination of a small number of muscular synergies (or motor primitives) organized by spinal neuronal populations.

At the same time, some evidence suggests also an important cortical contribution. For instance, cortical stimulation elicits a synergistic pattern of activities [70] while single neuronal response encodes the activity of a relatively small number of functionally related muscles [71]. In this vein, an integrative proposal suggests that recurrent activity propagating between the motor cortex and muscles could maintain accurate and discrete representations of muscle synergies [72]. All in all, synergies refer to covarying groups of muscles that remain in a fixed relationship during action control [73], thus providing the fundamental building blocks of motor control organization.

At the action observation level, this modular organization is rarely considered. In fact, the classical use of CSE recorded from few muscles may not be the most accurate way to explore action observation effects. In addition to that, CSE is highly variable across time [74, 75] and hugely dependent on cortical states [76] and on spontaneous cortical oscillatory dynamics [77, 78].

As a consequence, a more robust marker of multi-joint action observation effects may be given by TMS-evoked kinematics (i.e. motor evoked kinematics, MEKs). This assumption is based on the principles of redundancy and invariance during motor execution [68, 79, 80] and it takes into account the fact that the control of movements relies upon the composition of intracortical, corticospinal, spinal and peripheral influences [81] which in turn regulate the temporal-spatial coordination of multiple agonist and antagonist muscles. At rest, MEKs replicate the modular organization of hand function [82] and, by reflecting the functional output of the motor system, MEKs may offer a reliable measure also during action observation [38, 48, 83].

A recent study has compared the respective robustness of MEPs and MEKs in a classical action observation task [64]. In this study, a single TMS was delivered over the observers' motor cortex at two timings of two reaching-grasping actions (precision vs. power grip). We recorded both MEPs on 4 hand/arm muscles and arm MEKs for 8 hand elevation angles. We repeated the same protocol twice, and we showed a significant time-dependent grip-specific MEPs and MEKs modulation. However, MEKs, differently from MEPs, exhibit a consistent significant modulation across sessions (Fig. 2.1).

As predicted by considerations about motor redundancy in action execution, MEKs data in action observation tasks might be more reliable than MEPs. Beside obvious methodological considerations about the reliability and replicability of results, a theoretical point can be raised here. Indeed, these results are highly suggestive of the fact that the modularity employed to solve the complexity of motor control may as well be used by the observing brain. In this latter case, modularity may be used to map between visual appearance of actions (kinematics) and action goals, thus bypassing the need to estimate a point-by-point muscle-level description of actions.

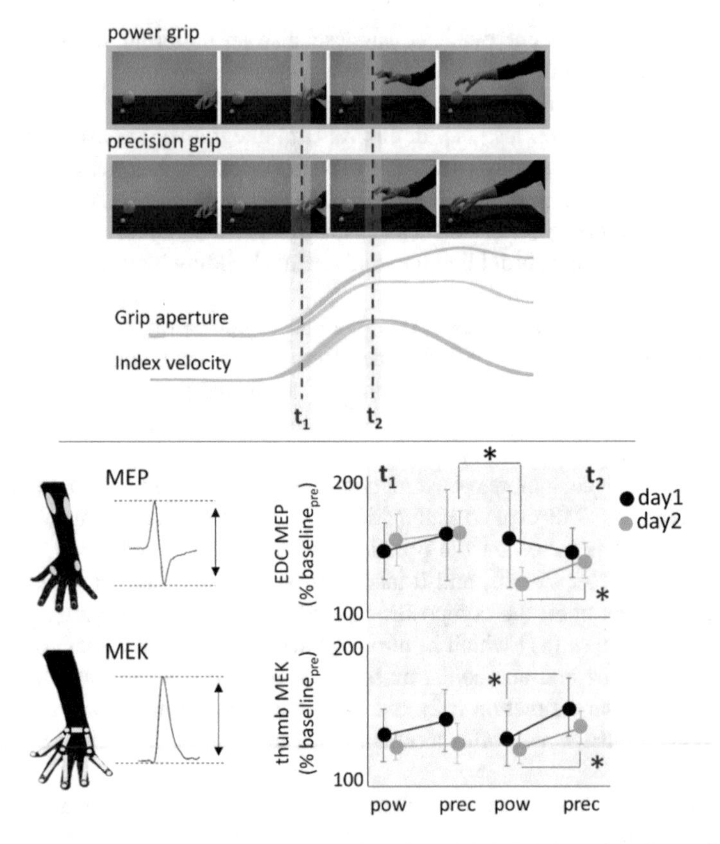

Fig. 2.1 Illustration of the experimental protocol used to study MEKs in action observation [64]. Upper panel. Four representative frames of the two displayed movies (power and precision grip) and associated kinematic (grip aperture and index velocity). Timing t_1 and t_2 are represented by black dotted vertical lines. Lower panel. On the left, typical recording for MEPs and MEKs. On the right, mean and standard error of the extensor digitorum muscle (EDC) MEPs and thumb MEKs expressed as a % of the average of the baseline, separately for the two sessions (day 1, day 2), timing (t_1, t_2), and grasp type (precision (prec), power (pow)). Significant differences are represented by an asterisk ($p < 0.05$). Adapted from [64]

2.5 Action Inhibition in the Presence of Errors: Own Versus Other's

As for the case of motor redundancy, other central themes in motor control can effectively be exploited to investigate action observation effects. For example, a long tradition has accumulated on the neural responses to errors in execution and the associated motor reprogramming.

Activity in the motor system is largely influenced by preparatory and inhibitory processes associated with action selection and reprogramming [84, 85]. Indeed, the motor system quickly inhibits an ongoing action plan to select and execute an appropriate alternative movement [86–88] or to prevent the unwanted release of a wrong response [89].

Many studies have investigated the neural substrates of behavioral inhibition by using tasks that require stopping an action [90–92]. A rapid suppression of activity can be observed both during execution, likely reflecting a pause in motor output [93, 94] or during action preparation [95, 96]. Hence, the motor system is inhibited not only when a movement needs to be aborted, but also when it is in the process of specifying the future action. In this sense monitoring processes are essential to detect unexpected events or mistakes and thus to efficiently update the planned action [97, 98]. Although motor inhibition is considered as a fundamental part of action execution similar processes that might occur in the observer's brain are poorly understood.

In fact, if errors occur during action execution, these are usually unexpected by the observer and it is unlikely that we can efficiently predict them [99]. Therefore, an error during the unfolding of an action should create a surprise signal in the observer [100]. Hence, the role played by motor activity in action observation, as well as its similarity with action execution processes, should be clarified by investigating the observation of action mistakes.

In this regard, action error detection requires sensing subtle kinematic violations in the observed action [101–103], reflecting a matching process between observed actions onto corresponding stored representations of the same actions. It is currently believed that others' actions cues are compared to stored internal motor models of the same action [104]. Two alternative models could explain how this comparison takes place in the AON (Action Observation Network): The direct matching hypothesis and the predictive coding approach.

The direct matching hypothesis [1, 23] is based on the idea that action observation automatically activates the neuronal population that represents this action in the observer's premotor-parietal brain network, mapping the visual representation of the observed action onto a motor representation of the same action. These motor activations are triggered by action observation and are iteratively refined during the temporal deployment of the observed action. As the action unfolds, more and more data is integrated into the emergence of a specific motor activation pattern that matches the one that has been implemented by the actor. On this basis, action outcomes become accessible to the observer as if he was himself acting. In this context, an error is a signal that do not fit with the representation of the observed action. According to this account, observation of an error should activate the same inhibitory mechanisms at play during error execution [105].

Differently, the predictive-coding approach suggests that "reading" other's actions stem from an empirical Bayesian inference process, in which top-down expectations (e.g. goal) allow the prediction of lower levels of action representation (e.g. motor commands; [106]). Predicted motor commands are compared with observed kinematics to generate a prediction error that is further propagated across neural processing levels to update information according to the actual outcome. In this perspective, the computation of an error between predicted and currently perceived movements, translates into an increase of activity in the action observation network. In other words, the direct matching hypothesis suggests that motor activation during action observation represents the reactivation of a specific motor pattern, while predictive coding suggests that the same activity represents the discrepancy between a bottom-up and a top-down process.

Results from a recent study [107] investigating the temporal dynamics of modulation of the AON during an action error lend some initial support towards the last hypothesis. In this study, the temporally fine-grained balance of motor excitation and inhibition was investigated at three time points (120, 180, 240 ms after action error). CSE, short intracortical inhibition (sICI), and intracortical facilitation (ICF) were measured during the observation of a sequence of pictures depicting either correct or erroneous actions. The authors used two different type of errors: procedural-execution errors (wrong passage of the rope) and control errors (in which the rope suddenly appears cut in two segments).

As described earlier, CSE provides an instantaneous read-out of the state of the motor system while sICI and ICF reflect the activity of distinct intracortical inhibitory and excitatory circuits [108]. Results show an early (120 ms) reduction of inhibition (sICI) for the observation of a motor execution error, while the control error elicited a similar effect but with a longer latency (240 ms) (Fig. 2.2). These results suggest that the neural mechanisms involved in detecting action execution errors mainly consist in the modulation of intracortical inhibitory circuits. According to these results when an action error is detected, a decrease in inhibition rather than an increase is present. This is the opposite of what we would expect from a complete functional match between action execution and action observation processes but is congruent with the predictive coding hypothesis [109].

The early effect is associated to the presentation of a motor execution error. This delay of 120 ms is consistent with the time required to activate the motor system during graspable object presentation [110]. The late effect instead, is triggered by errors requiring access to strategic and abstract reasoning regarding the feasibility of the action plan, that only later translates into similar neurophysiological modulation [111]. A similar biphasic modulation has also been shown for corticospinal excitability during action observation [38].

In conclusion, according to the predictive coding account the brain uses all available information to continuously predict forthcoming events and reduce sensory uncertainty by dynamically formulating perceptual hypotheses [112]. In this context, the main function carried out by the AON could be the minimization of the sensory prediction error (i.e., Bayesian-like inferences are generated and dynamically compared to the incoming sensory information) based on visually perceived actions. In this view, motor activation during action observation do not merely reflect an automatic resonance mechanism but rather the interplay between an internal prediction and the incoming signals.

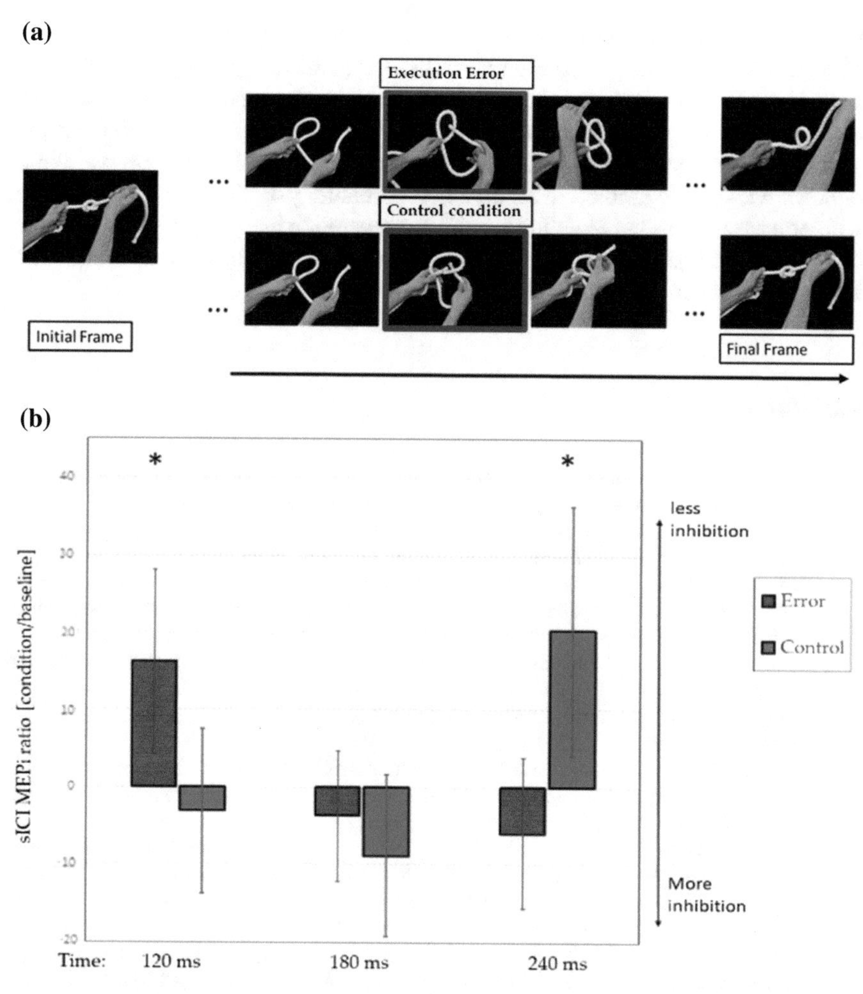

Fig. 2.2 Motor inhibition while observing action errors. In panel **a**, each row represents the timeline of a knotting action. The red square highlights the frame associated to the motor error (wrong insertion of the rope) while the blue square the control error (in which the rope suddenly appears cut in two segments). The TMS pulses were given at three time points (120, 180, 240 ms) after the errors. In panel **b**, intracortical inhibition results. The normalized modulation of inhibitionas function of the TMS timing shows the early effect for the execution error. Adapted from [107]

2.6 Conclusions

The present chapter moves from the observation that multiple inconsistencies characterize our understanding of how motor activations impact others' action discrimination. Our claim is that, the motor control literature could inject fresh new ideas in the study of the neurophysiological mechanisms of action perception. In this regard, here we presented two examples of how core principles of motor control have helped the formulation of new testable hypothesis. At the same time though, the importance of such a cross-domain fertilization is not limited to the design of new experiments. In fact, one clear implication relates to the basic claim surrounding the interpretation of mirror-like activity: the functional similarity between action execution and perception. In this regard, limiting the investigation to the spatio-temporal overlap of brain activity would amount to modern neo-phrenology. Rather, building a parallel with processes recruited in action planning, preparation and execution might eventually show the neural mechanisms that are functionally similar as well as those that highlight the obvious and central differences.

References

1. Rizzolatti, G., & Craighero, L. (2004). The mirror-neuron system. *Annual Review of Neuroscience, 27,* 169–192. https://doi.org/10.1146/annurev.neuro.27.070203.144230.
2. di Pellegrino, G., Fadiga, L., Fogassi, L., et al. (1992). Understanding motor events: A neurophysiological study. *Experimental Brain Research, 91,* 176–180.
3. Fogassi, L., Ferrari, P. F., Gesierich, B., et al. (2005). Parietal lobe: From action organization to intention understanding. *Science, 80*(308), 662–667. https://doi.org/10.1126/science.1106138.
4. Gallese, V., Fadiga, L., Fogassi, L., & Rizzolatti, G. (2002). Action representation and the inferior parietal lobule. In *Common mechanisms in perception and action* (pp. 334–355).
5. Iacoboni, M. (2005). Neural mechanisms of imitation. *Current Opinion in Neurobiology, 15,* 632–637. https://doi.org/10.1016/j.conb.2005.10.010.
6. Rizzolatti, G., & Sinigaglia, C. (2010). The functional role of the parieto-frontal mirror circuit: Interpretations and misinterpretations. *Nature Reviews Neuroscience, 11,* 264–274. https://doi.org/10.1021/am4002502.
7. Rizzolatti, G., & Sinigaglia, C. (2016). The mirror mechanism: A basic principle of brain function. *Nature Reviews Neuroscience, 17,* 757–765. https://doi.org/10.1038/nrn.2016.135.
8. Flanagan, J. R., & Johansson, R. S. (2003). Action plans used in action observation. *Letter to Nature, 424,* 769–771. https://doi.org/10.1038/nature01861.
9. Kilner, J. M., Vargas, C., Duval, S., et al. (2004). Motor activation prior to observation of a predicted movement. *Nature Neuroscience, 7,* 1299–1301. https://doi.org/10.1038/nn1355.
10. Schütz-Bosbach, S., & Prinz, W. (2007). Prospective coding in event representation. *Cognitive Processing, 8,* 93–102. https://doi.org/10.1007/s10339-007-0167-x.
11. Dinstein, I. (2008). Human cortex: Reflections of mirror neurons. *Current Biology, 18,* 956–959. https://doi.org/10.1016/j.cub.2008.09.007.
12. Lingnau, A., Gesierich, B., & Caramazza, A. (2009). Asymmetric fMRI adaptation reveals no evidence. *PNAS, 106,* 9925–9930.
13. Turella, L., Pierno, A. C., Tubaldi, F., & Castiello, U. (2009). Mirror neurons in humans: Consisting or confounding evidence? *Brain and Language, 108,* 10–21. https://doi.org/10.1016/j.bandl.2007.11.002.
14. Fadiga, L., Craighero, L., & Olivier, E. (2005). Human motor cortex excitability during the perception of others' action. *Current Opinion in Neurobiology, 15,* 213–218. https://doi.org/10.1016/j.conb.2005.03.013.
15. Naish, K. R., Houston-Price, C., Bremner, A. J., & Holmes, N. P. (2014). Effects of action observation on corticospinal excitability: Muscle specificity, direction, and timing of the mirror response. *Neuropsychologia, 64,* 331–348. https://doi.org/10.1016/j.neuropsychologia.2014.09.034.

16. Fox, N. A., Yoo, K. H., Bowman, L. C., et al. (2016). Assessing human mirror activity With EEG mu rhythm: A meta-analysis. *Psychological Bulletin, 142,* 291–313. https://doi.org/10.1037/bul0000031.
17. Hardwick, R. M., Caspers, S., Eickhoff, S. B., & Swinnen, S. P. (2018). Neural correlates of action: Comparing meta-analyses of imagery, observation, and execution. *Neuroscience and Biobehavioral Reviews, 94,* 31–44. https://doi.org/10.1016/j.neubiorev.2018.08.003.
18. Mukamel, R., Ekstrom, A. D., Kaplan, J., et al. (2010). Single-neuron responses in humans during execution and observation of actions. *Current Biology, 20,* 750–756. https://doi.org/10.1016/j.cub.2010.02.045.
19. Decety, J., Grèzes, J., Costes, N., et al. (1997). Brain activity during observation of actions. Influence of action content and subject's strategy. *Brain, 120,* 1763–1777. https://doi.org/10.1093/brain/120.10.1763.
20. Iacoboni, M., Woods, R. P., Brass, M., et al. (1999). Cortical mechanisms of human imitation. *Science (New Series), 286,* 2526–2528. https://doi.org/10.1038/020493a0.
21. Rizzolatti, G., Fadiga, L., Matelli, M., et al. (1996). Localization of grasp representations in humans by PET: 1. Observation versus execution. *Experimental Brain Research, 111,* 246–252. https://doi.org/10.1007/BF00227301.
22. Petrides, M. (2005). Lateral prefrontal cortex: Architectonic and functional organization. *Philosophical Transactions of the Royal Society B, 360,* 781–795. https://doi.org/10.1098/rstb.2005.1631.
23. Rizzolatti, G., Fogassi, L., & Gallese, V. (2001). Neurophysiological mechanisms underlying the understanding and imitation of action. *Nature Reviews Neuroscience, 2,* 1–10. https://doi.org/10.1038/35090060.
24. Pineda, J. A. (2008). Sensorimotor cortex as a critical component of an "extended" mirror neuron system: Does it solve the development, correspondence, and control problems in mirroring? *Behavioral and Brain Functions, 4,* 1–16. https://doi.org/10.1186/1744-9081-4-47.
25. Babiloni, C., Babiloni, F., Carducci, F., et al. (2002). Human cortical Electroencephalography (EEG) rhythms during the observation of simple aimless movements: A high-resolution EEG study. *Neuroimage, 17,* 559–572.
26. Cochin, S., Barthelemy, C., Lejeune, B., et al. (1998). Perception of motion and qEEG activity in human adults. *Electroencephalography and Clinical Neurophysiology, 107,* 287–295.
27. Lapenta, O. M., Ferrari, E., Boggio, P. S., et al. (2018). Motor system recruitment during action observation: No correlation between mu-rhythm desynchronization and corticospinal excitability. *PLoS ONE, 13,* 1–15. https://doi.org/10.1371/journal.pone.0207476.
28. Sebastiani, V., de Pasquale, F., Costantini, M., et al. (2014). Being an agent or an observer: Different spectral dynamics revealed by MEG. *Neuroimage, 102,* 717–728. https://doi.org/10.1016/j.neuroimage.2014.08.031.
29. Fadiga, L., Fogassi, L., Pavesi, G., & Rizzolatti, G. (1995). Motor facilitation during action observation: A magnetic stimulation study. *Journal of Neurophysiology, 73,* 2608–2611.
30. D'Ausilio, A., Bartoli, E., & Maffongelli, L. (2015). Grasping synergies: A motor-control approach to the mirror neuron mechanism. *Physics of Life Reviews, 12,* 91–103. https://doi.org/10.1016/j.plrev.2014.11.002.
31. Hallett, M. (2007). Transcranial magnetic stimulation: A primer. *Neuron, 55,* 187–199. https://doi.org/10.1016/j.neuron.2007.06.026.
32. Rothwell, J. C. (1997). Techniques and mechanisms of action of transcranial stimulation of the human motor cortex. *Journal of Neuroscience Methods, 74,* 113–122. https://doi.org/10.1016/S0165-0270(97)02242-5.
33. Fadiga, L., Buccino, G., Craighero, L., et al. (1998). Corticospinal excitability is specifically modulated by motor imagery: A magnetic stimulation study. *Neuropsychologia, 37,* 147–158. https://doi.org/10.1016/S0028-3932(98)00089-X.
34. Fourkas, A. D., Avenanti, A., Urgesi, C., & Aglioti, S. M. (2006). Corticospinal facilitation during first and third person imagery. *Experimental Brain Research, 168,* 143–151. https://doi.org/10.1007/s00221-005-0076-0.

35. Urgesi, C., Maieron, M., Avenanti, A., et al. (2010). Simulating the future of actions in the human corticospinal system. *Cerebral Cortex, 20,* 2511–2521. https://doi.org/10.1093/cercor/bhp292.
36. Hoshiyama, M., Koyama, S., Kitamura, Y., et al. (1996). Effects of judgement process on motor evoked potentials in Go/No-go hand movement task. *Neuroscience Research, 24,* 427–430. https://doi.org/10.1016/0168-0102(95)01013-0.
37. Michelet, T., Duncan, G. H., & Cisek, P. (2010). Response competition in the primary motor cortex: Corticospinal excitability reflects response replacement during simple decisions. *Journal of Neurophysiology, 104,* 119–127. https://doi.org/10.1152/jn.00819.2009.
38. Barchiesi, G., & Cattaneo, L. (2013). Early and late motor responses to action observation. *Social Cognitive and Affective Neuroscience, 8,* 711–719. https://doi.org/10.1093/scan/nss049.
39. Prabhu, G., Voss, M., Brochier, T., et al. (2007). Excitability of human motor cortex inputs prior to grasp. *Journal of Physiology, 581,* 189–201. https://doi.org/10.1113/jphysiol.2006.123356.
40. Gangitano, M., Mottaghy, F. M., & Pascual-Leone, A. (2001). Phase-specific modulation of cortical motor output during movement observation. *NeuroReport, 12,* 1489–1492. https://doi.org/10.1097/00001756-200105250-00038.
41. Borroni, P., Montagna, M., Cerri, G., & Baldissera, F. (2005). Cyclic time course of motor excitability modulation during the observation of a cyclic hand movement. *Brain Research, 1065,* 115–124. https://doi.org/10.1016/j.brainres.2005.10.034.
42. Cavallo, A., Becchio, C., Sartori, L., et al. (2012). Grasping with tools: Corticospinal excitability reflects observed hand movements. *Cerebral Cortex, 22,* 710–716. https://doi.org/10.1093/cercor/bhr157.
43. Alaerts, K., Senot, P., Swinnen, S. P., et al. (2010). Force requirements of observed object lifting are encoded by the observer's motor system: A TMS study. *European Journal of Neuroscience, 31,* 1144–1153. https://doi.org/10.1111/j.1460-9568.2010.07124.x.
44. Senot, P., D'Ausilio, A., Franca, M., et al. (2011). Effect of weight-related labels on corticospinal excitability during observation of grasping: A TMS study. *Experimental Brain Research, 211,* 161–167. https://doi.org/10.1007/s00221-011-2635-x.
45. Cattaneo, L., Caruana, F., Jezzini, A., & Rizzolatti, G. (2009). Representation of goal and movements without overt motor behavior in the human motor cortex: A transcranial magnetic stimulation study. *Journal of Neuroscience, 29,* 11134–11138. https://doi.org/10.1523/JNEUROSCI.2605-09.2009.
46. Cattaneo, L., Maule, F., Barchiesi, G., & Rizzolatti, G. (2013). The motor system resonates to the distal goal of observed actions: Testing the inverse pliers paradigm in an ecological setting. *Experimental Brain Research, 231,* 37–49. https://doi.org/10.1007/s00221-013-3664-4.
47. Borroni, P., & Baldissera, F. (2008). Activation of motor pathways during observation and execution of hand movements. *Social Neuroscience, 3,* 276–288. https://doi.org/10.1080/17470910701515269.
48. Finisguerra, A., Maffongelli, L., Bassolino, M., et al. (2015). Generalization of motor resonance during the observation of hand, mouth, and eye movements. *Journal of Neurophysiology, 114,* 2295–2304. https://doi.org/10.1152/jn.00433.2015.
49. Senna, I., Bolognini, N., & Maravita, A. (2014). Grasping with the foot: Goal and motor expertise in action observation. *Human Brain Mapping, 35,* 1750–1760. https://doi.org/10.1002/hbm.22289.
50. Betti, S., Castiello, U., Guerra, S., & Sartori, L. (2017). Overt orienting of spatial attention and corticospinal excitability during action observation are unrelated. *PLoS ONE, 12,* 1–22. https://doi.org/10.1371/journal.pone.0173114.
51. Amoruso, L., & Urgesi, C. (2016). Contextual modulation of motor resonance during the observation of everyday actions. *Neuroimage, 134,* 74–84. https://doi.org/10.1016/j.neuroimage.2016.03.060.
52. Cavallo, A., Bucchioni, G., Castiello, U., & Becchio, C. (2013). Goal or movement? Action representation within the primary motor cortex. *European Journal of Neuroscience, 38,* 3507–3512. https://doi.org/10.1111/ejn.12343.

53. Sartori, L., Betti, S., Chinellato, E., & Castiello, U. (2015). The multiform motor cortical output: Kinematic, predictive and response coding. *Cortex, 70,* 169–178. https://doi.org/10.1016/j.cortex.2015.01.019.
54. Catmur, C., Gillmeister, H., Bird, G., et al. (2008). Through the looking glass: Counter-mirror activation following incompatible sensorimotor learning. *European Journal of Neuroscience, 28,* 1208–1215. https://doi.org/10.1111/j.1460-9568.2008.06419.x.
55. Catmur, C., Walsh, V., & Heyes, C. (2007). Sensorimotor learning configures the human mirror system. *Current Biology, 17,* 1527–1531. https://doi.org/10.1016/j.cub.2007.08.006.
56. Bizzi, E., D'avella, A., Saltiel, P., & Tresch, M. (2002). Modular organization of spinal motor systems. *Neurosci, 8,* 437–442. https://doi.org/10.1177/107385802236969.
57. Bernstein, N. A. (1967). *The coordination and regulation of movements*. Oxford: Pergamon Press.
58. Hilt, P. M., Berret, B., Papaxanthis, C., et al. (2016). Evidence for subjective values guiding posture and movement coordination in a free-endpoint whole-body reaching task. *Science Report, 6,* 23868. https://doi.org/10.1038/srep23868.
59. Grasso, R., Bianchi, L., & Lacquaniti, F. (1998). Motor patterns for human gait: Backward versus forward locomotion. *Journal of Neurophysiology, 80,* 1868–1885. https://doi.org/10.1007/s002210050274.
60. Levin, O., Wenderoth, N., Steyvers, M., & Swinnen, S. P. (2003). Directional invariance during loading-related modulations of muscle activity: Evidence for motor equivalence. *Experimental Brain Research, 148,* 62–76. https://doi.org/10.1007/s00221-002-1277-4.
61. Hilt, P. M., Cardellicchio, P., Dolfini, E., Pozzo, T., Fadiga, L. & D'Ausilio, A. (2020). Motor recruitment during action observation: Effect of interindividual differences in action strategy. *Cerebral Cortex*.
62. Romani, M., Cesari, P., Urgesi, C., et al. (2005). Motor facilitation of the human cortico-spinal system during observation of bio-mechanically impossible movements. *Neuroimage, 26,* 755–763. https://doi.org/10.1016/j.neuroimage.2005.02.027.
63. Hannah, R., Rocchi, L., & Rothwell, J. C. (2018). Observing without acting: A balance of excitation and suppression in the human corticospinal pathway? *Frontiers in Neuroscience, 12,* 1–10. https://doi.org/10.3389/fnins.2018.00347.
64. Hilt, P. M., Bartoli, E., Ferrari, E., et al. (2017). Action observation effects reflect the modular organization of the human motor system. *Cortex, 95,* 104–118. https://doi.org/10.1016/j.cortex.2017.07.020.
65. Palmer, C. E., Bunday, K. L., Davare, M., & Kilner, J. M. (2016). A causal role for primary motor cortex in perception of observed actions. *Journal of Cognitive Neuroscience*, 1–9. https://doi.org/10.1162/jocn_a_01015.
66. Bizzi, E., Cheung, V. C. K., Avella, A., et al. (2008). Combining modules for movement. *Brain Research Reviews, 57,* 125–133. https://doi.org/10.1016/j.brainresrev.2007.08.004.
67. Bizzi, E., Mussa-Ivaldi, F. A., & Giszter, S. F. (1991). Computations underlying the execution of movement: A biological perspective. *Science, 80*(25), 287–291.
68. Flash, T., & Hochner, B. (2005). Motor primitives in vertebrates and invertebrates. *Current Opinion in Neurobiology, 15,* 660–666. https://doi.org/10.1016/j.conb.2005.10.011.
69. Hilt, P. M., Delis, I., Pozzo, T., & Berret, B. (2018). Space-by-time modular decomposition effectively describes whole-body muscle activity during upright reaching in various directions. *Frontiers in Computational Neuroscience*.
70. Overduin, S. A., D'Avella, A., Carmena, J. M., & Bizzi, E. (2012). Microstimulation activates a handful of muscle synergies. *Neuron, 76,* 1071–1077. https://doi.org/10.1016/j.neuron.2012.10.018.
71. Holdefer, R. N., & Miller, L. E. (2002). Primary motor cortical neurons encode functional muscle synergies. *Experimental Brain Research, 146,* 233–243. https://doi.org/10.1007/s00221-002-1166-x.
72. Aumann, T. D., & Prut, Y. (2015). Do sensorimotor β-oscillations maintain muscle synergy representations in primary motor cortex? *Trends in Neurosciences, 38,* 77–85.

73. Hart, C. B., & Giszter, S. F. (2010). A neural basis for motor primitives in the spinal cord. *Journal of Neuroscience, 30,* 1322–1336. https://doi.org/10.1523/JNEUROSCI.5894-08.2010.
74. Kiers, L., Cros, D., Chiappa, K. H., & Fang, J. (1993). Variability of motor potentials evoked by transcranial magnetic stimulation. *Electroencephalography and Clinical Neurophysiology, 89,* 415–423.
75. Schmidt, S., Cichy, R. M., Kraft, A., et al. (2009). Clinical neurophysiology an initial transient-state and reliable measures of corticospinal excitability in TMS studies. *Clinical Neurophysiology, 120,* 987–993. https://doi.org/10.1016/j.clinph.2009.02.164.
76. Klein-Flügge, M. C., Nobbs, D., Pitcher, J. B., & Bestmann, S. (2013). Variability of human corticospinal excitability tracks the state of action preparation. *Journal of Neuroscience, 33,* 5564–5572. https://doi.org/10.1523/JNEUROSCI.2448-12.2013.
77. Van Elswijk, G., Maij, F., Schoffelen, J., et al. (2010). Corticospinal Beta-band synchronization entails rhythmic gain modulation. *Journal of Neuroscience, 30,* 4481–4488. https://doi.org/10.1523/JNEUROSCI.2794-09.2010.
78. Keil, J., Timm, J., Sanmiguel, I., et al. (2014). Cortical brain states and corticospinal synchronization influence TMS-evoked motor potentials. *Journal of Neurophysiology, 111,* 513–519. https://doi.org/10.1152/jn.00387.2013.
79. Guigon, E., Baraduc, P., & Desmurget, M. (2007). Computational motor control: Redundancy and invariance. *Journal of Neurophysiology, 97,* 331–347. https://doi.org/10.1152/jn.00290.2006.
80. Sporns, O., & Edelman, G. M. (1993). Solving Bernstein's problem: A proposal for the development of coordinated movement by selection. *Child Development, 64,* 960–981. https://doi.org/10.1111/j.1467-8624.1993.tb04182.x.
81. Fetz, E. E., Perlmutter, S. I., Prut, Y., et al. (2002). Roles of primate spinal interneurons in preparation and execution of voluntary hand movement. *Brain Research Reviews, 40,* 53–65. https://doi.org/10.1016/S0165-0173(02)00188-1.
82. Gentner, R., & Classen, J. (2006). Modular organization of finger movements by the human central nervous system. *Neuron, 52,* 731–742. https://doi.org/10.1016/j.neuron.2006.09.038.
83. D'Ausilio, A., Maffongelli, L., Bartoli, E., et al. (2014). Listening to speech recruits specific tongue motor synergies as revealed by transcranial magnetic stimulation and tissue-Doppler ultrasound imaging. *Philosophical Transactions of the Royal Society B, 369,* 20130418. https://doi.org/10.1098/rstb.2013.0418.
84. Bestmann, S., Harrison, L. M., Blankenburg, F., et al. (2008). Influence of uncertainty and surprise on human corticospinal excitability during preparation for action. *Current Biology, 18,* 775–780. https://doi.org/10.1016/j.cub.2008.04.051.
85. Stinear, C. M., Coxon, J. P., & Byblow, W. D. (2009). Primary motor cortex and movement prevention: Where stop meets go. *Neuroscience and Biobehavioral Reviews, 33,* 662–673.
86. Bestmann, S., & Duque, J. (2016). Transcranial magnetic stimulation: Decomposing the processes underlying action preparation. *Neuroscientist, 22,* 392–405. https://doi.org/10.1177/1073858415592594.
87. Neubert, F. X., Mars, R. B., Olivier, E., & Rushworth, M. F. S. (2011). Modulation of short intra-cortical inhibition during action reprogramming. *Experimental Brain Research, 211,* 265–276. https://doi.org/10.1007/s00221-011-2682-3.
88. Cardellicchio, P., Dolfini, E., Hilt, P. M., Fadiga, L. & D'Ausilio, A. (2020). Motor cortical inhibition during concurrent action execution and action observation. *NeuroImage, 208,* 116445.
89. Ficarella, S. C., & Battelli, L. (2019). Motor preparation for action inhibition: A review of single pulse TMS studies using the Go/NoGo paradigm. *Frontiers in Psychology, 10.* https://doi.org/10.3389/fpsyg.2019.00340.
90. Hilt, P. M., & Cardellicchio, P. (2018). Attentional bias on motor control: Is motor inhibition influenced by attentional reorienting? *Psychological Research,* 1–9. https://doi.org/10.1007/s00426-018-0998-3.

91. Seeley, T. D., Visscher, K. P., Schlegel, T., et al. (2012). Stop signals provide cross inhibition in collective decision making by honeybee swarms. *Science, 80*(335), 108–111. https://doi.org/10.1126/science.1210361.
92. Wessel, J. R., Jenkinson, N., Brittain, J. S., et al. (2016) Surprise disrupts cognition via a fronto-basal ganglia suppressive mechanism. *Nature Communications, 7*. https://doi.org/10.1038/ncomms11195.
93. Aron, A. R., Wessel, J. R., Voets, S. H. E. M., et al. (2016). Surprise disrupts cognition via a fronto-basal ganglia suppressive mechanism. *Nature Communications, 7,* 1–10. https://doi.org/10.1038/ncomms11195.
94. Wessel, J. R., & Aron, A. R. (2017). On the globality of motor suppression: Unexpected events and their influence on behavior and cognition. *Neuron, 93,* 259–280.
95. Duque, J., Labruna, L., Verset, S., et al. (2012). Dissociating the role of prefrontal and premotor cortices in controlling inhibitory mechanisms during motor preparation. *Journal of Neuroscience, 32,* 806–816. https://doi.org/10.1523/jneurosci.4299-12.2012.
96. Greenhouse, I., Sias, A., Labruna, L., & Ivry, R. B. (2015). Nonspecific inhibition of the motor system during response preparation. *Journal of Neuroscience, 35,* 10675–10684. https://doi.org/10.1523/jneurosci.1436-15.2015.
97. Friston, K. J. (2010). Is the free-energy principle neurocentric? *Nature Reviews Neuroscience,* 2–4. https://doi.org/10.1038/nrn2787-c2.
98. Summerfield, C., & De Lange, F. P. (2014). Expectation in perceptual decision making: Neural and computational mechanisms. *Nature Reviews Neuroscience, 15,* 745–756. https://doi.org/10.1038/nrn3838.
99. Schiffer, A.-M., Krause, K. H., & Schubotz, R. I. (2014). Surprisingly correct: Unexpectedness of observed actions activates the medial prefrontal cortex. *Human Brain Mapping, 35,* 1615–1629. https://doi.org/10.1002/hbm.22277.
100. Schiffer, A.-M., & Schubotz, R. I. (2011). Caudate nucleus signals for breaches of expectation in a movement observation paradigm. *Frontiers in Human Neuroscience, 5,* 1–12. https://doi.org/10.3389/fnhum.2011.00038.
101. Bond, C. F., Omar, A., Pitre, U., et al. (1992). Fishy-looking liars: Deception judgment from expectancy violation. *Journal of Personality and Social Psychology, 63,* 969–977. https://doi.org/10.1037/0022-3514.63.6.969.
102. Frank, M. G., & Ekman, P. (1997). The ability to detect deceit generalizes across different types of high-stake lies. *Journal of Personality and Social Psychology, 72,* 1429–1439. https://doi.org/10.1037/0022-3514.72.6.1429.
103. Sebanz, N., & Shiffrar, M. (2009). Detecting deception in a bluffing body: The role of expertise. *Psychonomic Bulletin & Review, 16,* 170–175. https://doi.org/10.3758/PBR.16.1.170.
104. Wolpert, D. M., Doya, K., & Kawato, M. (2003). A unifying computational framework for motor control and social interaction. *Philosophical Transactions of the Royal Society of London, 358,* 593–602. https://doi.org/10.1098/rstb.2002.1238.
105. Buch, E. R., Mars, R. B., Boorman, E. D., & Rushworth, M. F. S. (2010). A network centered on ventral premotor cortex exerts both facilitatory and inhibitory control over primary motor cortex during action reprogramming. *Journal of Neuroscience, 30,* 1395–1401. https://doi.org/10.1523/JNEUROSCI.4882-09.2010.
106. Kilner, J. M., Friston, K. J., & Frith, C. D. (2007). The mirror-neuron system: A Bayesian perspective. *NeuroReport, 18,* 619–623. https://doi.org/10.1097/WNR.0b013e3281139ed0.
107. Cardellicchio, P., Hilt, P. M., Olivier, E., et al. (2018). Early modulation of intra-cortical inhibition during the observation of action mistakes. *Science Report, 8,* 1784. https://doi.org/10.1038/s41598-018-20245-z.
108. Kujirai, T., Caramia, M. D., Rothwell, J. C., et al. (1993). Corticocortical inhibition in human motor cortex. *Journal of Physiology, 471,* 501–519.
109. Friston, K. J. (2005). A theory of cortical responses. *Philosophical Transactions of the Royal Society B, 360,* 815–836. https://doi.org/10.1098/rstb.2005.1622.
110. Franca, M., Turella, L., Canto, R., et al. (2012). Corticospinal facilitation during observation of graspable objects: A transcranial magnetic stimulation study. *PLoS One*, *7*. https://doi.org/10.1371/journal.pone.0049025.

111. Andersen, R. A., & Cui, H. (2009). Review intention, action planning, and decision making in parietal-frontal circuits. *Neuron, 63,* 568–583. https://doi.org/10.1016/j.neuron.2009.08.028.
112. Donnarumma, F., Costantini, M., Ambrosini, E., et al. (2017). Action perception as hypothesis testing. *Cortex, 89,* 45–60. https://doi.org/10.1016/j.cortex.2017.01.016.

Chapter 3
Beyond Automatic Motor Mapping: New Insights into Top-Down Modulations on Action Perception

Alessandra Finisguerra, Lucia Amoruso, and Cosimo Urgesi

Abstract Our ability to recognize other people's actions is central to everyday life. Transcranial magnetic stimulation (TMS) studies have shown that observing an action induces activity in the observer's motor system that replicates the muscle selectivity, direction and temporal profile of the observed movement when executed. This motor resonant activity has long been presumed to reflect an inner, automatic replica of the observed movement. However, recent empirical evidence has challenged this view by showing that motor resonance can be tuned to high-level features, such as the overarching goal and intention of the observed action, while being simultaneously influenced by top-down contextual factors. Interestingly, current predictive coding models provide a mechanistic account to explain how action recognition is achieved, stressing the role of prior expectations and the interplay of bottom-up and top-down processes in terms of matching and mismatching predictions across hierarchical levels of action representation. In this chapter, we first provide an overview of seminal TMS findings that point to the characterization of motor resonance as an automatic fine-grained simulation of the observed movement. Second, we discuss more recent sources of evidence supporting the notion of motor resonance as a flexible phenomenon, stressing the role of top-down modulations during action perception.

Alessandra Finisguerra and Lucia Amoruso contributed equally.

A. Finisguerra (✉) · C. Urgesi
Scientific Institute, IRCCS E. Medea, Pasian di Prato, Udine, Italy
e-mail: alessandra.finisguerra@lanostrafamiglia.it

L. Amoruso
Basque Center on Cognition, Brain and Language, San Sebastian, Spain

IKERBASQUE, Basque Foundation for Science, Bilbao, Spain

C. Urgesi
Laboratory of Cognitive Neuroscience, Department of Languages, Literatures, Communication, Education and Society, University of Udine, Udine, Italy

N. Noceti et al. (eds.), *Modelling Human Motion*,
https://doi.org/10.1007/978-3-030-46732-6_3

3.1 Introduction

The ability to understand others' intentions via observing their actions in naturalistic contexts constitutes the bedrock of social cognition. Indeed, what we can afford from others' actions is more than the way in which they are executed (i.e., the movement kinematics pattern). Observing others' movements allows the understanding of their intentions and mental states and drives the observer to behave in a supportive, interactive or defensive way. Even when very poor visual information of moving persons is provided, the visual perception of biological motion [42] per se has been shown to allow understanding their mental states[19] and to increase the activation of those areas, in an observer's brain, that are considered to be crucial for action understanding [62, 73]. Importantly, minimal contextual cue can affect the way in which a motion pattern is perceived [71]. On this view, movements are never perceived in isolation but rather context-embedded, with objects, actors and the relationships amongst them "gluing together" into a unified representation.

Action understanding is considered to be supported by the activation of the so-called mirror neuron system (MNS; [59]), a collection of fronto-parietal regions which become active during both observation and execution of similar actions. While primarily discovered in monkeys [56], MNS-like activity has also been shown to be present in humans, from newborns to the elderly [48, 46]. Infants are indeed able to recognize and respond to social signals since birth. Moreover, the evidence that neonatal imitation is present from birth has been considered as reflecting the presence of a mechanism that matches the observed facial gestures with the internal motor representation of the same action, thus highlighting a mirror-like coupling between the observed action and the motor code to produce the same action [65]. Importantly, different techniques spanning form single-cell recording in implanted patients [54] to electro- and magneto-encephalographic [13, 35, 57], neuroimaging [52] and non-invasive brain stimulation methods [15, 25, 37, 75] allow to investigate the occurrence and the functional meaning of mirror-like responses during action observation, in a more direct way. Here, we focus on studies measuring motor-evoked potentials (MEPs) induced with transcranial magnetic stimulation (TMS) in peripheral muscles, which reflect the level of corticospinal excitability (CSE). Using this method, several studies have shown higher CSE, thought to reflect higher activation in the primary motor cortex (M1), in response to observed human actions, as compared to action-unrelated control conditions [24]. We refer to this action-specific CSE increase as motor resonance as it reflects an index of mirror-like activity in the observer's motor system during action observation.

Briefly, motor resonance can be featured by, at least, three core elements: muscle-specificity, direction and timing of the modulation. First, muscle-specificity during action observation implies a change in the activation of the cortical representation of the muscles that are specifically involved in the execution of the observed action [25, 67]. Second, the direction of the modulation consists of an increase or decrease in CSE during action observation. The direction could mirror, or not, the modulation of muscle activation during action execution [60, 76] as well as reflect the final balance of simulative and inhibitory mechanisms that are necessary for selecting or withdrawing unwanted imitation responses [78]. Lastly, the timing of the modulation depends upon the delay at which motor resonance occurs, with respect to action observation onset. In this vein, a CSE modulation occurring immediately after the observation of an action is taken as a marker of the automatic simulation and the faithful covert replica of the observed movement. Time-locked modulations can be assessed by recording MEPs at different time points during action observation [5, 18, 51].

Based on the presence of somatotopic, direction-specific and time-locked effects, motor resonance has been traditionally conceived as an inner replica of the observed movements [55], thus reflecting the automatic mapping of what, according to the hierarchical model of action representation [33], have been defined as *low-level* motor aspects. However, recent evidence from TMS studies suggests that motor resonance may be actually a more flexible phenomenon than previously thought. Briefly, these studies show that the more we know about the interacting partner or about the scenario in which an action is taking place, the more the likelihood of understanding the meaning of the observed action. Furthermore, when movements are observed embedded in richer contexts (i.e., where the information at hand allows representing the goal or the intention behind the observed action), the observed kinematic seems to be mirrored in a less specific extent. Thus, motor resonance becomes prone to top-down modulations and switches across different representational levels.

Before considering how these high-level factors can drive action understanding, it is necessary to understand how actions can be described and hierarchically represented, and thus, what these factors are. According to Hamilton and Grafton [33], actions can be described and thus understood [45] at four different levels of a hierarchy. Starting from the bottom, these levels are as follows: (i) the muscle, which codes for the pattern of muscular activity required to execute the action; (ii) the kinematics, which maps the movements of the effectors in time and space; (iii) the goal, which includes the short-term transitive or intransitive aim; and (iv) the intention, which includes the long-term purpose behind the action. By considering these levels of action representation, generative models of action prediction [45] postulate a flow of predictions about what sensory consequences (i.e., the low action representation levels) are the most likely given the hypothesized goal and intention. For example, let us consider that we are observing someone grasping a mug by its handle (see Fig. 3.1a). Given a prior expectation about the intention underlying the observed action (e.g., to drink), we can predict, on the basis of our own motor representations, the proximal goal (e.g., bringing the mug toward the mouth). Moving further in the hierarchy, based on the goal information, we can estimate the upcoming movement kinematics (e.g., trajectory of the hand toward the object and their interaction). Lastly, the lowest level would involve the synergic muscular activation required to attain the movement. From an action-oriented predictive coding approach [45], given an observed action, the MNS allows intention comprehension by generating for each hierarchical level top-down predictions (priors) about lower levels of action representation. When the priors and the incoming sensory information match each other, the overarching intention becomes clear; otherwise, a feedback is sent back to higher

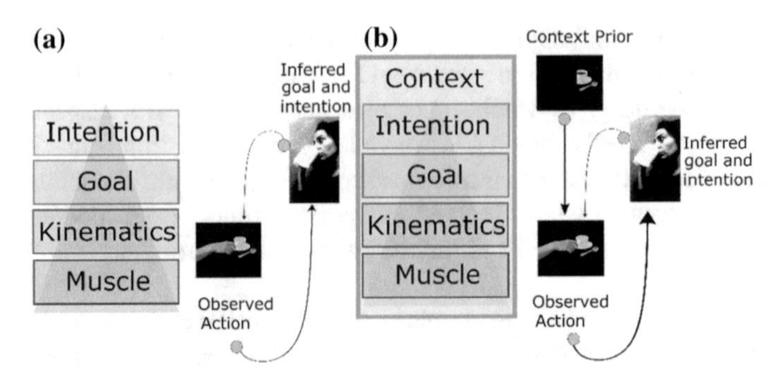

Fig. 3.1 **a** Intention understanding within a hierarchical model of action representation. **b** Extension of the hierarchical model of action understanding including the fifth level of contextual prior

levels to adjust the initial prediction and minimize the error signal. Interestingly, it has been proposed that, in addition to these well-known four levels, context can be seen as a fifth top-down level guiding action comprehension under situations of perceptual uncertainty [44], namely, when similar perceptual kinematics are at the service of different intentions (see Fig. 3.1b). A well-known thought experiment to exemplify this situation is the one provided by Dr. Jekyll and Mr. Hyde [40]. Let us suppose that Dr. Watson witnesses this (or these) character(s) reaching and grasping a scalpel. By observing the way in which the scalpel is grasped, would it be possible for Dr. Watson to recognize which intention underlies the observed action? In other words, the person grasping the scalpel is Dr. Jekyll trying to cure or Mr. Hyde intending to kill? According to current predictive coding views, in these ambiguous cases, contextual cues would aid intention recognition. For instance, if Dr. Watson witnesses this action in an operating theater, it is more likely that Dr. Jekyll's intention is "to cure" [9]. On this view, contextual knowledge would aid action recognition by signaling which intentions are more likely to drive upcoming actions given the information present in the environment, forming the basis to estimate lower level aspects of action representation (i.e., kinematics).

In this chapter, we show how in some cases motor resonance corresponds to the inner simulation of the observed movement, while, in other cases, top-down modulation makes motor resonance a less faithful replica of the observed action. More specifically, taking into account the hierarchical model of action representation, we start analyzing seminal studies on mirror responses during action observation finding a more or less strict similarity between CSE modulation and the low-level motor aspects of the observed action. In this view, the role of expertise in modulating motor resonance is also considered. Then, going up across the hierarchy, we move to examine those studies testing how this low-level mirroring could be affected by other high-level factors, such as expectations, or dissociated from the higher representational levels (e.g., the goal and the intention of an action). Finally, we present the most representative and recent studies showing the occurrence of top-down contextual effects, in which factors corresponding to internal and external milieus influence the way in which actions are perceived and interpreted.

3.2 Kinematics Representation

After Fadiga et al. [25]'s seminal study showing motor resonance during action observation in humans, Gangitano et al. [30] reported a muscle-specific increase in CSE for finger opening movements which reflected the increased amount of aperture coherently with the observed movement phase. Accordingly, phase and muscle-specific modulations were further shown during the observation of either transitive movements [53] as well as during the observation of intransitive cyclic flexion and extension movements of the wrist [12]. Notably, the finding of CSE modulations for intransitive actions challenged the initial view, grounded on monkey studies, that the MNS primarily encodes the action goal rather than the observed movements [29, 74]. Beside muscle-specificity, motor resonance has been shown to be sensitive to less salient changes in kinematic signals, such as muscle contraction and force requirements during action execution. A series of studies considered the effects of observing object-lifting actions on motor resonance. For instance, Alaerts et al. [2] found that the amount of facilitation for action observation was congruent with the degree of muscular involvement in action execution. Indeed, while keeping constant the observed action (i.e., grasping-and-lifting-the-object) but changing the object weight, greater facilitation for heavy than light object lifting was found. Moreover, the observation of either precision or power grasp-to-lift actions resulted in a weight-dependent muscle-specific motor resonance. In a similar vein, enhanced muscle-specific motor resonance for abduction finger movements was influenced by onlooker's hand orientation [49] and position [76] congruently with the maximal activation of the same muscles during action execution under different postures. While favoring the low-level hypothesis, according to which motor resonance is a covert mimicry of the observed movement, these findings opened the debate about which action level is motor resonance coding for.

A first break into the wall of the "covert mimicry hypothesis" came from studies investigating the degree of correspondence between the observed movement and the motor representation in the observer. First of all, Gangitano et al. [31] found that, while CSE facilitation of finger muscles peaked at the maximal finger aperture of a grip observed in its typical time course, it was rather suppressed when the maximal finger aperture unexpectedly occurred soon after the observation of a closed hand (but see [32]). Furthermore, somatotopic modulation of CSE was found during the observation of not only possible but also impossible fingers movements [60], which could not be executed by the observer (and thus could not be mimicked), but could be matched with the possible version for direction of movement in space. Finally, somatotopic CSE facilitation was obtained from finger muscles not only during observation of an actual movements but also during observation of a static image of a hand implying a grasping movement [76]. Critically, maximal CSE facilitation was not obtained when the hand was in a movement phase with maximal contraction of the finger muscles but rather at the initial phase of the movement [77]. This suggested that CSE facilitation does not merely map what is being shown to the observer, but it represents what can be anticipated by the observer on the basis of the observed information. In favor of this predictive view of motor resonance are also the findings of a recent study [66] showing that CSE of a mouth muscle is facilitated during observation of a grasping-to-drink versus grasping-to-poor action already at the contact of the hand with the bottle, thus before any mouth action could be observed. Thus, motor resonance allows representing the future deployment of an action or even the expected action in a chain, going beyond what has been shown to the observer. In fact, an even greater CSE facilitation during observation of implied action images was obtained after interfering with the activity of a visual area devoted to the analysis of biological motion [10], suggesting that the less visual information is available, the more the observer's motor system needs to fill-in missing information.

The use of motor representations to disambiguate action perception was also demonstrated by the investigation of the role prior experience and/or familiarity with the observed action in modulating motor resonance [1, 14]. For example, Aglioti et al. [1] found a greater facilitation of finger muscles for the observation of the initial phases of "out" as compared to "in" shots to a basket; this, however, occurred only for basketball players and not for players of other sports or for sports journalists, who may have visual experience but not direct motor experience basketball actions. Similar results were obtained for the observation of soccer penalty kicks comparing field-players and goalkeepers: the first ones, having direct motor experience with penalty kicks, showed greater motor resonance than the latter ones, having direct visual but not motor experience, particularly when observing fainted kicks [70]. In a different domain, Candidi et al. presented expert pianists and naïve controls with videos displaying a professional pianist that could perform fingering errors while playing musical scales. Despite the fact that non-pianist controls had previously received a visual training to recognize the errors in the videos, only piano experts showed a somatotopic modulation in the *abductor pollicis brevis*, the muscle that was actually involved in the execution of the piano fingering errors. Specifically, they observed increased CSE 300 ms after error onset. Overall, the findings suggest that prior motor but not visual experience is necessary to induce motor resonance. While demonstrating that prior motor experience with the observed action provides a fine-grained simulative error monitoring system to evaluate others' movements, these studies suggest that high-level information (i.e., movement correctness) can influence motor resonance. In this respect, in spite of the supposed innateness of MNS, its plasticity allows being sensitive to previous experience and enables anticipatory motor representations.

3.3 Goal Representation

The aforementioned studies did not allow dissociating the contribution of kinematics versus goal encoding, when this last was present, leaving the controversy about the level of action representation unsolved. In this regard, Cattaneo et al. [16] designed a paradigm to dissociate the contribution of kinematics versus goal encoding. Participants observed a hand manipulating normal or reverse players without an evident goal or, alternatively, with the goal to grasp an object. Crucially, during the observation of grasping action performed with these pliers, the action goal level (grasping an object) could be isolated from finger closing or opening movements necessary to achieve that goal. Indeed, grasping the object with the normal plier would imply a closing movement of the hand, while performing the same action by using the reverse plier would require opening movement of the hand to operate pliers. Besides these conditions, the activation of the *Opponens Pollicis*, a muscle involved in thumb opposition for finger closing movement, was measured during the observation of similar movement of the pliers but devoid of a goal [16]. When no goal was present, higher activation of the *Opponens Pollicis* was found during observation of closing than opening hand movements, suggesting that motor resonance was mirroring the hand movement. Conversely, when a goal was present, motor resonance no longer reflected the observed hand movement but rather the motor goal. Indeed, higher activation of the *Opponens Pollicis* was observed during the observation of opening hand movements (and thumb extension) required to attain the goal of closing the reverse pliers. While suggesting an incorporation of the tool into the body representation and the possibility for motor resonance to be shaped accordingly, results suggested that high-level features, such as the goal of an action, beside low-level kinematics, modulate motor resonance.

Although divergent findings to this conclusion have been reported [17], the integrated contribution of kinematics and goal coding depending on the information about the action goal has been widely supported. Accordingly, Mc Cabe et al. [51] found that observing the grasping of small or big objects induced a kinematic-specific modulation of CSE (in terms of time course and muscle involvement), which was coupled by a goal-specific modulation only when the goal (i.e., to be grasped object) could be inferred from the initial part of the movement. Conversely, when the motor goal was ambiguous (i.e., switched online between objects), CSE modulation mirrored low-level kinematics only. In this study, the amount of visual information provided and the time at which motor resonance was recorded was crucial in biasing the modulation toward either lower- or higher-level modulation of motor resonance. Thus, when the goal was present and the information provided allows representing the action at higher-level, motor resonance switches from low to the high representational level. In the absence of this information, motor facilitation mimics the observed kinematics (see also [4]).

In line with this view are also the results from a study by Betti et al. [11]. In this study, participants observed videoclips showing an index finger kicking a ball into the goal with (symbolic kick) or without (finger kick) wearing a miniaturized soccer shoe; they also observed finger movements without any contextual information (biological movement). Findings revealed a muscle-specific CSE facilitation for an index finger muscle during the observation of all conditions, thus reflecting a low-level mapping of the kinematic profile. Importantly, CSE from a leg muscle was facilitated, until being similar to the finger muscle activation, only when participants observed the symbolic action, which mimicked a leg action even if performed with the finger. This pointed to a simultaneous representation of symbolic action and movement kinematics in the observer's motor system. Differently stated, when semantic representation of the observed action is possible, motor resonance switches from low- to high-level features mapping, showing a generalization between muscles. Likewise, Finisguerra et al. [28] reported that the increased CSE recorded from a forearm flexor muscle (that is involved in closing the hand) during the observation of intransitive closing rather than opening movements of the hand generalized across different effectors (i.e., eyelid, mouth) observed during execution of similar closing versus opening movements [28]. It is likely that the common action goal, which could be inferred from the observed set of stimuli, contributed to this high-level mapping of action meaning. Additional evidence comes from Senna et al. [63]'s study, in which familiarity with an observed action elicited a shift from lower- to higher-level motor mapping. Specifically, participants viewed typical hand actions (i.e., grasping a pencil) and typical foot actions (i.e., pressing a pedal) that could be performed by either a hand or a foot effector, resulting in familiar or unfamiliar actions when performed with typical or atypical effector, respectively. Observing unfamiliar actions resulted into an effector-specific modulation of hand and foot muscle CSE. Conversely, during the observation of familiar actions, CSE modulation of the muscle involved in the represented action generalized across both effectors. This evidence hence suggests that actions can be coded either in a somatotopic low-level or in a goal high-level fashion, depending on the familiarity with the observed action.

3.4 Intention Level Representation

As discussed above, observing object-lifting actions modulated motor resonance according to the object weight and the role of the recorded muscle in action execution [2]. Despite this modulation might resemble the gradually increasing activation of these muscles in action execution, these findings only partially support the low-level coding hypothesis. To be more specific, in this study [2], participants were asked to observe lifting actions performed upon different objects, whose weight could be easily inferred from their appearance. For example, participants could observe lifting actions toward an empty, half-full, or full bottle, or toward a brass balance weight or a ribbon cable, thus toward objects that clearly had different weight. For this reason, they did not allow clarifying whether the force-related effects on motor resonance were driven by the observation of the low-level kinematics of lifting actions per se or by high-level expectations triggered by the intrinsic properties of the lifted objects. In a follow-up study [3], the role of different visual cues (i.e., kinematic profile, hand contraction and intrinsic object properties) contributing to motor resonance during object lifting was separately tested. In this study, a kinematic profile without other confounding effects was provided through the lifting movement of a hand wearing a glove. Information about hand contraction void of other visual cues was obtained by showing static images of a hand exerting maximal or minimal isometric contraction upon the same object. The role of intrinsic object properties was tested by showing lifting action toward heavy object having the appearance of a heavy or a light object; or toward light object with the appearance of a light or a heavy one. Even if the weight information carried by movement kinematics and hand contraction modulated motor resonance in finger and forearm muscles, when movement kinematics conflicted with object appearance, the weight-dependent modulation of motor resonance was reduced. While partially supporting a low-level coding of observed action, we interpret these findings as evidence for a contribution of observers' expectations (e.g., triggered by object properties) in interfering with the weight-dependent modulation of motor resonance. Other findings support this view. For instance, in Senot et al. [64], participants were asked to observe lift-to-place actions performed upon: (i) a transparent bottle that could be full of sand or empty; (ii) an opaque bottle that could be full of sand or empty; (iii) an opaque bottle full of sand that could be labeled as "heavy" or as "light." In this study, the observation of lift-to-place actions directed to heavy or light objects led to a weight-dependent activation of an index finger muscle, regardless of the object intrinsic properties when the content of a bottle was visible or hidden from view, suggesting that movement kinematics was enough to modulate CSE. However, when explicit congruent or incongruent semantic cues about object weight were provided by verbal labels, the weight-dependent modulation ceased, suggesting the role exerted by expectations on motor resonance phenomena.

The relevance of low- and high-order factors in shaping motor resonance was further confirmed by a set of subsequent studies dealing with object-weight discrimination [27, 69]. In Tidoni et al. [69], motor resonance modulations were assessed during the observation of reach-to-lift actions directed to either light or heavy objects that could be performed with a genuine or a deceptive intention. Participants were asked to observe these videoclips and to indicate whether the actors of the videoclips were moving with a genuine or a deceptive intention. Even if motor resonance was greater for heavy than light object grasping, thus replicating the weight-dependent modulation of motor resonance described in previous study [2], the authors also found that observing deceptive actions facilitated an index finger muscle CSE more than observing genuine actions, regardless of object weight. While showing that motor resonance increased during the observation of deceptive action regardless from the object weight, results from this study still did not allow ascertaining whether low-level features (i.e., the altered kinematic pattern that is required to deceive the observer about the object weight) or high-level aspects (i.e., deceptive intention) could explain the increased CSE for deceptive action. Therefore, in a subsequent study, Finisguerra et al. [27] sought to test whether the modulation of the observer's motor system during observation of deceptive actions [69] mirrors the mapping of kinematic adaptation required to fool an observer (low-level mapping) or the decoding of actor's intention (high-level coding). To this aim, participants were asked to predict the weight of cubes lifted by actors who received truthful information on the object weight and provided (1) truthful (truthful actions) or (2) deceptive (deceptive actions) cues to the observers (thus in keeping with the Tidoni et al.'s manipulation) or (3) who received fooling information and were asked to provide truthful cues (deceived actions). This way, actor's intentions and kinematic adaptations were independently manipulated. While the observation of deceptive actions facilitated CSE in a muscle-independent fashion, observing kinematic alterations determined by genuine intentions induced a

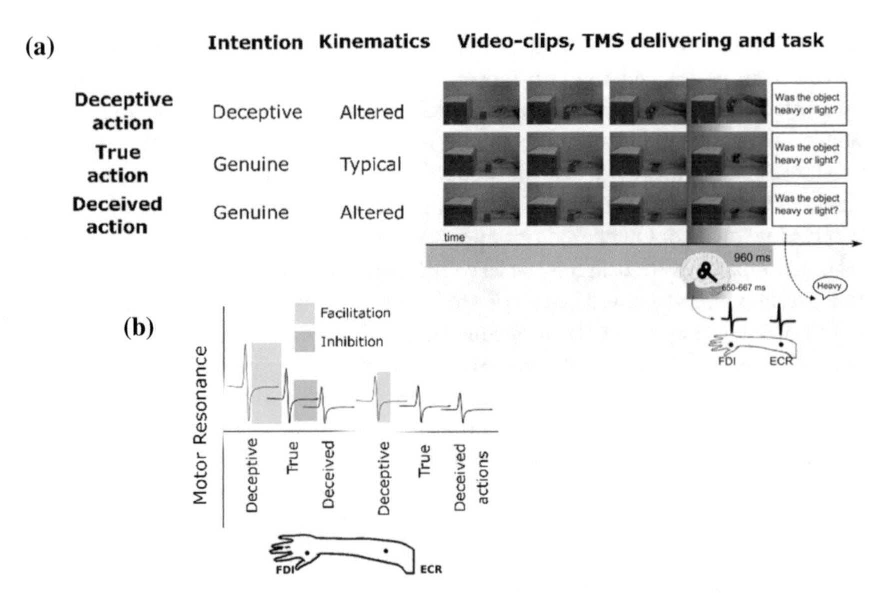

Fig. 3.2 **a** Schematic representation of different action observation conditions and of the paradigm to investigate the role of deceptive intention and kinematics alterations in the motor mapping of deceptive actions. **b** Facilitatory or inhibitory effects on motor resonance during the observation of actions driven by deceptive or genuine intentions with kinematics adaptations with respect to genuine actions without kinematics alterations (Adapted from [27])

muscle-specific CSE inhibition that resembled the pattern of muscle activation during AE in the same condition. Overall, both low-level and high-level features were mirrored into the observer's motor system in a dissociable fashion (see Fig. 3.2). Thus, these findings showed that MR mirrors force-related modulation only when no additional information about the object intrinsic and extrinsic properties in the observed scene was present. As soon as a conflict between the observed action and the expectations due to internal representations of the object properties occurs, or when the expectations about object properties are diverted through deceptive intention, high-level rather than low-level features shape motor resonance.

3.5 Contextual Modulations

As mentioned in the introduction of this chapter, it has been proposed that, in addition to the well-known four levels (i.e., muscle, kinematics, goal and intention) in which an action can be represented [33], context can serve as a fifth top-down level guiding action comprehension under situations of perceptual ambiguity [44, 45]. On this view, context-based information would aid action recognition by signaling which intentions are more likely to drive upcoming actions. On this view, the intention that is inferred from action observation now depends also upon prior information received from the contextual level.

Mounting evidence from TMS studies suggests that motor resonance can be modulated by a wide range of high-level contextual factors. Indeed, human actions do not occur in isolation but rather embedded in internal (i.e., observer's values, temperament, personality traits and previous experience) and external contexts (i.e., surrounding objects and their affordances, constraints and opportunities provided by environmental cues).

On one hand, recent studies reporting top-down modulations involving internal factors have suggested that they play a critical role during action observation. For instance, individuals with higher score in the harm avoidance trait (i.e., a personality trait characterized by excessive worrying and being fearful) exhibited lower motor resonance during the observation of immoral as compared to neutral actions (i.e., stealing a wallet vs. picking up a notepaper, respectively) containing similar movement kinematics [47]. In another study, Craighero and Mele [20] have recently reported that the observation of an agent performing an action with negative (i.e., unpleasant) consequences on a third person results in decreased motor resonance as compared to the observation of actions underpinning positive and neutral intentions with equal kinematics. Nevertheless, this result was independent of the personality traits of the participants. Overall, these studies suggest that individual's ethical values modulate action coding, with immoral and/or a negative intentions leading to a suppression of the ongoing motor simulation.

The observer's current state also plays a critical role during action observation. For instance, Hogeveen and Obhi [39] found that, during the observation of human and robotic actions (i.e., a human hand or grabber reaching tool squeezing a ball, respectively), participants previously involved in a social interaction with the experimenter, showed increased CSE for the observation of human actions as compared to robotic ones. This effect was absent in those individuals not previously engaged in the social interaction, with human and robotic actions triggering similar levels of CSE. In a more recent study, Hogeveen et al. [38] reported that CSE triggered by the observation of a hand squeezing a ball becomes differentially modulated after participants being exposed to a low- or a high-power induction priming procedure. During the power priming procedure, participants were asked to write an essay documenting a low-, neutral-, high-, power experience, thus recalling a memory in which someone else had power over the observer or in which the observer had power over someone else. Participants in the high-power group showed less motor resonance facilitation relative to those in the low-power group, suggesting that people in positions of power display reduced interpersonal sensitivity and diminished processing of social input. In a similar vein, even the hierarchical status of an observer in the virtual world modulates his/her motor resonance to observed actions, with followers on social networks showing greater motor resonance than the individuals who are more followed [26]. All in all, both studies indicate that action perception is modulated by prior social interactions.

On the other hand, parallel top-down modulations have been observed when considering external contextual factors. In a series of studies, Amoruso et al. [5, 7]

explored the role of contextual information in modulating action coding at low levels of representation (i.e., muscle and kinematics). CSE was measured, while participants were asked to observe an actor model performing everyday actions embedded in congruent, incongruent or ambiguous contexts and to recognize actor's intention. Context-action congruency was manipulated in terms of the compatibility between grasping movement kinematics and the action setting. For instance, within a breakfast scenario (i.e., a cup full of coffee), the actor could grasp the cup by its handle with a precision grip (congruent condition) or with a whole-hand grip from the top (incongruent condition). Finally, in ambiguous contexts, the cup was half-full of coffee, and thus, both types of grip were equally plausible. Videos were interrupted before action ending, and participants were requested to predict action unfolding. Specifically, two possible descriptors (i.e., to drink and to move) were presented, and participants had to select which was the actor's more likely intention, given contextual and kinematics information. As compared to the ambiguous condition, the congruence between the movements and the context increased CSE at early stages (~250 ms after action onset), while incongruence between them resulted into a later inhibition (>400 ms) for the index finger muscle, which is known to be involved in the observation and execution of reaching-to-grasping movements (see Fig. 3.3). Crucially, the different time course and direction (i.e., facilitation vs. inhibition) of

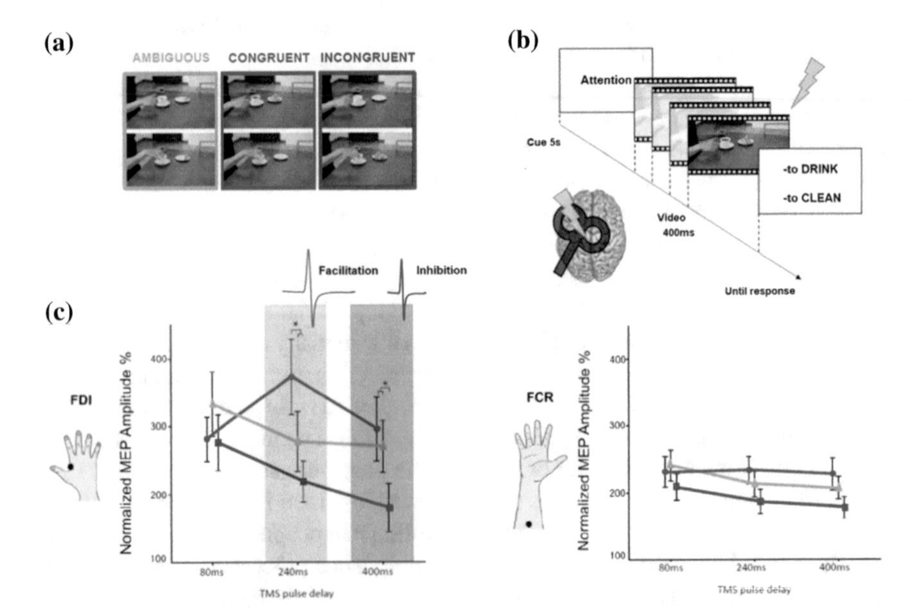

Fig. 3.3 **a** Schematic representation of actions embedded in congruent, incongruent and ambiguous contexts; **b** Experimental paradigm to investigate the role of contextual congruency on the motor mapping of observed action kinematics. **c** Early facilitatory effects on motor resonance during the observation of actions embedded in a congruent context and later inhibitory effects on motor resonance during the observation of actions embedded in an incongruent context (Adapted from [5])

the observed effects suggest that they stem from partially independent mechanisms, with the early facilitation directly involving simulative motor resonance through the classical action observation network, and the later inhibition recruiting structures outside of this network conveying information about the intention estimated from the context. Additional evidence from a role of top-down contextual modulation on motor resonance comes from two recent studies. In the first one, Riach et al. [58] used a similar logic but introduced a baseline condition in which actions were observed without a context. Similar to Amoruso et al. [5]' findings, observation of actions within congruent contexts (i.e., pinching a sponge in a kitchen background) facilitated CSE as compared to baseline. However, no CSE modulations resulted from observing the actions in incongruent contexts. In a second study by Cretu et al. [21], participants observed either full or occluded videos of an actor grasping and lifting a jar using a precision or a whole-hand grip. Color cues preceded observation trials and were manipulated in term of their informativeness in predicting the upcoming action. Overall, the authors found that even in the absence of movement kinematics information (i.e., occluded condition), contextual reliable cues were sufficient to trigger a muscle-specific response in the observer. Nevertheless, when presenting both sources of information together (i.e., kinematics and context), CSE facilitation became stronger than when either source was presented alone. These findings support the view that bottom-up simulative motor mapping triggered by observed kinematics and top-down contextual information interact in the observer's motor system.

Regarding the inhibitory effects on CSE reported for contextual conflicting information [5, 7], similar findings were reported by Janssen et al. [41]. They showed that incongruence between an action specified by a prior symbolic cue (i.e., an arrow indicating the requirement of a whole-hand grip) and the observed action (i.e., movement implying a precision grip) resulted in a reduction of motor resonance for the observed action, with CSE replicating the motor pattern of the action specified by the prior cue. Likewise, Mattiassi et al. [50] found that the observation of hand movements preceded by an incongruent motor-related masked prime (e.g., a different hand movement) led to a comparable drop in motor resonance.

Last but not least, another aspect that has been shown to modulate motor resonance is the social nature of the context in which actions are observed. In the study by Sartori and colleagues [61], the authors recorded MEPs, while participants observed action sequences that could call for a complementary response from the observer or not. For instance, participants watched videos of a model grasping a thermos using a whole-hand grip to pour coffee into three mugs located next to her. Afterward, the model could pour coffee into a fourth mug located in perspective, closer to the participant (social condition) or returned to the initial position (non-social condition). Crucially, the coffee mugs could either depict a handle or not, and hence, a person intending to pick it up would need to use a precision or a whole-hand grip, respectively. Interestingly, greater increase of CSE was found for social than for non-social contexts. In those cases where the social requests demanded the use of a whole-hand grip (i.e., a mug without handle), increased CSE in the *abductor digiti minimi*, a muscle involved in this type of grasping, was observed. Conversely, its activity decreased when the request demanded the use of a precision grip (i.e., a mug with a handle). Furthermore, this effect took place during the last part of the video, namely, when the complementary social request could potentially occur, while at the beginning of the video, CSE mimicked model's kinematics. These findings point to the fact that social contexts can induce a modulation in the observer's motor system (i.e., shifting from emulation to reciprocity) that is consistent with the intention to accept the request (i.e., grasping the mug) rather than with the tendency to resonate with the observed action (i.e., grasping the thermos).

Overall, these studies provide striking evidence for top-down effects in action perception, underscoring the involvement of a wide range of internal and external contextual factors in modulating this process and overall supporting a flexible view of the motor resonance phenomenon [36, 43].

3.6 Conclusions

Motor resonance has long been presumed to reflect an automatic inner replica of observed movements. In the present chapter, in light of recent empirical evidence supporting the interplay between high- and low-level aspects of action coding, we provide a broader view on this phenomenon accounting for the top-down influences on action perception. The overview of findings discussed here indicates that when experimental designs allow dissociating between the kinematics and the goal/intention levels or when naturalistic information about the context in which actions occur is available, a shift from low- to high-level mapping becomes evident in the observer's motor system, leading to a dissociation between the observed kinematics and the observer's motor activations.

Furthermore, when considering the timing of these top-down modulations, it seems that they arise from around ~250 ms post-movement onset but not before. This is in line with a recent two-stage model proposed by Ubaldi et al. [72] suggesting that a fast bottom-up process mediated by the dorsal action observation network would produce an early automatic simulative response before 250 ms, while top-down processes would be mediated by a slower system relying on the prefrontal cortex, reflected in later CSE modulations occurring 300 ms after movement onset. Future studies need to take advantage of the high-temporal resolution of TMS to investigate the deployment of motor resonance in the observer's motor system and to unravel the possibly multiple levels of coding of others' behaviors.

All in all, the major conclusion of this chapter is that motor resonance is more dynamic and flexible than typically thought and that different sources of top-down influences can impact on action perception, determining the way in which others' behavior is interpreted. At a mechanistic level, these effects can be broadly grouped in terms of the agreement across levels of action representation, with CSE facilitation representing compatible information across levels (i.e., movement kinematics and contextual cues hinting to the same motor intention) and CSE suppression the incompatible one. Thus, the status of CSE modulation seems to ultimately reflect the highest-level representation of the final intention of another person. When no other information is available (as in the laboratory setting of most earlier studies), this representation can only stem from the mapping of the observed movement kinematics. When, however, the ultimate intention can be clarified from other information regarding the target object and the surrounding environment or about the observer's and the model's internal status, or their interaction, motor resonance codes for the highest-level available aspect of others' behavior. A suppression of motor resonance occurs, however, when the ultimate intention cannot be disambiguated from multiple, conflicting sources of information.

In this perspective, the preserved [22] or the reduced [23, 68] evidence of motor resonance during action observation in individuals with social perception disorders (e.g., autism) can reflect whether or not these individuals can represent or not the ultimate intention of others' behavior, rather than simply echoing the inability to map their movement kinematics (see also [6, 8, 34]).

References

1. Aglioti, S. M., Cesari, P., Romani, M., & Urgesi, C. (2008). Action anticipation and motor resonance in elite basketball players. *Nature Neuroscience, 11*(9), 1109–1116.
2. Alaerts, K., Senot, P., Swinnen, S. P., Craighero, L., Wenderoth, N., & Fadiga, L. (2010). Force requirements of observed object lifting are encoded by the observer's motor system: A TMS study. *European Journal of Neuroscience, 31*(6), 1144–1153. https://doi.org/10.1111/j.1460-9568.2010.07124.x.
3. Alaerts, K., Swinnen, S. P., & Wenderoth, N. (2010). Observing how others lift light or heavy objects: Which visual cues mediate the encoding of muscular force in the primary motor cortex? *Neuropsychologia, 48*(7), 2082–2090. https://doi.org/10.1016/j.neuropsychologia.2010.03.029.
4. Amoruso, L., & Finisguerra, A. (2019). Low or high-level motor coding? The role of stimulus complexity. *Frontiers in Human Neuroscience (Motor Neuroscience)*. https://doi.org/10.3389/fnhum.2019.00332.
5. Amoruso, L., Finisguerra, A., & Urgesi, C. (2016). Tracking the time course of top-down contextual effects on motor responses during action comprehension. *The Journal of Neuroscience: The Official Journal of the Society for Neuroscience, 36*(46), 11590–11600. https://doi.org/10.1523/JNEUROSCI.4340-15.2016.
6. Amoruso, L., Finisguerra, A., & Urgesi, C. (2018). Autistic traits predict poor integration between top-down contextual expectations and movement kinematics during action observation. *Scientific Reports, 8*(1), 16208. https://doi.org/10.1038/s41598-018-33827-8.
7. Amoruso, L., & Urgesi, C. (2016). Contextual modulation of motor resonance during the observation of everyday actions. *NeuroImage, 134,* 74–84. https://doi.org/10.1016/j.neuroimage.2016.03.060.
8. Amoruso, L., Narzisi, A., Pinzino, M., Finisguerra, A., Billeci, L., Calderoni, S. et al. (2019). Contextual priors do not modulate action prediction in children with autism. *Proceedings of the Royal Society B: Biological Sciences, 286*(1908), 20191319. https://doi.org/10.1098/rspb.2019.1319.
9. Ansuini, C., Cavallo, A., Bertone, C., & Becchio, C. (2015). Intentions in the brain: The unveiling of mister hyde. *The Neuroscientist, 21*(2), 126–135. https://doi.org/10.1177/1073858414533827.
10. Avenanti, A., Annella, L., Candidi, M., Urgesi, C., & Aglioti, S. M. (2013). Compensatory plasticity in the action observation network: Virtual lesions of STS enhance anticipatory simulation of seen actions. *Cerebral Cortex, 23*(3), 570–580. https://doi.org/10.1093/cercor/bhs040.

11. Betti, S., Castiello, U., & Sartori, L. (2015). Kick with the finger: Symbolic actions shape motor cortex excitability. *European Journal of Neuroscience, 42*(10), 2860–2866. https://doi.org/10.1111/ejn.13067.
12. Borroni, P., Montagna, M., Cerri, G., & Baldissera, F. (2005). Cyclic time course of motor excitability modulation during the observation of a cyclic hand movement. *Brain Research, 1065*(1–2), 115–124. https://doi.org/10.1016/j.brainres.2005.10.034.
13. Bufalari, I., Aprile, T., Avenanti, A., Di Russo, F., & Aglioti, S. M. (2007). Empathy for pain and touch in the human somatosensory cortex. *Cerebral Cortex (New York, N.Y.: 1991), 17*(11), 2553–2561. https://doi.org/10.1093/cercor/bhl161.
14. Candidi, M., Sacheli, L. M., Mega, I., & Aglioti, S. M. (2014). Somatotopic mapping of piano fingering errors in sensorimotor experts: TMS Studies in pianists and visually trained musically naives. *Cerebral Cortex, 24*(2), 435–443. https://doi.org/10.1093/cercor/bhs325.
15. Candidi, M., Urgesi, C., Ionta, S., & Aglioti, S. M. (2008). Virtual lesion of ventral premotor cortex impairs visual perception of biomechanically possible but not impossible actions. *Social Neuroscience, 3*(3–4), 388–400. https://doi.org/10.1080/17470910701676269.
16. Cattaneo, L., Caruana, F., Jezzini, A., & Rizzolatti, G. (2009). Representation of goal and movements without overt motor behavior in the human motor cortex: A transcranial magnetic stimulation study. *Journal of Neuroscience, 29*(36), 11134–11138. https://doi.org/10.1523/JNEUROSCI.2605-09.2009.
17. Cavallo, A., Becchio, C., Sartori, L., Bucchioni, G., & Castiello, U. (2012). Grasping with tools: Corticospinal excitability reflects observed hand movements. *Cerebral Cortex, 22*(3), 710–716. https://doi.org/10.1093/cercor/bhr157.
18. Cavallo, A., Heyes, C., Becchio, C., Bird, G., & Catmur, C. (2014). Timecourse of mirror and counter-mirror effects measured with transcranial magnetic stimulation. *Social Cognitive and Affective Neuroscience, 9*(8), 1082–1088. https://doi.org/10.1093/scan/nst085.
19. Clarke, T. J., Bradshaw, M. F., Field, D. T., Hampson, S. E., & Rose, D. (2005). The perception of emotion from body movement in point-light displays of interpersonal dialogue. *Perception, 34*(10), 1171–1180. https://doi.org/10.1068/p5203.
20. Craighero, L., & Mele, S. (2018). Equal kinematics and visual context but different purposes: Observer's moral rules modulate motor resonance. *Cortex; a Journal Devoted to the Study of the Nervous System and Behavior, 104,* 1–11. https://doi.org/10.1016/j.cortex.2018.03.032.
21. Cretu, A. L., Ruddy, K., Germann, M., & Wenderoth, N. (2019). Uncertainty in contextual and kinematic cues jointly modulates motor resonance in primary motor cortex. *Journal of Neurophysiology, 121*(4), 1451–1464. https://doi.org/10.1152/jn.00655.2018.
22. Enticott, P. G., Kennedy, H. A., Rinehart, N. J., Bradshaw, J. L., Tonge, B. J., Daskalakis, Z. J., & Fitzgerald, P. B. (2013). Interpersonal motor resonance in autism spectrum disorder: Evidence against a global "mirror system" deficit. *Frontiers in Human Neuroscience, 7.* https://doi.org/10.3389/fnhum.2013.00218.
23. Enticott, P. G., Kennedy, H. A., Rinehart, N. J., Tonge, B. J., Bradshaw, J. L., Taffe, J. R. et al. (2012). Mirror neuron activity associated with social impairments but not age in autism spectrum disorder. *Biological Psychiatry, 71*(5), 427–433. https://doi.org/10.1016/j.biopsych.2011.09.001.
24. Fadiga, L., Craighero, L., & Olivier, E. (2005). Human motor cortex excitability during the perception of others' action. *Current Opinion in Neurobiology, 15*(2), 213–218. https://doi.org/10.1016/j.conb.2005.03.013.
25. Fadiga, L., Fogassi, L., Pavesi, G., & Rizzolatti, G. (1995). Motor facilitation during action observation: A magnetic stimulation study. *Journal of Neurophysiology, 73*(6), 2608–2611.
26. Farwaha, S., & Obhi, S. S. (2019). Differential motor facilitation during action observation in followers and leaders on instagram. *Frontiers in Human Neuroscience, 13,* 67. https://doi.org/10.3389/fnhum.2019.00067.
27. Finisguerra, A., Amoruso, L., Makris, S., & Urgesi, C. (2018). Dissociated representations of deceptive intentions and kinematic adaptations in the observer's motor system. *Cerebral Cortex, 28*(1), 33–47. https://doi.org/10.1093/cercor/bhw346.

28. Finisguerra, A., Maffongelli, L., Bassolino, M., Jacono, M., Pozzo, T., & D'Ausilio, A. (2015). Generalization of motor resonance during the observation of hand, mouth, and eye movements. *Journal of Neurophysiology, 114*(4), 2295–2304. https://doi.org/10.1152/jn.00433.2015.
29. Gallese, V., Fadiga, L., Fogassi, L., & Rizzolatti, G. (1996). Action recognition in the premotor cortex. *Brain: A Journal of Neurology, 119 (Pt 2)*, 593–609.
30. Gangitano, M., Mottaghy, F. M., & Pascual-Leone, A. (2001). Phase-specific modulation of cortical motor output during movement observation. *NeuroReport, 12*(7), 1489–1492.
31. Gangitano, M., Mottaghy, F. M., & Pascual-Leone, A. (2004). Modulation of premotor mirror neuron activity during observation of unpredictable grasping movements. *The European Journal of Neuroscience, 20*(8), 2193–2202. https://doi.org/10.1111/j.1460-9568.2004.03655.x.
32. Gueugneau, N., Mc Cabe, S. I., Villalta, J. I., Grafton, S. T., & Della-Maggiore, V. (2015). Direct mapping rather than motor prediction subserves modulation of corticospinal excitability during observation of actions in real time. *Journal of Neurophysiology, 113*(10), 3700–3707. https://doi.org/10.1152/jn.00416.2014.
33. Hamilton, A. F. D. C., & Grafton, S. (2007). *The motor hierarchy: From kinematics to goals and intentions* (Haggard, P., Rosetti, Y., Kawato, M.).
34. Hamilton, A. F. de C. (2013). Reflecting on the mirror neuron system in autism: A systematic review of current theories. *Developmental Cognitive Neuroscience, 3*, 91–105. https://doi.org/10.1016/j.dcn.2012.09.008.
35. Hari, R., Forss, N., Avikainen, S., Kirveskari, E., Salenius, S., & Rizzolatti, G. (1998). Activation of human primary motor cortex during action observation: A neuromagnetic study. *Proceedings of the National Academy of Sciences of the United States of America, 95*(25), 15061–15065. https://doi.org/10.1073/pnas.95.25.15061.
36. Heyes, C. (2010). Where do mirror neurons come from? *Neuroscience and Biobehavioral Reviews, 34*(4), 575–583. https://doi.org/10.1016/j.neubiorev.2009.11.007.
37. Hilt, P. M., Bartoli, E., Ferrari, E., Jacono, M., Fadiga, L., & D'Ausilio, A. (2017). Action observation effects reflect the modular organization of the human motor system. *Cortex, 95*, 104–118. https://doi.org/10.1016/j.cortex.2017.07.020.
38. Hogeveen, J., Inzlicht, M., & Obhi, S. S. (2014). Power changes how the brain responds to others. *Journal of Experimental Psychology: General, 143*(2), 755–762. https://doi.org/10.1037/a0033477.
39. Hogeveen, J., & Obhi, S. S. (2012). Social interaction enhances motor resonance for observed human actions. *Journal of Neuroscience, 32*(17), 5984–5989. https://doi.org/10.1523/JNEUROSCI.5938-11.2012.
40. Jacob, P., & Jeannerod, M. (2005). The motor theory of social cognition: A critique. *Trends in Cognitive Sciences, 9*(1), 21–25. https://doi.org/10.1016/j.tics.2004.11.003.
41. Janssen, L., Steenbergen, B., & Carson, R. G. (2015). Anticipatory planning reveals segmentation of cortical motor output during action observation. *Cerebral Cortex, 25*(1), 192–201. https://doi.org/10.1093/cercor/bht220.
42. Johansson, G. (1973). Visual perception of biological motion and a model for its analysis. *Perception and Psychophysics, 14*(2), 201–211. https://doi.org/10.3758/BF03212378.
43. Keysers, C., & Gazzola, V. (2014). Hebbian learning and predictive mirror neurons for actions, sensations and emotions. *Philosophical Transactions of the Royal Society of London. Series B, Biological Sciences, 369*(1644), 20130175. https://doi.org/10.1098/rstb.2013.0175.
44. Kilner, J. M. (2011). More than one pathway to action understanding. *Trends in Cognitive Sciences, 15*(8), 352–357. https://doi.org/10.1016/j.tics.2011.06.005.
45. Kilner, J. M., Friston, K. J., & Frith, C. D. (2007). Predictive coding: An account of the mirror neuron system. *Cognitive Processing, 8*(3), 159–166. https://doi.org/10.1007/s10339-007-0170-2.
46. Lepage, J.-F., & Théoret, H. (2007). The mirror neuron system: Grasping others? Actions from birth? *Developmental Science, 10*(5), 513–523. https://doi.org/10.1111/j.1467-7687.2007.00631.x.

47. Liuzza, M. T., Candidi, M., Sforza, A. L., & Aglioti, S. M. (2015). Harm avoiders suppress motor resonance to observed immoral actions. *Social Cognitive and Affective Neuroscience, 10*(1), 72–77. https://doi.org/10.1093/scan/nsu025.
48. Léonard, G., & Tremblay, F. (2007). Corticomotor facilitation associated with observation, imagery and imitation of hand actions: A comparative study in young and old adults. *Experimental Brain Research, 177*(2), 167–175. https://doi.org/10.1007/s00221-006-0657-6.
49. Maeda, F., Kleiner-Fisman, G., & Pascual-Leone, A. (2002). Motor facilitation while observing hand actions: Specificity of the effect and role of observer's orientation. *Journal of Neurophysiology, 87*(3), 1329–1335. https://doi.org/10.1152/jn.00773.2000.
50. Mattiassi, A. D. A., Mele, S., Ticini, L. F., & Urgesi, C. (2014). Conscious and unconscious representations of observed actions in the human motor system. *Journal of Cognitive Neuroscience, 26*(9), 2028–2041. https://doi.org/10.1162/jocn_a_00619.
51. Mc Cabe, S. I., Villalta, J. I., Saunier, G., Grafton, S. T., & Della-Maggiore, V. (2015). The relative influence of goal and kinematics on corticospinal excitability depends on the information provided to the observer. *Cerebral Cortex, 25*(8), 2229–2237. https://doi.org/10.1093/cercor/bhu029.
52. Molenberghs, P., Cunnington, R., & Mattingley, J. B. (2012). Brain regions with mirror properties: A meta-analysis of 125 human fMRI studies. *Neuroscience and Biobehavioral Reviews, 36*(1), 341–349. https://doi.org/10.1016/j.neubiorev.2011.07.004.
53. Montagna, M., Cerri, G., Borroni, P., & Baldissera, F. (2005). Excitability changes in human corticospinal projections to muscles moving hand and fingers while viewing a reaching and grasping action. *The European Journal of Neuroscience, 22*(6), 1513–1520. https://doi.org/10.1111/j.1460-9568.2005.04336.x.
54. Mukamel, R., Ekstrom, A. D., Kaplan, J., Iacoboni, M., & Fried, I. (2010). Single-neuron responses in humans during execution and observation of actions. *Current Biology: CB, 20*(8), 750–756. https://doi.org/10.1016/j.cub.2010.02.045.
55. Naish, K. R., Houston-Price, C., Bremner, A. J., & Holmes, N. P. (2014). Effects of action observation on corticospinal excitability: Muscle specificity, direction, and timing of the mirror response. *Neuropsychologia, 64C,* 331–348. https://doi.org/10.1016/j.neuropsychologia.2014.09.034.
56. di Pellegrino, G., Fadiga, L., Fogassi, L., Gallese, V., & Rizzolatti, G. (1992). Understanding motor events: A neurophysiological study. *Experimental Brain Research, 91*(1), 176–180. https://doi.org/10.1007/BF00230027.
57. Pineda, J. A. (2005). The functional significance of mu rhythms: Translating "seeing" and "hearing" into "doing". *Brain Research. Brain Research Reviews, 50*(1), 57–68. https://doi.org/10.1016/j.brainresrev.2005.04.005.
58. Riach, M., Holmes, P. S., Franklin, Z. C., & Wright, D. J. (2018). Observation of an action with a congruent contextual background facilitates corticospinal excitability: A combined TMS and eye-tracking experiment. *Neuropsychologia, 119,* 157–164. https://doi.org/10.1016/j.neuropsychologia.2018.08.002.
59. Rizzolatti, G., Fadiga, L., Gallese, V., & Fogassi, L. (1996). Premotor cortex and the recognition of motor actions. *Cognitive Brain Research, 3*(2), 131–141. https://doi.org/10.1016/0926-6410(95)00038-0.
60. Romani, M., Cesari, P., Urgesi, C., Facchini, S., & Aglioti, S. M. (2005). Motor facilitation of the human cortico-spinal system during observation of bio-mechanically impossible movements. *Neuroimage, 26*(3), 755–763. https://doi.org/10.1016/j.neuroimage.2005.02.027.
61. Sartori, L., Bucchioni, G., & Castiello, U. (2013). When emulation becomes reciprocity. *Social Cognitive and Affective Neuroscience, 8*(6), 662–669. https://doi.org/10.1093/scan/nss044.
62. Saygin, A. P., Wilson, S. M., Hagler, D. J., Bates, E., & Sereno, M. I. (2004). Point-light biological motion perception activates human premotor cortex. *The Journal of Neuroscience: The Official Journal of the Society for Neuroscience, 24*(27), 6181–6188. https://doi.org/10.1523/JNEUROSCI.0504-04.2004.
63. Senna, I., Bolognini, N. & Maravita, A. (2014). Grasping with the foot: Goal and motor expertise in action observation. *Hum Brain Mapp, 35*(4), 1750–1760. https://doi.org/10.1002/hbm.22289

64. Senot, P., D'Ausilio, A., Franca, M., Caselli, L., Craighero, L., & Fadiga, L. (2011). Effect of weight-related labels on corticospinal excitability during observation of grasping: A TMS study. *Experimental Brain Research, 211*(1), 161–167. https://doi.org/10.1007/s00221-011-2635-x.
65. Simpson, E. A., Murray, L., Paukner, A., & Ferrari, P. F. (2014). The mirror neuron system as revealed through neonatal imitation: Presence from birth, predictive power and evidence of plasticity. *Philosophical Transactions of the Royal Society B: Biological Sciences, 369*(1644), 20130289. https://doi.org/10.1098/rstb.2013.0289.
66. Soriano, M., Cavallo, A., D'Ausilio, A., Becchio, C., & Fadiga, L. (2018). Movement kinematics drive chain selection toward intention detection. *Proceedings of the National Academy of Sciences of the United States of America, 115*(41), 10452–10457. https://doi.org/10.1073/pnas.1809825115.
67. Strafella, A. P., & Paus, T. (2000). Modulation of cortical excitability during action observation: A transcranial magnetic stimulation study. *NeuroReport, 11*(10), 2289–2292.
68. Théoret, H., Halligan, E., Kobayashi, M., Fregni, F., Tager-Flusberg, H., & Pascual-Leone, A. (2005). Impaired motor facilitation during action observation in individuals with autism spectrum disorder. *Current Biology: CB, 15*(3), R84–R85. https://doi.org/10.1016/j.cub.2005.01.022.
69. Tidoni, E., Borgomaneri, S., di Pellegrino, G., & Avenanti, A. (2013). Action simulation plays a critical role in deceptive action recognition. *Journal of Neuroscience, 33*(2), 611–623. https://doi.org/10.1523/JNEUROSCI.2228-11.2013.
70. Tomeo, E., Cesari, P., Aglioti, S. M., & Urgesi, C. (2013). Fooling the kickers but not the goalkeepers: Behavioral and neurophysiological correlates of fake action detection in soccer. *Cerebral Cortex, 23*(11), 2765–2778. https://doi.org/10.1093/cercor/bhs279.
71. Tremoulet, P. D., & Feldman, J. (2000). Perception of animacy from the motion of a single object. *Perception, 29*(8), 943–951. https://doi.org/10.1068/p3101.
72. Ubaldi, S., Barchiesi, G., & Cattaneo, L. (2015). Bottom-up and top-down visuomotor responses to action observation. *Cerebral Cortex, 25*(4), 1032–1041. https://doi.org/10.1093/cercor/bht295.
73. Ulloa, E. R., & Pineda, J. A. (2007). Recognition of point-light biological motion: Mu rhythms and mirror neuron activity. *Behavioural Brain Research, 183*(2), 188–194. https://doi.org/10.1016/j.bbr.2007.06.007.
74. Umiltà, M. A., Kohler, E., Gallese, V., Fogassi, L., Fadiga, L., Keysers, C., et al. (2001). I know what you are doing. *Neuron, 31*(1), 155–165. https://doi.org/10.1016/S0896-6273(01)00337-3.
75. Urgesi, C., Candidi, M., Ionta, S., & Aglioti, S. M. (2007). Representation of body identity and body actions in extrastriate body area and ventral premotor cortex. *Nature Neuroscience, 10*(1), 30–31. https://doi.org/10.1038/nn1815.
76. Urgesi, C., Moro, V., Candidi, M., & Aglioti, S. M. (2006). Mapping implied body actions in the human motor system. *Journal of Neuroscience, 26*(30), 7942–7949. https://doi.org/10.1523/JNEUROSCI.1289-06.2006.
77. Urgesi, C., Maieron, M., Avenanti, A., Tidoni, E., Fabbro, F., & Aglioti, S. M. (2010). Simulating the future of actions in the human corticospinal system. *Cerebral Cortex (New York, N.Y.: 1991), 20*(11), 2511–2521. https://doi.org/10.1093/cercor/bhp292.
78. Vigneswaran, G., Philipp, R., Lemon, R. N., & Kraskov, A. (2013). M1 corticospinal mirror neurons and their role in movement suppression during action observation. *Current Biology: CB, 23*(3), 236–243. https://doi.org/10.1016/j.cub.2012.12.006.

Chapter 4
The Visual Perception of Biological Motion in Adults

Paul Hemeren and Yves Rybarczyk

Abstract This chapter presents research about the roles of different levels of visual processing and motor control on our ability to perceive biological motion produced by humans and by robots. The levels of visual processing addressed include high-level semantic processing of action prototypes based on global features as well as lower-level local processing based on kinematic features. A further important aspect concerns the interaction between these two levels of processing and the interaction between our own movement patterns and their impact on our visual perception of biological motion. The authors' results from their research describe the conditions under which semantic and kinematic features influence one another in our understanding of human actions. In addition, results are presented to illustrate the claim that motor control and different levels of the visual perception of biological motion have clear consequences for human–robot interaction. Understanding the movement of robots is greatly facilitated by the movement that is consistent with the psychophysical constraints of Fitts' law, minimum jerk and the two-thirds power law.

4.1 Introduction

It is no small secret that human vision is highly sensitive to the motion patterns created by the movement of other individuals (e.g. [7, 63, 70]). This sensitivity, however, is not restricted to motion patterns as such. When we see the movements of others, we do not merely see the independent movement of hands, arms, feet and legs and movement of the torso. Instead, we are able to quickly and accurately identify many motion patterns as meaningful actions (running, jumping, throwing, crawling, etc.) [28]. Gunnar Johansson [33, 34] clearly demonstrated an effective method for investigating the sensitivity of human action perception by using the point-light technique and creating point-light displays (PLDs) of human actions (Fig. 4.1). By

P. Hemeren (✉)
University of Skövde, Skövde, Sweden
e-mail: paul.hemeren@his.se

Y. Rybarczyk
Dalarna University, Falun, Sweden

N. Noceti et al. (eds.), *Modelling Human Motion*,
https://doi.org/10.1007/978-3-030-46732-6_4

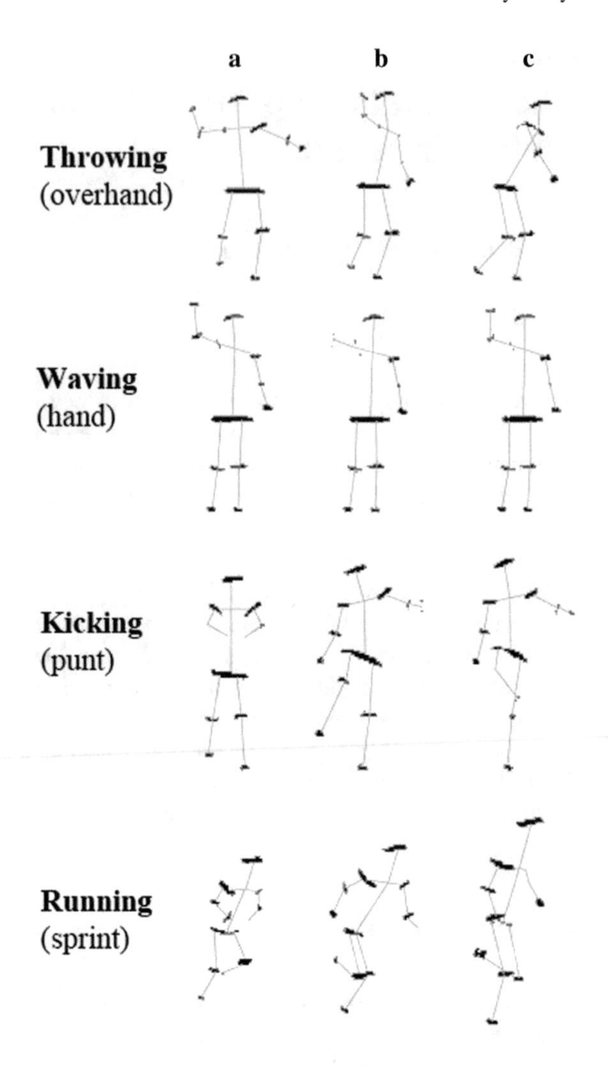

Fig. 4.1 Examples of the patch-light stimuli used in Experiments 1 and 2. The letters (**a**), (**b**) and (**c**) refer to the beginning, middle and end frames, respectively. The lines connecting the patches are for illustrative purposes only and were not included in the experiments

placing small lights or reflective patches on the joints of a human actor dressed in dark clothing and filming various actions, Johansson could isolate the motion (kinematic) information associated with the different actions by adjusting the contrast of the filmed sequences so that only the points of light were visible to human observers. When just one (static) frame from one of these motion sequences was presented to observers, they were unable to discern any meaningful representation. When, however, consecutive frames were displayed to produce (apparent) motion, observers could immediately see the portrayed action, walking or a couple dancing a Swedish folk dance. Johansson [34] states, "that as little as a tenth of a second (the time needed to project two motion-picture frames) is often enough to enable a naïve observer to identify a familiar biological motion". These early results from Johansson's research have created both a controlled methodological and empirical basis for systematically studying action perception and action categorization.

The first part of the chapter presents two categories (global and local) of the critical visual features in our perception of human biological motion (BM) presented as PLDs and how these features contribute to different levels of visual and cognitive processing in relation to action category prototypes and action segmentation. This emphasis on the use of PLDs is based on the vast research that uses PLDs to systematically study the influence of different visual features on BM perception. This should not be understood as ruling out the contribution of a more extensive visual context that includes human body form, other interactive agents and objects. A further step (though not included in this chapter) would then be to scale up the stimulus complexity by including more visually natural human movement situations in order to assess the impact of this additional information on the visual perception of human movements and actions. (See for example Hemeren [28, Sect. 4.3] and Yovel and O'Toole [70] for a discussion of the differences in using visually natural human movement conditions.) The global visual features refer to action concepts and categories while local visual features refer to the specific movement parameters of points in the PLDs.

A hypothesis defended by the authors is the fact that the sensitivity to BM could be explained by the motor properties of the observer [23]. The second part of the chapter describes several results that support the hypothesis. Many characteristics of human motor control can be interpreted as a signature of BM and, consequently, may facilitate the discrimination of a biological stimulus by a human observer. Among them, we can cite the Fitts' law, the minimum jerk and the power law, which seem to alter BM perception if their implementation in a PLD or humanoid robot is violated [13, 25, 69]. Besides improving our understanding on the psychological mechanics that underly the perception of BM, these findings provide us with the keys to design machines (e.g. cobots) for which the actions and intentions will be easily decodable by a human operator [52].

4.2 Action Categories and Biological Motion: Prototypes and Graded Structure

The ability to categorize human actions is a fundamental cognitive function. Given the predictive nature of human cognition, we need to understand the physical and social consequences of our actions and the actions of others. Not only do we see certain things as CUPS, BOOKS, DOGS, CARS, APPLES, etc., but we also see various patterns of movement as RUNNING, WALKING, JUMPING, THROWING, etc. Furthermore, the ability to recognize actions and events would seem to be a basic cognitive function given the fact that we live in an environment that is largely dynamic with respect to our own movements and interactions with objects and people.

On a more general level, concepts and categorization have been referred to as the "glue" [46] and "building blocks" [22] of human cognition. See also Harnad [27].

An important "glue-like" property of natural actions concerns their spatiotemporal form in movement kinematics, and this form can be used to group different actions together under the same category or to distinguish actions from one another. The structure of action categories appears to be similar to object categories where prototypes and exemplar typicality indicate a graded (radial) structure based on the perceptual similarity among exemplars of action categories. Previous findings from Dittrich [16] and Giese and Lappe [21] support the idea of action prototypes and accompanying typicality gradients. Results from Sparrow et al. [59] showed that moving stick figures of forearm flexion movements were categorized according to action prototype templates.

Two experiments from Hemeren [28] directly tested this hypothesis about the graded structure of action categories by instructing participants to view five different action exemplars from each of four different action categories (KICKING, RUNNING, THROWING and WAVING).

In the first experiment, twenty-four native Swedish-speaking students judged the typicality of each action in relation to a presented category label. The main results for the typicality ratings for the matching trials (Table 4.1) showed that a significant three-tiered difference between the five exemplars was found for KICKING and RUNNING, while only a two-tiered difference was found for THROWING and WAVING. Similar to object categories, these results suggest that for a restricted domain and number, typicality ratings for actions show graded structure. This is consistent with the robustness of typicality effects for a broad range of categorical domains [46]. A more revealing finding would show a typicality gradient together with verification judgements, which was the purpose of creating the second experiment. If the typicality ratings in the previous experiment and the verification reaction times in this experiment are a function of the same process, i.e. judging the similarity between an exemplar and a category prototype, then they should be highly correlated.

Table 4.1 Mean typicality ratings (Typ.) and standard deviations (SD) for action exemplars in relation to an action category label. Scale is from 0 to 8

	Action category label							
	Kicking	Typ.	Running	Typ.	Throwing	Typ.	Waving	Typ.
Exemplar	Soccer	7.7 (0.6)	Sprint	7.9 (0.5)	Overhand	7.2 (1.2)	Hand	7.8 (0.6)
	Punt	7.6 (0.8)	Skip	5.0 (2.1)	Throw-in	6.8 (1.1)	Both arms	5.3 (1.8)
	Toe-kick	7.3 (0.9)	Backward	4.3 (2.1)	Side arm	6.6 (1.4)	Get back	5.2 (1.7)
	Karate	6.6 (1.5)	Sideways	4.3 (2.1)	Underhand	5.9 (1.5)	Come here	4.6 (2.1)
	Heel-kick	4.8 (1.9)	In place	4.0 (2.2)	Side toss	5.8 (1.2)	Arm	4.3 (2.0)

If category verification of an action exemplar is carried out by accessing an action prototype for the action category label, then highly typical actions should be verified faster than less typical actions. This should result in the typicality-RT effect. Along similar lines, less typical actions may also share more properties (spatiotemporal pattern) with a prototype from a contrast category, which will result in longer verification times and/or more verification "errors".

Twenty-one native Swedish-speaking students participated in Experiment 2. Following the presentation of the category label, participants were instructed to verify whether the action exemplar belonged to the presented action category.

As a test of the relationship between the obtained typicality ratings from the previous experiment and the verification times in this experiment, these two measures were used in a correlation analysis. The results for this typicality-RT effect (Fig. 4.2) showed a significant (and strong) correlation between rated typicality and verification reaction time, $r = -0.82$, [$F(1, 18) = 35.64$, $p < 0.0001$]. Indeed, typicality seems to be an excellent predictor of the time it takes to verify category membership for the actions used in this study. The more typical an action exemplar is rated in relation to a "correct" category label, the less time it takes to correctly verify category membership.

These results indicate that recognizing and understanding the actions of others are due, at least in part, to having access to action meaning in the form of knowledge about action categories, i.e. groups of similar kinematic patterns of human motion. This also suggests that participants used high-level categorical knowledge in judging the typicality of action exemplars in relation to category labels and when given a speeded category verification task. The results also show an effect of perceptual relatedness indicating access to the spatiotemporal visual shape of actions presented as point-light displays. This information can be used to judge the typicality of action exemplars and to make judgements of category verification in relation to previously presented action category labels. Results from both experiments demonstrate a radial

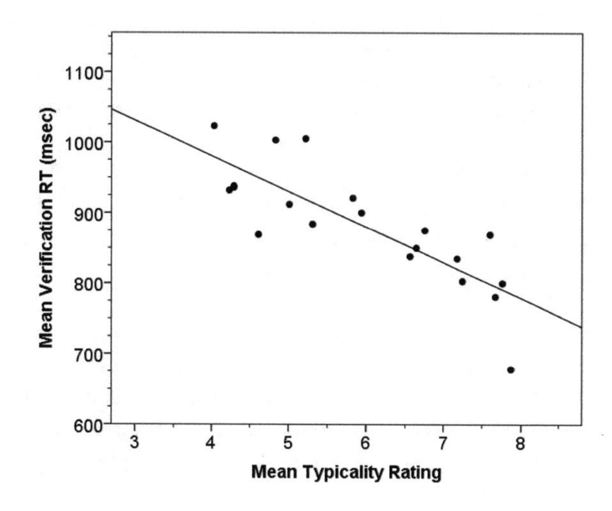

Fig. 4.2 Scatterplot for the relation between mean typicality ratings from Exp. 1 and mean verification reaction times from Exp. 2

structure for action concepts which is also expressed as a high correlation between judgements of typicality and category verification (typicality-RT effect).

4.3 Orientation Specificity

When actions presented as PLDs are viewed in an upright orientation, the depicted actions are easily recognized. They seem to pop-out. There is phenomenally direct access to a global, semantic level of representation. As Johansson [33] pointed out, "... we have found that it seems to be a highly mechanical, automatic type of visual data treatment that is most important". In a study that specifically tested the extent to which human movement/actions pop-out, Mayer et al. [44] found that the visual search slopes for human targets were affected by the number of distractors, which does not indicate pop-out. However, even though people do not pop-out, their movement was detected easier than movements of machines, which means that there is an established finding that human vision is especially sensitive to human biological motion.

Visual sensitivity to human biological motion can also be assessed by investigating visual access to high-level information about the categorical structure and semantic meaning of human body movement. An effective method for disturbing the categorical/semantic processing of biological motion is to turn them upside down, i.e. invert them. However, if the PLDs are inverted, people have a greater difficulty to see the action portrayed in the movement of the points of light. This effect represents one of the most replicated findings in BM perception. There is a wealth of converging behavioural and neuroscientific results that demonstrate impaired recognition, identification, detection and priming when displays of biological motion are viewed in an inverted orientation (e.g. [1, 3, 10, 16, 26, 49, 56, 60, 62, 64]).

In one of the earliest studies demonstrating the inversion effect, Sumi [60] let subjects view inverted walking and running sequences. The majority of participants who reported seeing a human figure failed to see it as inverted. They reported arm movements for the legs and leg movements for the arms. Other responses included non-human elastic forms indicative of non-rigid motion and mechanical changes. These results suggest that human perception of biological motion is sensitive to the image plane orientation of the displays. This effect is also found for faces and is referred to as the Thatcher effect [45]. Of particular importance is the fact that participants apparently were able to see local motion in terms of the motion of arms or legs but failed to get "the whole gestalt."

The theoretical significance of this inversion effect has to do with the fact that inverted displays contain the same hierarchical structural information as upright displays. The same local pair-wise relations and their relations to a principal axis of organization occur in inverted and upright displays. The performance differences between perceiving upright and inverted displays therefore seem to indicate different processing mechanisms. By systematically investigating performance differences under varying experimental conditions, we may gain further insight into our understanding of the factors that influence our keen ability to perceive the actions of others.

4.4 Global and Local Visual Features in Action Segmentation

To investigate the potential differences in action understanding due to access to different levels/kinds of information, Hemeren and Thill [29] conducted a study to document the effects of high- and low-level visual features on action segmentation where the actions were examples of hand/arm actions (Table 4.2). The focus on action segmentation reflects much of the research dealing with how to create complex actions by using combinations of action primitives, i.e. action parts (e.g. [47, 51, 61]). For the visual recognition of hand/arm actions, the visual system in most cases requires access to limb position, velocity, and acceleration. The method for determining the information used to derive motor primitives was to engage participants in an action segmentation task and then assess the degree of agreement between the kinematics of each action and the segmentation behaviour of the participants. This result was then compared to a similar segmentation task where participants not only had access to the kinematics of each action, but also had access to the semantic-level category information for each action. The purpose was to investigate whether or not segmentation was driven primarily by the kinematics of the action, as opposed to high-level top-down information about the action and the object used in the action.

Twelve hand/arm actions involved interaction with an object. The recorded actions shown as PLDs contained the precise reference points of finger, hand and arm positions of the person performing the actions (Fig. 4.3).

Table 4.2 Number of correct verbal descriptions of the actions

Viewing condition	Action					
	Cut with scissors	Lift dumb-bell	Open door	Pour from bottle	Saw wood	Spray with spray bottle
Pictures	12	12	12	12	12	12
No pictures	4	1	1	2	2	1
Viewing condition	**Action**					
	Drink from mug	Open a can and drink	Play tower of Hanoi	Turn pages in book	Unscrew bottle cap	Write on board
Pictures	12	12	12	12	11	12
No pictures	1	1	1	0	1	3

Fig. 4.3 Black-on-white (for clarity) point-light frame from a hand/arm action

Two different conditions were created in order to test the difference between high-level (semantic-level) category processing and low-level kinematic feature processing in an action segmentation task. Twenty-four participants were randomly assigned to one of the two viewing conditions. For the picture condition (semantic-level processing condition), twelve participants first viewed a picture of the object that was used the action. After viewing the object, the participants viewed the upright PLD all the way through once. They were then asked to describe what action the person was performing in the PLD. The participants then proceeded to the segmentation task where they were instructed to mark breakpoints in the action sequences that constituted the transitions between different segments in the action sequences.

In the inverted + no picture condition, each participant viewed a mirror-inverted PLD of the upright actions. No pictures of the objects were shown to the participants in this group, which together with the inverted version of each PLD should create limited access to any semantic-level recognition of the action. The purpose of the difference between these two groups was to have different access to semantic category labels for the actions but maintains the same access to the kinematic variables. Both upright and inverted PLDs contain the same kinematic variables. The placements of the segmentation task breakpoints by the participants were then analysed in relation to the kinematic variables of velocity, change in direction and acceleration for the point of the movement of the wrist.

The main question to assess in the analysis concerns the extent to which differential access to the semantic categories might influence the segmentation behaviour of the participants in relation to having access to the same kinematic variables. The results in Table 4.2 show that participants in the upright + picture condition were able to describe the different actions, whereas in the inverted + no picture condition, participants generally lacked (though not completely) access to any correct semantic category description of the PLDs.

The fact that participants can see and describe the movements of body parts but fail to identify the higher-level semantic meaning of the actions is similar to association agnosia for objects where patients can see the parts of objects but fail to identify the object as such [19]. There is no strictly visual deficit as such but rather an inability to recognize the object. When participants viewed the inverted point-light actions, they were able to visually discern the relevant body parts and segment the actions on the basis of changes in the direction of movement and velocity. What seemed to be missing was epistemic visual perception [32].

A critical question, given the different viewing conditions between the two groups, concerns the extent to which their segmentation behaviour differed when viewing the actions. In order to address this question, a density function of the marks placed along the timeline by each group of participants was computed for every action (see Hemeren and Thill [29] for details). Linear correlation coefficients were then calculated for the positioning of the segmentation breakpoints for each action according to the different groups.

According to the results in Table 4.3, there was significant segmentation agreement between the two groups, despite different access to semantic category information.

There was however a large difference between the highest (unscrew a bottle cap) and lowest (tower of Hanoi) correlation coefficient, which suggests different levels of agreement for the different actions. In order to visualize this difference, the plots of the density functions for the segmentation marks for the two groups and for two actions (*drink from mug* and *tower of Hanoi*) are presented in Fig. 4.4. For the bottom density function profile (tower of Hanoi), the main difference between the groups seems to concern whether or not to mark the recurring grasping–moving–releasing movements involved in the action. For the picture group, where participants identified the action as solving the tower of Hanoi puzzle, there was a much greater tendency to segment the movements that moved each disc on the tower. For the inverted + no picture group, there was much less segmentation that marked those recurring movements.

Table 4.3 Correlation coefficients (Pearson r) for the relationship between the density functions for the picture and no picture-inverted groups

Action	r (n)	Action	r (n)
1. Cut with scissors	0.72 (650)	7. Drink from a mug	0.86 (477)
2. Lift a dumb-bell	0.71 (435)	8. Open a can and take a drink	0.50 (564)
3. Open a door	0.60 (236)	9. Solve the tower of Hanoi	0.18 (634)
4. Pour from a bottle	0.60 (343)	10. Turn pages in a book	0.58 (584)
5. Saw wood	0.46 (457)	11. Unscrew a bottle cap	0.93 (404)
6. Spray from a spray bottle	0.66 (349)	12. Write on whiteboard	0.67 (670)

All coefficients are significant at the 0.001 level
n = number of observations of velocity and mark density function over the time course of the action, i.e. number of frames for each action

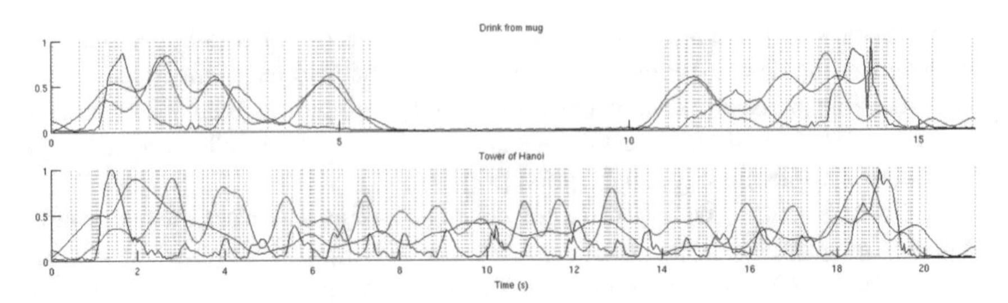

Fig. 4.4 Relationship between density functions for the segmentation marks for the picture and no picture-inverted groups for two actions. Black = velocity. Red = picture group. Blue = no picture group. Dotted vertical lines indicate marks placed by participants in the picture (red) and no picture group (blue), respectively

The main finding here is that when participants were given the task of segmenting hand/arm actions presented as point-light displays, segmentation was largely based on the kinematics, i.e. the velocity and acceleration of the wrist, regardless of whether or not participants had access to higher-level information about the action. If access to high-level information about the identification of the action, e.g. drinking was impaired by inverting the point-light displays, the kinematic information remained a salient source of information on which to base action segmentation. If participants had access to the high-level information, they still tended to rely on the kinematics of the hand/arm actions for determining where to place segmentation marks.

4.5 Influence of Movement Control on Motion Perception

A further central issue based on the previous studies concerns the relationship between the visual perception of actions and the potential effect of movement control. The kinematic variables and the movement of different body parts together create a high number of degrees of freedom that characterizes biological motion and contributes to people having to develop strategies to reduce the complexity of recognizing complex actions as well as motor control. This section presents several demonstrations of the strategies that constrain the control of movement that also seems to influence the perception of motion.

4.6 Physical and Functional Constraints

4.6.1 Musculoskeletal Constraints

The important starting point here concerns the anatomy of the cervical spine which constrains head rotations in order to maintain the skull in a horizontal position during the completion of an action. This transformation of the head into a stabilized referential frame facilitates the integration of vestibular and visual information to support the coordination of human movement [4, 15]. Another characteristic of the skeleton and muscles that simplifies the biomechanics is a limitation of allowed movements. The movement of limbs is constrained by anatomical characteristics which have been selected for their adaptive benefit. Thus, the musculoskeletal architecture of the living organisms induces a reduction of the possible movements, which simplifies the control of the action.

4.6.2 Kinematic Constraints

Individuals can grasp an object whatever their approximate position is in relation to this object. They just have to adjust the extension of their upper limbs according to the relative position of the object from the human body. Fortunately, they do not have to conscientiously calculate the angle between the limb segments to reach the target. Kinematic constraints between the segments enable a simplification of the degrees of freedom of the pointing movement [38]. The joint angle that links two segments (e.g. arm and forearm) is automatically adjusted according to the length of the movement. Thanks to this property, the individual has to focus on the control of the trajectory of the hand, only [58]. This is a basic problem addressed in robotics and which is known as the calculation of the inverse kinematics. It consists of applying kinematic equations to determine the joint parameters that provide a desired position of the robot end effector.

The programming of industrial robots that execute repetitive and precise tasks is inspired by this model. In the first stage, the programmer moves individually the robotic arm segments and records some key configurations that correspond to specific positions of the end effector. Then, in the second stage, the program is built by properly ordering the sequence of configurations according to the characteristic of the task to be completed. As in the control of the human movement, the programming of the robot's movement is simplified and transparent, because the programmer just worries about a single global parameter (i.e. position of the grasper) and not the multiple local variables of the action (i.e. joint angles).

4.6.3 Psychophysical Constraints

Fitts' Law

Fitts' law is a well-established principle of the relation between speed and accuracy of the biological movement [20]. The law states that the fastest movement time (MT) as possible between two targets relies on both the amplitude of the movement (A) and the width of the target (W), as described by Eq. 4.1.

$$\mathrm{MT} = a + b\left(\log_2\left(2\frac{A}{W}\right)\right) \tag{4.1}$$

In the equation, a and b are constants that depend on the characteristic of the effector and are determined empirically by linear regression. According to this equation, the larger the movement distance and the smaller the target is, the longer MT will take. Since the MT is determined by two variables, the second argument of the formula is usually identified as the index of difficulty (ID), and its calculation is obtained through Eq. 4.2.

$$\mathrm{ID} = \log_2\left(2\frac{A}{W}\right) \tag{4.2}$$

Fitts' law is a very robust characteristic of human motor performance, which can be verified in different contexts and kinds of effector [50].

Minimum Jerk

The motor program seeks for an optimization of the cost of the movement that involves a minimum amount of energy [67]. The result is a movement trajectory with the minimum jerk. This property can be defined by a cost function (CF) proportional to the mean-squared jerk, which is the derivative of the acceleration (Eq. 4.3).

$$\mathrm{CF} = \frac{1}{2}\int_{t1}^{t2}\left[\left(\frac{\mathrm{d}^3x}{\mathrm{d}t^3}\right)^2 + \left(\frac{\mathrm{d}^3y}{\mathrm{d}t^3}\right)^2\right]\mathrm{d}t \tag{4.3}$$

In this equation, x and y are the horizontal and vertical components of the motion, respectively. This suggests that the movement will be smoother when the CF will be minimized. Experiments show that the natural movements of subjects can be precisely predicted (trajectory and velocity) from this model [18]. The minimum jerk is characterized by a bell-shaped velocity profile, in which the movement speed increases progressively, reaches a peak near the midpoint and then decreases slowly. This absence of abrupt changes seems to support the execution of a smooth motion [30].

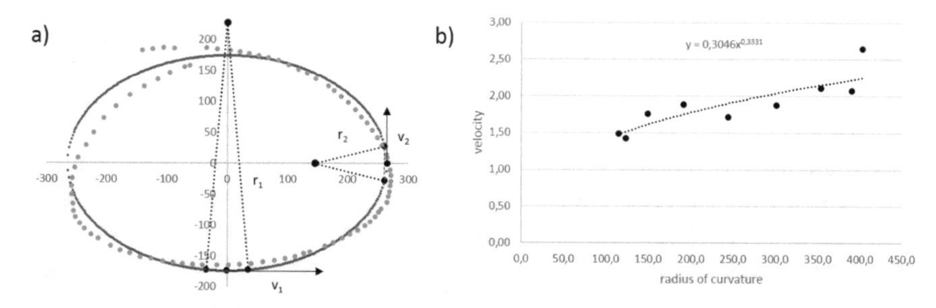

Fig. 4.5 Illustration of the modulation of the motion velocity dictated by the power law (panel a). The blue and red dots represent the actual elliptic movement performed by a subject and the ellipse that better fits the geometry of the trajectory, respectively. As indicated by the magnitude of the velocity vectors (v_1 and v_2), the larger the instantaneous radius of curvature (r_1 and r_2), the higher is the instantaneous motion speed. A sampling of each π/16 rad shows that the relationship between the velocity and radius of the curvature obeys indeed a two-thirds power law (panel b)

Two-Thirds Power Law

Another fundamental motor behaviour is the relationship between the velocity and the curvature of the biological movements, which is known as the two-thirds power law [39, 68]. This law states that the angular velocity of the end effector is proportional to the two-thirds root of its curvature or, equivalently, that the instantaneous tangential velocity (v_t) is proportional to the third root of the radius of curvature (r_t), as described in Eq. 4.4. In other words, it means that the velocity of the movement decreases in the highly curved parts of the trajectory and increases when the trajectory becomes straighter (Fig. 4.5).

$$v_t = k r_t^{-1/3} \tag{4.4}$$

There is a controversy regarding the origins and the violations of the 2/3 power law during the execution of biological movements [55, 67, 68, 72]. On the one hand, some studies tend to demonstrate that the power law is a signature of the central nervous system [31, 67, 71], because it seems to be independent of the dynamic properties of the limbs. This law is indeed observed in a wide variety of activities such as drawing [39], walking [66] and smooth pursuit eye [14]. On the other hand, different studies defend a biomechanical [24] or, even, an artefactual explanation [41, 42].

4.7 Evidences of the Impact of Motor Properties on Motion Perception

Several studies indicate that the physical and functional constraints described above tend to influence the perception of motion. For example, the smooth pursuit eye is a low-speed system (100° per second) in comparison with the ocular saccade (800° per second). Consequently, individuals such as predators have to anticipate the movement of the prey to be able to catch it. The easiest solution to predict the next position of the prey is to use an internal model of the movement that simulates the trajectory of the animal [5].

A study on the observation of individuals performing a tapping task tends also to confirm the linkage between a motor program and motion perception. Grosjean et al. [25] tested the hypothesis of the influence of Fitts' law on action perception. The authors asked participants to estimate if the speed of the observed movement was appropriate to complete the action without missing the targets. The result shows that perceived MTs are linearly correlated with the ID of the actions, which confirms that Fitts' law applies to motion perception. In 1995, a study carried out by Decety and Jeannerod [12] demonstrated that Fitts' law also holds for motor imagery. Both experiments constitute important evidence to support the theory that the motor properties constrain the perception of motion (lived and imagined).

A few number of experiments have addressed the question of the effect of the minimum jerk on visual perception. A neuropsychological study shows that autistic people have a reduced sensitivity to minimum jerk, which suggests an abnormal processing of biological motion in patients with this condition [9]. Another interesting and controversial work was carried out by Bisio et al. [6]. These authors aimed to identify which features of the biological motion would be responsible for the impact on the perception. Four parameters were analysed: the appearance (nature of the observed agent), the intention (action performed with or without a goal), the kinematics (velocity profile of the movement) and the geometry (smooth versus jerk trajectory). The experimental paradigm was the motor resonance (or motor contagion) of the observed movement on the action performed by the observer. The results indicate that the movement of the observer is influenced by the perception of biological motion regardless of the nature of the agent (human and humanoid robot), its intention and the curvilinearity of the movement. The only condition in which the motor contagion disappears is when the bell-shaped speed profile of the observed action is violated. Such a finding constitutes another example suggesting that the motor repertoire constrains the perception of motion. The main feature that seems to influence the observer is the kinematics of the movement and not the shape of the trajectory. In that sense, the outcome of this study is a bit paradoxical if we consider that the bell-shaped velocity profile of a movement that minimizes the jerk enables a smooth geometry of the trajectory of the end effector.

More studies focused on the influence of the 2/3 power law on the perceptual judgement in the absence of motor performance. In that respect, it was demonstrated that this kinematic principle affects motion imagery [48], which provides additional evidence of the perception–action coupling. This finding is supported by experiments that analyse the human capability to predict the course of an artefact that moves according to a biological or non-biological kinematic. For instance, Pozzo et al. [51] studied the human's performance to estimate the final location of a moving dot that has the last part of its trajectory masked. The results show that the precision of the prediction is significantly worst when the movement of the dot violates the kinematic laws of arm-pointing gesture. This outcome suggests that motion inference does not only depend on a visual information extrapolation based on pure physical characteristics of the movement. On the contrary, the observer seems to rely on an internal motor representation of the action to predict the future position of the stimulus. Moreover, the fact that the movement obeys the 2/3 power law seems to constitute a critical feature in the perceptual anticipation over the ongoing event [35].

It is also demonstrated that the power law has a direct effect on the velocity perception of an artefact [40, 69]. An individual exposed to a spot moving at constant velocity or according to the 2/3 power law tends to consider the movement of the latter stimulus as more uniform than the former. This conspicuous and robust illusion represents another behavioural evidence of the coupling between motor and perceptual processes. Salomon et al. [54] confirm the visual illusion of velocity constancy when the movement of the artefact follows the 2/3 power law. In addition, they demonstrate that the individuals tend to judge the 2/3 power law movement as more natural than a non-biological kinematic. They also tested the discrimination of a stimulus following its biological versus non-biological movement. The result shows that the participants require less time to discriminate a motion that violates the 2/3 power law. This outcome could be interpreted in terms of the attraction of the attention caused by the biological movement [57, 65], which would distract the subject from the task.

Even more striking is the fact that training individuals to perform movements that violate the 2/3 power law increases their sensibility to visual motions violating this law [2]. This result suggests that the impact of the motor program on visual perception would not be fully innate and could be transformed by a learning effect. The combined outcome of these studies supports the hypothesis that motor control underlies the visual perception of motion [8, 11].

4.8 Consequences on Human–Robot Interaction

The previous considerations have serious impacts on the collaboration between human beings and robots, particularly in industry (e.g. co-worker robots) and in services (e.g. caregiver robots). A successful cooperation requires an intuitive interaction between the two entities that enables the individual to predict the intentions of the machine. This is the reason why several studies have analysed the human perception of robot actions. The experimental paradigm is generally based on the reaction of the human being during the observation of a robot that produces a biological versus non-biological movement. Most of these studies agree on the fact that people prefer an interaction with robots that implement human-like behaviours [36, 53].

For instance, Huber et al. [30] tested the effect of the biological motion (minimum jerk of the end effector) on a handover interaction task between a robot and a human. The results show that the reaction time is significantly reduced for a biological (bell-shaped) than a non-biological (trapezoidal) speed profile. In addition, participants tend to feel safer in the former than in the latter condition. De Momi et al. [13] confirmed that human beings are able to distinguish biologically from non-biologically inspired robot movements. A machine learning approach based on a neural network algorithm was used to train a robotic arm on human actions. Then, the biological characteristic of the movement was tested by verifying that the robot indeed replicated the minimum jerk, the bell-shaped speed profile and the 2/3 power law. The experimental analysis on participant data shows that the observers discriminate between a machine-like and human-like robot behaviour. As observed by Bisio et al. [6], this finding suggests that the kinematic factors are sufficient to classify a movement as biological or non-biological.

This subjective perception is supported by the motor response of the observer too. Kupferberg et al. [37] showed that a motor interference occurs when a humanoid robot executes an incongruent movement in relation to the action of the observer (e.g. movements performed in different plans). However, this interference happens only when the machine moves according to the minimum-jerk principle, which suggests that the speed profile would facilitate the perception of a robot as an interactive partner. The effect of the naturalness of the movement seems to reflect a very deep process, because even after familiarization to non-biological kinematics an observer cannot predict unnatural actions of a robot as good as natural ones [17].

The facilitating effect of a robot replicating biological motions on the perception of the intentions of the machine is also confirmed in the context of physical interaction [43]. The force applied on the end effector of a robot by a human operator is significantly reduced if the robotic arm moves according to the 2/3 power law. In line with the findings of Beets et al. [2], an extensive training on non-biological profiles of velocity would permit the individual to lower the applied force, but without reaching the performance obtained with the natural pattern. It was also recently demonstrated that the best motor performance of a teleoperator is produced when the 2/3 power law is implemented in the remotely controlled robot [52].

All these experimental evidences support the assumption that the implementation of biological motion in the robot way of working would benefit the comprehension of the intention of the robot and consequently the cooperation between the human being and the machine. From a fundamental perspective, the findings obtained in robotics tend to demonstrate that movement can be processed as biological by the human brain, even if it is not produced by a living organism, provided that the artefact's motion simulates a kinematic property of biological motion.

References

1. Ahlström, V., Blake, R., & Ahlström, U. (1997). Perception of biological motion. *Perception, 26,* 1539–1548.
2. Beets, I. A. M., Rösler, R., & Fiehler, K. (2010). Nonvisual motor learning improves visual motion perception: Evidence from violating the two-thirds power law. *Journal of Neurophysiology, 10*(3), 1612–1624.
3. Bertenthal, B. I., & Pinto, J. (1994). Global processing of biological motions. *Psychological Science, 5,* 221–225.

4. Berthoz, A. (1991). Reference frames for the perception and control of movement. In *Brain and Space* (pp. 82–111). Oxford, UK: Oxford University Press.
5. Berthoz, A. (2000). *The brain's sense of movement*. Cambridge, MA, USA: Harvard University Press.
6. Bisio, A., Sciutti, A., Nori, F., Metta, G., Fadiga, L., Sandini, G., & Pozzo, T. (2014). Motor contagion during human–human and human–robot interaction. *PloS One, 9*(8).
7. Blake, R., & Shiffrar, M. (2007). Perception of human motion. *Annual Review of Psychology, 58.*
8. Casile, A., & Giese, M. A. (2006). Nonvisual motor training influences biological motion perception. *Current Biology, 16*(1), 69–74.
9. Cook, J., Saygin, A. P., Swain, R., & Blakemore, S. J. (2009). Reduced sensitivity to minimum-jerk biological motion in autism spectrum conditions. *Neuropsychologia, 47,* 3275–3278.
10. Daems, A., & Verfaillie, K. (1999). Viewpoint-dependent priming effects in the perception of human actions and body postures. *Visual Cognition, 6*(6), 665–693.
11. Dayan, E., Casile, A., Levit-Binnun, N., Giese, M. A., Hendler, T., & Flash, T. (2007). Neural representations of kinematic laws of motion: Evidence for action-perception coupling. *Proceedings of the National Academy of Sciences, 104*(51), 20582–20587.
12. Decety, J., & Jeannerod, M. (1995). Mentally simulated movements in virtual reality: Does Fitts' law hold in motor imagery? *Behavioral Brain Research, 72,* 127–134.
13. De Momi, E., Kranendonk, L., Valenti, M., Enayati, N., & Ferrigno, G. (2016). A neural network-based approach for trajectory planning in robot–human handover tasks. *Frontiers in Robotics and AI, 3,* 34.
14. de Sperati, C., & Viviani, P. (1997). The relationship between curvature and speed in two-dimensional smooth pursuit eye movements. *Journal of Neuroscience, 17,* 3932–3945.
15. De Waele, C., Graf, W., Berthoz, A., & Clarac, F. (1988). *Vestibular control of skeleton geometry. Posture and Gait* (pp. 423–432). Amsterdam, The Netherlands: Elsevier.
16. Dittrich, W. H. (1993). Action categories and the perception of biological motion. *Perception, 22,* 15–22.
17. Dragan, A. & Srinivasa, S. (2014). Familiarization to robot motion. In *Proceedings of the 2014 ACM/IEEE International Conference on Human–Robot Interaction* (pp. 366–373), Bielefeld, Germany, March.
18. Edelman, S., & Flash, T. (1987). A model of handwriting. *Biological Cybernetics, 57,* 25–36.
19. Farah, M. J. (2004). *Visual agnosia*. Cambridge, MA: MIT Press.
20. Fitts, P. M. (1954). The information capacity of the human motor system in controlling the amplitude of movement. *Journal of Experimental Psychology, 47,* 381–391.
21. Giese, M. A., & Lappe, M. (2002). Measurement of generalization fields for the recognition of biological motion. *Vision Research, 38,* 1847–1858.
22. Goldstone, R. L., Kersten, A., & Carvalho, P. F. (2018). Categorization and concepts. *Stevens' Handbook of Experimental Psychology and Cognitive Neuroscience, 3,* 1–43.
23. Gentsch, A., Weber, A., Synofzik, M., Vosgerau, G., & Schütz-Bosbach, S. (2016). Towards a common framework of grounded action cognition: Relating motor control, perception and cognition. *Cognition, 146,* 81–89.
24. Gribble, P. L., & Ostry, D. J. (1996). Origins of the power law relation between movement velocity and curvature: Modeling the effects of muscle mechanics and limb dynamics. *Journal of Neurophysiology, 76*(5), 2853–2860.
25. Grosjean, M., Shiffrar, M., & Knoblich, G. (2007). Fitts's law holds for action perception. *Psychological Science, 18*(2), 95–99.
26. Grossman, E. D., & Blake, R. (2001). Brain activity evoked by inverted and imagined biological motion. *Vision Research, 41,* 1475–1482.
27. Harnad, S. (2017). To cognize is to categorize: Cognition is categorization. In *Handbook of categorization in cognitive science* (pp. 21–54). Amsterdam: Elsevier.
28. Hemeren, P. E. (2008). Mind in action. *Lund University Cognitive Studies, 140.*
29. Hemeren, P. E., & Thill, S. (2011). Deriving motor primitives through action segmentation. *Frontiers in Psychology, 1,* 243.

30. Huber, M., Rickert, M., Knoll, A., Brandt, T., & Glasauer, S. (2008). Human–robot interaction in handing-over tasks. In: *Proceedings of the 17th IEEE International Symposium on Robot and Human Interactive Communication* (pp. 107–112), Munich, Germany, August.
31. Huh, D., & Sejnowski, T. J. (2015). Spectrum of power laws for curved hand movements. *Proceedings of the National Academy of Sciences of the United States of America, 112,* 3950–3958.
32. Jeannerod, M., & Jacob, P. (2005). Visual cognition: A new look at the two-visual systems model. *Neuropsychologia, 43,* 301–312.
33. Johansson, G. (1973). Visual perception of biological motion and a model for its analysis. *Perception and Psychophysics, 14*(2), 201–211.
34. Johansson, G. (1975). Visual Motion Perception. *Scientific American*, (June), pp 76–88.
35. Kandel, S., Orliaguet, J. P., & Viviani, P. (2000). Perceptual anticipation in handwriting: The role of implicit motor competence. *Perception and Psychophysics, 62*(4), 706–716.
36. Kilner, J., Hamilton, A. F. D. C., & Blakemore, S. (2007). Interference effect of observed human movement on action is due to velocity profile of biological motion. *Social Neuroscience, 2*(3), 158–166.
37. Kupferberg, A., Glasauer, S., Huber, M., Rickert, M., Knoll, A., & Brandt, T. (2011). Biological movement increases acceptance of humanoid robots as human partners in motor interaction. *Experimental Brain Research, 26*(4), 339–345.
38. Lacquaniti, F., Soechting, J., & Terzuolo, C. (1986). Path constraints on point to point arm movements in three dimensional space. *Neuroscience, 17,* 313–324.
39. Lacquaniti, F., Terzuolo, C., & Viviani, P. (1983). The law relating the kinematic and figural aspects of drawing movements. *Acta Psychologica, 54,* 115–130.
40. Levit-Binnun, N., Schechtman, E., & Flash, T. (2006). On the similarity between the perception and production of elliptical trajectories. *Experimental Brain Research, 172*(4), 533–555.
41. Marken, R. S., & Shaffer, D. M. (2017). The power law of movement: an example of a behavioral illusion. *Experimental Brain Research, 235*(6), 1835–1842.
42. Marken, R. S., & Shaffer, D. M. (2018). The power law as behavioral illusion: Reappraising the reappraisals. *Experimental Brain Research, 236*(5), 1537–1544.
43. Maurice, P., Huber, M., Hogan, N., & Sternad, D. (2018). Velocity-curvature patterns limit human–robot physical interaction. *IEEE Robotics and Automation Letters, 3*(1), 249–256.
44. Mayer, K. M., Vuong, Q. C., & Thornton, I. M. (2015). Do people "pop out"? *PLoS ONE, 10*(10), e0139618.
45. Mirenzi, A., & Hiris, E. (2011). The Thatcher effect in biological motion. *Perception, 40*(10), 1257–1260.
46. Murphy, G. L. (2002). *The big book of concepts*. Cambridge: Bradford Books, MIT Press.
47. Mussa-Ivaldi, F.A., & Bizzi, E. (2000). Motor learning through the combination of primitives. *Philosophical Transactions of the Royal Society of London B, 355*, 1755–1769.
48. Papaxanthis, C., Paizis, C., White, O., Pozzo, T., & Stucchi, N. (2012). The relation between geometry and time in mental actions. *PLoS ONE, 7*(11), e51191. https://doi.org/10.1371/journal.pone.0051161.
49. Pavlova, M., & Sokolov, A. (2000). Orientation specificity in biological motion perception. *Perception and Psychophysics, 62*(5), 889–899.
50. Plamondon, R., & Alimi, A. M. (1997). Speed/accuracy trade-offs in target-directed movements. *Behavioral and Brain Sciences, 20,* 279–349.
51. Pozzo, T., Papaxanthis, C., Petit, J. L., Schweighofer, N., & Stucchi, N. (2006). Kinematic features of movement tunes perception and action coupling. *Behavioral Brain Research, 169,* 75–82.
52. Rybarczyk, Y., & Carvalho, D. (2019). Bioinspired implementation and assessment of a remote-controlled robot. *Applied Bionics and Biomechanics*, Article ID 8575607. https://doi.org/10.1155/2019/8575607.
53. Rybarczyk, Y., & Mestre, D. (2012). Effect of temporal organization of the visuo-locomotor coupling on the predictive steering. *Frontiers in Psychology, 3*, Article ID 239. https://doi.org/10.3389/fpsyg.2012.00239.

54. Salomon, R., Goldstein, A., Vuillaume, L., Faivre, N., Hassin, R. R., & Blanke, O. (2016). Enhanced discriminability for nonbiological motion violating the two-thirds power law. *Journal of Vision, 16*(8), 1–12.
55. Schaal, S., & Sternad, D. (2001). Origins and violations of the 2/3 power law in rhythmic three-dimensional arm movements. *Experimental Brain Research, 136*(1), 60–72.
56. Shiffrar, M., & Pinto, J. (2002). The visual analysis of bodily motion. In W. Prinz & B. Hommel (Eds.), *Common mechanisms in perception and action: Attention and performance* (Vol. 19, pp. 381–399). Oxford: Oxford University Press.
57. Simion, F., Regolin, L., & Bulf, H. (2008). A predisposition for biological motion in the newborn baby. *Proceedings of the National Academy of Sciences, 105*(2), 809–813.
58. Soechting, J. F., & Flanders, M. (1989). Sensorimotor representations for pointings to targets in three-dimensional space. *Journal of Neurophysiology, 62,* 582–594.
59. Sparrow, W. A., Shinkfield, A. J., Day, R. H., Hollitt, S., & Jolley, D. (2002). Visual perception of movement kinematics and the acquisition of "action prototypes". *Motor Control, 6*(2), 146–165.
60. Sumi, S. (1984). Upside-down presentation of the Johansson moving light pattern. *Perception, 13,* 283–286.
61. Thoroughman, K. A., & Shadmehr, R. (2000). Learning of action through adaptive combination of motor primitives. *Nature, 407,* 742–747.
62. Troje, N. F. (2003). Reference frames for orientation anisotropies in face recognition and biological-motion perception. *Perception, 32,* 201–210.
63. Tversky, B. (2019). *Mind in motion: How action shapes thought*. London: Hachette.
64. Ueda, H., Yamamoto, K., & Watanabe, K. (2018). Contribution of global and local biological motion information to speed perception and discrimination. *Journal of Vision, 18*(3), 2.
65. Vallortigara, G., Regolin, L., & Marconato, F. (2005). Visually inexperienced chicks exhibit spontaneous preferences for biological motion patterns. *PLoS Biology, 3*(7), e208.
66. Vieilledent, S., Kerlirzin, Y., Dalbera, S., & Berthoz, A. (2001). Relationship between velocity and curvature of a human locomotor trajectory. *Neuroscience Letters, 305,* 65–69.
67. Viviani, P., & Flash, T. (1995). Minimum-jerk, two-thirds power law, and isochrony: converging approaches to movement planning. *Journal of Experimental Psychology: Human Perception and Performance, 21*(1), 32–53.
68. Viviani, P., & Schneider, R. (1991). A developmental study of the relationship between geometry and kinematics in drawing movements. *Journal of Experimental Psychology: Human Perception and Performance, 17*(1), 198–218.
69. Viviani, P., & Stucchi, N. (1992). Biological movements look uniform: evidence of motor-perceptual interactions. *Journal of Experimental Psychology: Human Perception and Performance, 18*(3), 603–623.
70. Yovel, G., & O'Toole, A. J. (2016). Recognizing people in motion. *Trends in Cognitive Sciences, 20*(5), 383–395.
71. Zago, M., Lacquaniti, F., & Gomez-Marin, A. (2016). The speed-curvature power law in Drosophila larval locomotion. *Biology Letters, 12*(10), Article ID 20160597.
72. Zago, M., Matic, A., Flash, T., Gomez-Marin, A., & Lacquaniti, F. (2018). The speed-curvature power law of movements: A reappraisal. *Experimental Brain Research, 236*(1), 69–82.

Chapter 5
The Development of Action Perception

Janny Christina Stapel

Abstract When newly born into this world, there is an overwhelming multitude of things to learn, ranging from learning to speak to learning how to solve a mathematical equation. Amidst this abundance, action perception is developing already in the first months of life. Why would learning about others' actions be among the first items to acquire? What is the relevance of action perception for young infants? Part of the answer probably lies in the strong dependence on others. Newborn human infants need caretakers even for fulfilling their basic needs. Weak neck muscles make it hard for them to lift up their head, and most of their movements come across as uncoordinated. Clearly, getting themselves a drink or dressing themselves is not part of their repertoire. Their reliance on their caregivers makes these caregivers and their actions important for the young infant. Seeing that the caregiver responds to their calls can already reduce some of the stress that comes with being so dependent. As such, it is helpful for an infant to learn to distinguish different actions of the caregiver. Not only are the caregivers' actions focused on the infant's physical needs, but also on helping the infant to regulate her emotions. Parents typically comfort a baby by softly rocking them, and by talking and smiling to them. Social interaction between caregiver and infant thus starts immediately after birth, and these interactions help them to bond. In the context of social interaction, it is useful to be able to distinguish a smile from a frown. Interpreting the facial actions of others is vital to successful communication. Moreover, in the period in which infants are still very limited in their own actions, observing others' actions forms a main resource for learning about the world. Making sense of others' actions is therefore of central importance already during early development.

J. C. Stapel (✉)
Donders Institute for Brain, Cognition, and Behaviour, Radboud University, Nijmegen, The Netherlands
e-mail: j.c.stapel@donders.ru.nl

N. Noceti et al. (eds.), *Modelling Human Motion*,
https://doi.org/10.1007/978-3-030-46732-6_5

5.1 Action Perception: What It Is

Before elaborating on the development of action perception, it might be good to provide a definition of action perception and set boundaries that clarify what falls under the heading action perception and what not.

The term *action perception* is composed of two separate terms: *action* and *perception*. *Action* is commonly defined to be 'the goal-directed movements of an agent'. This definition of action begs the question what goal-directedness is. Although many scholars in the action perception domain use the word goal-directedness, the assigned meaning differs between scholars. To make things worse, scholars sometimes flexibly switch between different meanings of the word. Unclear use of terminology makes it hard to grasp which findings should be considered together when crafting a theoretical account that captures the current body of empirical findings. Findings may seem incongruent with each other, but in fact may be tapping into different concepts. *Goal-directedness* for instance can refer to the directly observable property of a movement, namely that it has a specific endpoint in time and space. While cycling on a home-trainer is a continuous movement, putting down a cup is not. The movement of putting down a cup has a clear end location, and the movement stops at that end location. The goal of this action is the table, the visible end location of the cup. In this chapter, I will use the term *target* to refer to these unambiguous goals.

Goal-directedness can also refer to a potentially observable, potentially unobservable property of a movement grounded in the actor, namely the notion that the movements were intended by the actor. Unintentional movements can result from accidents or from external forces or a combination of the two. Like the different usages of the term *goal*, the term *intention* also has multiple meanings and is used in different ways by different scholars. Some use action intention to refer the next action: to pick up a ball in order to throw it [49]. Others use intention to refer to a more abstract plan: picking up a cup with the intention to drink from it. Drinking can be broken down into multiple smaller steps, multiple sub-actions, and in that sense, the word *intention* here still refers to the next step in the action sequence. It can also be used to point to a desired result that is not part of the action domain: to kick a ball in order to impress someone else.

There is no doubt whether movements can be performed with intentions from each of these different levels. As such, each of these levels can be used to describe movements. But are all of these various levels of intentions and thereby the various definitions of action useful when studying action *perception*? Arguably, if the task of the observer is to decipher why the actor is doing what she is doing, then the intentionality of the action is relevant for the observer. However, it may not always be clear to the observer whether the movements were performed intentionally or not. One could place movements on a continuum starting from highly functional without any communicative intent, to highly arbitrary with solely the intention to communicate. At the end of this spectrum are gestures, which may be unknown to the observer and hence it might be hard to grasp whether these movements were performed intentionally or not. Given that drawing a distinction between intentional and non-intentional movements will be arbitrary unless we have access to unambiguous reports from the actor, *action* will be defined in this chapter as *any type of human movements*. Other scholars extend the scope of action perception to actions of agents which are not necessarily human. However, an entity is often considered to be an agent if it performs intentional movements [63] and if for the current chapter the intentionality of the movements is not taken into account, we would be including any moving entity. This would make the term action perception very broad and hence not very useful. Most probably, the perception of a rolling ball is based on very different processes and is built on very different experiences than the perception of a hand grasping a cup. Therefore, action perception will here refer to the perception of human actions.

Perception is the second term in action perception and refers to the information-processing that follows sensation. While sensation is the process of receiving physical inputs through the senses, perception can be viewed as an interpretation step. The study of perception can be confined to a single sense, such as visual perception which only describes the information-processing originating from light falling on the retina. Perception can also be studied in a broader sense, as objects are often sensed through various sensory modalities simultaneously: A person holding a bell can see, feel and hear the bell at the same time. Whereas object perception is often multisensory, action perception is predominantly visual in nature. Most human movements do not produce sound unless there is friction with a surface (e.g., scratching), or only at specific points in the action which frequently coincides with the end point of the action (e.g., a cup touching the table when putting down a cup). The primarily visual nature of action perception is also reflected in the empirical study of action perception as most studies use videos or pictures as stimulus materials.

Action perception in its full-fletched version that adults exhibit is the higher-order cognitive interpretation of the movements of other people, namely the understanding of why the actor does what she does. Developmentally, this ability may be built up out of less complex forms of action perception. The most rudimentary form of action perception is the ability to *discriminate* different actions. When infants demonstrate that they perceive the differences between different actions, this means they can discriminate different actions from each other. Speech perception offers a nice analogy to action perception. One may be able to hear the difference between 'ball' and 'bell', but that does not necessarily imply that the person understands what these speech sounds refer to. Another prerequisite for understanding speech is the ability to *segment* the speech stream into meaningful parts like words or sentences. In similar fashion, action understanding requires the perceiver to segment the continuous stream of ongoing visual motion into meaningful chunks, and further into distinguishable actions and sub-actions. While action discrimination and segmentation concern the incoming stream of information, action *prediction* and *expectations* concern the future states of an action. Expectations about future states of the action may be broad and relatively unspecific. In that case, the expectation may help to guide attention, and this can lead to reduced response times in comparison to when the observer has no clue what to expect. Expectations may also be very narrow and specific. In that case, the observer might predict the exact time and location where the observed action will end. However, even if an observer has made a perfect prediction about how the observed action continued, this does not mean that she fully understood the action. An observer may for instance predict the timing of a click sound when watching someone turn the key of a door, but not understand that this is the only and necessary way to unlock the door, and that the person did this with the intention to enter the house. As such, action *understanding* can be seen as the most sophisticated form of action perception.

5.2 Methodologies for Studying Action Perception Development

Studying cognitive abilities in development is challenging, and studying action perception development is no exception to that. The challenges lie—among others—in infants' limited attention span, their inability to tell what they think or feel and them not understanding nor adhering to instructions. Intrigued by what goes on in the mind of those little ones, infant researchers have come up with a number of methods to study (action) perception development.

5.2.1 Looking Time

At the end of the 1950s and the start of the '60s, Robert Fantz devised a method that proved to be very useful for studying whether infants perceive the difference between visual stimuli. Two visual stimuli were presented to the infant in alternating fashion, while the infant's gaze direction was monitored. Looking longer to one of the two stimuli would indicate a preference for that stimulus, and hence, the method is called 'preferential looking paradigm.' The preferential looking paradigm is used widespread in various forms. A frequently used form is one in which stimuli are presented next to each other simultaneously. Typically, infants' looking behavior is recorded by a camera facing the infant. Afterward, coders score whether the infant looked to the left or to the right half of the screen. Another influential research method derived from the preferential looking paradigm is the so-called habituation paradigm. In the habituation paradigm, an infant is presented with the same stimulus over and over again until she loses interest. Then, a different stimulus is presented. If the infant regains attention to this novel stimulus, indicated by an increase in looking time, then the infant is thought to detect the difference between the initial and the novel stimulus.

Looking time measures have been proven to be highly effective in assessing action perception development. For instance, the well-known Woodward paradigm is based on a habituation procedure. In the Woodward paradigm, infants face two toys that are positioned next to each other, a ball and a bear. In the habituation phase, they observe an actor repeatedly reaching for the same toy. After habituation, the toys switch location. The actor then either reaches to the same toy as before, which requires the arm to take a different path, or reaches to the same location as before but now to a different toy. Five-month-olds are found to dishabituate stronger to the novel toy than to the novel path action, as evidence by longer looking times [139]. When provided with hands-on experience with the toys prior to the habituation procedure, even 3-month-olds display a preference for the novel toy action [123]. At the very least, these studies demonstrate that 3-months are capable of discriminating different actions.

Beyond indicating action discrimination, longer looks to the novel toy action are also taken as indication of surprise. In general, looking longer to novel stimulus A than to novel stimulus B is commonly interpreted as being more surprised to see A than B. If the infant is indeed surprised, then this means she must have formed an expectation about what was about to happen next. Looking time measures can thus be used to study action expectations.

Lastly, looking time measures have been employed to study action segmentation in infancy. Baldwin et al. [6] familiarized 10- to 11-month-old infants with videos of everyday action sequences, such as picking up a towel from the floor and placing it on a towel rack. In the test phase, infants saw versions of the same videos that were either paused right after a sub-action was completed, or shortly before the sub-action was completed. The infants looked longer at the videos that were interrupted shortly before compared to shortly after sub-actions were completed. These and follow-up results [115] indicate that at least from 9-months of age, infants are sensitive to action boundaries and seem able to segment a continuous action stream into functional sub-actions.

5.2.2 Eye-Tracking

Infants' gaze direction can only be assessed in a coarse-grained manner when relying on video recordings. The introduction of eye-trackers has made it possible to record infants' gaze in a more fine-grained manner both in spatial and temporal terms. This has allowed for major advances in the study of action perception development.

The high temporal resolution for instance enabled infancy researchers to uncover more about infants' expectations through registering saccadic reaction times. The saccadic reaction time (SRT) is the time between a trigger (for instance: stimulus onset) and the start of a saccadic eye movement to a specified location. SRTs are known to be shorter when covert attention has already shifted toward to the location where the eyes will land. For instance, adults are quicker at detecting a target at the left when they have first seen a left-pointing arrow [107]. Daum and Gredebäck [25] used this phenomenon called *priming* to investigate whether infants form expectations when they see an open hand. The researchers showed 3-, 5-, and 7-month-old infants static pictures of an open hand that was ready to grasp an object. After seeing the open hand, the 7-month-olds were quicker to look at an object appearing at the location congruent with the grasping hand than at an incongruent location. A large portion of the 5-month-olds showed a similar pattern of results: shorter SRTs for the congruent trials than for the incongruent trials. In the group of 7-month-olds, the advantage of congruent over incongruent trials was 83 ms, which underscores the necessity of a high temporal resolution of the measurements.

Furthermore, the combination of a relatively high spatial and a high temporal resolution allows infant researchers to study anticipatory eye movements. Anticipatory eye movements are eye movements that observers make to a location where they expect something to happen in the near future. Therefore, anticipatory eye movements are interpreted as a sign of prediction: apparently, the observer expects visual input at a particular time and a particular location. Directing gaze to a location where nothing happens yet may seem odd, but is actually very useful. In this way, the observer avoids missing the upcoming event on that location. Falck-Ytter et al. [33] demonstrated that infants also exhibit anticipatory eye movements. Their 12-month-old participants observed an actor reaching for a ball and placing the ball into a bucket. Gaze of the infants was recorded with an eye-tracker, and analysis of the gaze data revealed that the infants looked at the bucket before the actor's hand had arrived there. This revealed that infants, like adults, look ahead of others' actions.

5.2.3 *Neuroscientific Methods*

Another elegant but challenging way to study the development of action perception is measuring the brain activity of infants and young children while they watch others' actions. The most commonly used method in this domain is electroencephalography (EEG), and since recently, infant researchers also started to use functional near-infrared spectroscopy (fNIRS).

An electroencephalogram is a recording of the electrical activity of the brain captured by placing sensors (electrodes) on the skin of the head. This type of recording can be used to measure a broad range of action perception capacities, starting from action discrimination, segmentation, to the more future-oriented capacities as expectation and prediction. For example, researchers have investigated the expectations infants build up when observing actions by showing infants a series of action pictures that either ended in an expected manner or in an unexpected manner [18, 112]. By analyzing the time-locked electrical response of the brain (called event-related response, abbreviated: ERP) that is elicited in response to the last picture in the sequence, researchers can deduce whether the infants had expected the action to end differently or as observed. Specifically, the focus of such research is on the N400, which is the negative deflection in the ERP that is commonly observed in adults around 400 ms after stimulus onset when processing unexpected semantic information both in language [41] and in action perception [73, 120].

fNIRS measures how much near-infrared light is absorbed in different regions of the brain depending on the task the person performs. Local and temporal changes in near-infrared light absorption are thought to be related to changes in oxygenation of the underlying brain tissue—underlying as fNIRS is measured at the skin and not inside the skull or brain. Changing in oxygenation of brain regions is indicative of the involvement of that brain region in the task at hand.

fNIRS has the same potential as EEG for studying action perception development in the sense that it can capture the same action perception capacities. In comparison with EEG, fNIRS provides more certainty about the cerebral origin of the captured brain activity, but the downside is that the temporal resolution of the response is much poorer than of EEG. Lloyd-Fox and colleagues nicely demonstrated the added advantage of fNIRS over EEG by showing—with fNIRS—that neighboring but distinct fronto-temporal areas are activated when 5-month-olds are observing different actions, such as mouth, hand and eye movements [78].

Neuroscientific methods have the advantage over behavioral methods that the region or the type of activity found can be informative about the processes that take place when observing actions. Specifically, researchers have strived to discover the potential role of motor processes in action perception development. In EEG, the power in the mu-frequency band (in infants: activity overlying central sites peaking between 6 and 9 Hz) is known to decrease during action production and has also been found to decrease during action observation [85]. Reduction in the power of the mu-frequency band is therefore taken as an indication that motor processes are active during action perception in infants, resembling findings from studies on adults [96]. The downside of using mu-frequency activity as a marker for motor processes is that it can be easily confused with the alpha-frequency activity generated in occipital areas of the brain [130] which typically overlaps in its frequency range. EEG signals are spatially smeared out over a large area on the surface of the head due to the electrically insulating skull, which makes it hard to separate mu- from alpha-frequency activity. Activity picked up with fNIRS is more localized which makes fNIRS an interesting tool to capture motor processes. Shimada and Hiraki [118] demonstrated that it is indeed possible to register motor responses in action observation in infants. They found motor cortical responses when 6- to 7-month-olds observed live actions, corroborating the findings from EEG studies.

Lastly, neuroscientific methods are appealing as these methodologies do not require any overt behavior which implies that they are not confounded by motor skills. The lack of requirements on the motor side makes the neuroscientific methods very useful for across age comparisons. Adults, children, toddlers and infants can all be subjected to the same experimental procedures as long as the procedure is fitted to the infant population.

5.2.4 Active Responses

In stark contrast with the neuroscientific methods are methods that require action from the participant. Whereas in neuroscientific studies, data quality is best if the participant moves as little as possible, in active response studies, movements are a necessity, otherwise nothing can be measured. Depending on age and temperament, infants and small children need some time to get acquainted with the researcher and the setup to feel free enough to respond spontaneously. In such paradigms, the researcher needs to find a balance between engaging the individual participant—all children are different and require a different approach—and experimental control. Another downside of using active responses as the basis for the outcome measure of a study is that it is hard if not impossible to construct a task that can be performed across a wide age range. What is motorically feasible for the infant and what is capturing his or her attention changes within a few months.

Despite these hurdles, developmental researchers have successfully managed to carefully construct numerous active response paradigms to test infants' emerging action perception capacities. A reason for using an active rather than a passive response paradigm is that it allows the researcher to tap into a complex capacity, namely action understanding.

A good example of such a paradigm comes from Andrew Meltzoff, who tested the responses of 18-month-olds to successful and failed demonstrations of an object-directed action [87]. For instance, some of the toddlers observed a model holding a dumbbell and pulling off the wooden cubes at the ends of the dumbbell, whereas others observed a model trying to achieve the same effect, but failing as the model's hand slipped off one of the cubes. Interestingly, even without ever seeing the completed action, toddlers reproduced the successful version of the action. The frequency of re-enacting the intended actions was not different for the group that observed the successful than for the group that observed the failed actions. According to the author, this demonstrates that 18-month-olds already understand the intentions of others' actions. *What* infants imitate can thus provide information about infants' interpretation of the observed action.

The *frequency* or selectivity with which infants imitate others' actions can likewise reveal their understanding of the action. For instance, Carpenter and colleagues [17] compared whether 14- to 18-month-olds would imitate intentional actions as frequently as accidental actions. If the frequency would not be different in these situations, then that might imply that the infants merely mimic the movements of the model, whereas if infants selectively imitate intentional rather than accidental actions, then that might imply that infants understand that the accidental action did not work out as intended.

In some cases, children imitate only part of the modeled action. Researchers use this as an indication of what the children consider to be essential aspects of the modeled actions. A classic example comes from Bekkering and colleagues [10] who invited preschool children (4–6 year-olds) to do what the actor did. The children observed an actor point with his left hand to one of two dots in front of him on the table. If the pointing was contralateral (arm crossing body midline), then children frequently imitated by using the ipsilateral arm, but pointing to the correct dot. However, when there were no dots on the tabletop, but the actor would point using the same arm postures, children only infrequently chose the 'wrong' arm, as if now the purpose of the action had been to generate a posture rather than to achieve a goal ('point to a specific dot').

5.3 Types of Information Used for Action Perception in Development

Actions contain various types of information that infants, like adults, can use for identifying, segmenting, predicting and ultimately understanding observed actions. First, the action scene might contain manipulable *objects* which can play a role in object-directed actions. Second, these objects and other aspects of the scene may hint at the setting in which the action takes place. Typically, different actions take place in the *context* of an operating theater than in a class room. Third, the *movements* of the actor are an essential if not the most important aspect of the action. Fourthly, the actor's focus of attention, as revealed by the *gaze* direction of the actor, can form an important clue for understanding and predicting what the actor is doing.

The seminal work of Amanda Woodward clearly reveals that infants pay attention to the *objects* involved in other's actions already from a young age. As discussed above, if 5-month-old infants are habituated to a reaching action to the same target object, they subsequently dishabituate to 'same path, novel object' test trials, and not to the 'different path, same object' test trials. According to Woodward, observed actions are encoded in terms of their goal rather than the means. This effect is not found in 3-month-olds, unless they receive brief hands-on experience manipulating the same objects using special mittens equipped with Velcro on which the objects stick [123]. This may imply that in everyday life, which does not offer this specific 'sticky-mittens' experience, infants do not differentiate different object-directed actions at 3-months of age: Objects seem to become a more relevant part in the perception of others' actions once infants can manipulate objects themselves. The importance of target objects in encoding others' actions is echoed in the work of Bekkering and colleagues, in which children seemed to prioritize goals over means in imitating others' reaching-to-the-ear actions [10].

Objects can also play a different role in actions, namely provide context to the action. For instance, grasping a cup from a table that looks nicely prepared for having tea together may lead the observer to expect that the actor will drink from the cup, whereas if the table is a mess, the observer may expect that the actor wants to clean the table and that grasping is part of the cleaning action [62]. Whether infants use these type scenarios to interpret observed actions is not entirely clear yet. Much research, however, has been devoted to infants' perception of objects as obstacles along the actor's desired path. For instance, when habituated with a person jumping over a wall to get to another person, 14-month-olds look longer when the wall is removed to a repetition of the same jumping than to a novel walking action that leads to the same result (person A meeting person B). Sodian and her colleagues draw the conclusion from these data that infants expect others' to act efficiently [122]. A comparable study was conducted by Philips and Wellman [105] on visually more familiar actions namely reaching. They habituated 12-month-olds to curved reaches over a barrier. The barrier was removed in test trials, and infants looked longer to similar now unnecessarily curved reaches than to reaches straight to the target object.

These results illustrate that infants take objects into account in action perception also when the object is not the target of the action.

While objects are an essential part in object-directed actions, the *movements* of the actor reveal important clues about the actor's intentions and future events as well. For instance, when reaching for an object with the intent to throw it, the velocity profile of the reach is different from a reach that is performed with the intent to subsequently place the object into a bucket. This modulation of kinematics in action production can already be observed from 10-months of age [19, 49]. Velocity profiles of actions also differ based on the size of the target object. Smaller targets require more precision which impacts the average speed of the action, as precise movements require more time than less precise movements [35]. An eye-tracking study testing 9-, 12- and 15-month-old infants demonstrated that only the oldest age group was able to use the velocity profile of a reaching-and-aiming movement to predict the target of the aiming action [126]. The velocity profile hinted whether the actor aimed at a small or a large button which were positioned right next to each other. It is still an open question whether the age at which infants can productively use velocity as a cue for target prediction is action-specific, or whether this a general skill developing around 15 months of age.

Target predictions in these studies are often spatial predictions, focusing on where the action will end. Temporal predictions—specifying when something will happen—have been shown to be action-specific in infancy. Fourteen-month-olds were found to be more accurate in predicting the timing of reappearance of a shortly occluded crawler, than in predicting the timing of reappearance of a walker they observed [127]. There was no significant difference in prediction accuracy for walking and crawling at 18-months of age, suggesting that the walking actions were not intrinsically harder to predict, but rather that temporal prediction of walking develops later than the temporal prediction of crawling.

Before they can predict actions based on kinematics, infants develop the ability to discriminate actions with different kinematics. A classic way of testing sensitivity to aspects of observed kinematics involves presenting point-light displays (PLDs) in which only moving dots are presented against a uniform background. The dots represent the major joints of a person in motion. Many aspects can be 'read' from the dots, such as the identity of the actor [24, 79], the performed action [27] and the actor's emotions [6]. Fox and McDaniel were the first in 1982 [37] to report that 4- but not 2-month-olds prefer looking at a point-light figure of a walking person than at a set of randomly moving dots. At first, differentiation between these stimuli might be based on local grouping: 3-month-olds discriminate regular PLDs from PLDs in which the phase relationship between the points is perturbed, both when the figure is presented in the normal orientation and when presented upside-down [11]. At 5-months of age, infants only distinguish canonical point-light walkers from point-light walkers with perturbed phase relationships when the PLDs are presented upright, but no longer when the stimuli are inverted. This implies that around 5-months, upright human motion has become familiar and the motion is processed globally [108].

Within the realm of all possible movements that may feed action perception processes, head and eye movements form a special class. Head and eye movements frequently uncover what a person will do next. Position and angle of head and eyes are informative about upcoming actions because gaze direction can be derived from these two sources, and in most of our actions, vision is used to guide and control our actions. This even holds for actions we think we can do blindly, like walking [129]. In everyday manual actions, gaze is either ahead of the hand movements or supporting minutious hand movements [74]. As gaze often reveals the actor's target [36], it has the potential to serve as an important cue for action prediction. Infants gradually develop the ability to follow the gaze of their interaction partner. Scaife and Brunner [116] reported that 30% of their 2-month-old participants followed gaze, a number which rose to 100% in the 14-month-old participants. When positioned opposite to each other, the interaction partner may start looking at an object located behind the observer. Following gaze in such a situation, or distinguishing smaller differences in gaze angles develops in the second year of life [16]. Highest gaze following scores are found when the experimenter moves head and eyes in combination, but infants do also follow gaze if only the eyes of the experimenter move [16, 23]. Gaze following has the potential to guide action perception as the actor's gaze can point the observer to the relevant aspects of the action. However, recent work from Yu and Smith does not converge with this hypothesis [144, 145]. They allowed toddlers (11–24 months) to play with a few preselected toys together with one of their parents while gaze of both parent and child were recorded. Dyads frequently established joint visual attention, but did so by hand–eye interaction rather than through gaze following. That is, mutual gaze at the same object was established by one of partners picking up and manipulating a toy. Hand movements thus triggered eye movements of the interaction partner. In similar vein, 14-month-olds were found to predict the target of observed reaching actions based on the hand movements of the actor, despite the contingent head and eye movements that always preceded these reaching movements [71]. At 12-months, but not at 7- or 9-months, infants seem to encode the gaze movements of the actor as a goal-directed action [140]. That is, the 12-month-olds dishabituated to 'new target' but not to 'new location' events in a classic Woodward paradigm where the actor performed an eye movement to an object (replacing the typically used reaching movement). However, if the actor inspected the object with multiple gaze fixations, infants already at 9-months of age were able interpret the gaze behavior of the actor as goal-directed [65].

5.4 Potential Driving Factors for Action Perception Development

Over the course of the first years of life, humans develop the ability to discriminate and segment observed actions, form expectations and predictions about these actions, and ultimately, gain an understanding of what the other person is doing. But how are these abilities acquired? Part of the answer may be that infants possess a number of potentially hardwired preferences. The other part of the answer may lie in the experiences infants build up through interaction with their environment. However, to effectively use these experiences as building blocks for their action perception abilities, infants need at least a few basic skills. I will first address the preferences that might steer action perception development. Then I will discuss how experiences might support the development of action perception abilities and which capacities might be fundamental for action perception development.

5.4.1 Early Preferences

Infants have an early visual preference for face-like stimuli [95]. Studies with newborns suggest that this preference is inborn [48, 64, 86, 132]. The in-built preference for faces heightens the propensity that they will look at faces whenever these are around, which is very likely to be beneficial for development. A face is namely an important and rich stimulus, which conveys information about emotions [29], identity [136], direction of attention [75] and spoken language [143]. Being visually exposed to faces on a regular basis means that there are ample learning opportunities for detecting the subtle messages faces may send. For instance, infants may learn to detect the other's direction of attention, which is an important ingredient in the development of gaze following.

When looking at faces, infants spend a considerable time looking at the eyes. Haith and colleagues for instance report that between 3 and 11 weeks of age, participating infants looked ten times longer at the eyes than at the mouth even if the person was talking [50]. At 6-months of age, infants start looking longer at the mouth, though the eyes still form the major attraction of the face [60, 90]. It is around this age that infants start babbling [30, 99]. Developing a preference for looking at the mouth can feed into a perception-action loop in which perception may inspire action and vice versa. Visual speech can form an additional cue to disambiguate the acoustic speech stream, which is not purely redundant as categories of speech sounds often overlap in their acoustic features [22, 77]. In other words, some instantiations of speech sounds cannot be disambiguated without additional cues. Auditory speech perception is therefore trainable through audiovisual experience [54]. Infants direct their visual attention to the audiovisual cues provided by the mouth in the same developmental time frame as when they learn to produce speech sounds themselves [76], which may suggest that the infant brain makes use of the benefits visual speech offers. First, visual attention to the mouth region of a speaker's face increases from 4- to 8-months of age. At this stage, the preference for looking at the speaker's mouth is independent of the language spoken, be it native or non-native to the infant. From 8- to 12-months, the proportion of looks to the mouth decreases when the infant's native tongue is spoken, while the proportion of looks remains high when it concerns a foreign tongue. What drives the initial increase in interest for the mouth is not clear, it seems that the curiosity in mouth movements increases in the period in which the infant tries to explore and control its own vocalizations.

Together with a visual preference for faces, infants seem to have a preference for upright biological motion from early on. Newborn infants prefer looking at the point-light display of a walking hen over looking at random dot motion [119]. In the same way, newborns prefer a display of an upright walking hen over an inverted version of the same walking movements. The authors chose to display a walking hen rather than a walking person because this ruled out the possibility that the participating infants had visual experience with the observed motion. Furthermore, prior research had employed the same stimuli, testing the sensitivity of newly hatched chickens to biological motion [134], and reuse of the stimuli allowed cross-species comparisons. However, it cannot be excluded that these early preferences for upright biological motion are influenced by or are defacto an interest in motion patterns that are consistent with gravity [133]. A follow-up study from Bardi and colleagues [8] tested whether newly born human infants preferred upright biological motion over point-light stimuli with the same motion profile per dot, but spatially scrambled. The spatially scrambled dots contain the same acceleration and deceleration profiles as the regular point-light dots, behaving in congruence with the impact of gravity. The tested newborns did not display a spontaneous preference, which leaves room for an interpretation in terms of a gravity bias. At 3-months of age, a gravity bias can no longer explain the data, as infants spontaneously look longer at scrambled motion dot displays than at canonical upright point-light walkers [13]. Together, these motion-perception studies illustrate that biological motion is treated differently than other motion patterns already early in life.

Infants may also have a preference for manipulable objects. Longer looks to actions involving new objects compared actions involving new means, as found in studies employing the Woodward paradigm, have classically been interpreted as a sign that infants understand the actor–goal relationship. However, these data also hint at something else, namely that from 5-months of age, infants seem to display a preference for objects. Movements are typically more interesting than static objects, but still, infants look longer at scenes in which the movements are similar and the objects are different rather than at scenes in which the movements are different but the objects are the same [15, 130–142]. In comparable vein, from 6 months of age, infants look at a target object while a hand is reaching toward that object [68], suggesting that infants prefer to look at the object rather than at the hand motion. However, potentially it is not the object per se that attracts their interest, as 6-month-olds have also been found to look at the ear of a person when that person is bringing a phone to her ear, and likewise, they have been found to look at the mouth when a person is bringing a cup to the mouth [61]. Potentially, infants prefer to watch the most critical parts of an action, which typically takes place at the end of an action—think of a football landing in the goal—or at turning points—say a person reaches to a cup, then the grasp of the cup might reveal where the hand and cup might move next. Work from Land and Hayhoe [74] illustrates that in everyday activities, our eyes are drawn to the targets of our actions and to transition points. At these points, the eyes are most needed to support manual actions. It is also at these points that hand movements decelerate. It might be that infants tend to prioritize looking at these crucial turning points based on the kinematics of the action. In that sense, the action segmentation skills as found by Baldwin may emerge from a sensitivity to changes in velocity: If the action slows down, something interesting is about to happen. Given that segmentation skills thus far mainly have been tested in 10-month-old infants [7], it is an open question whether infants learn that deceleration often precedes interesting events, or that infants have an early preference for deceleration independent of rewards.

5.4.2 Capacities Fundamental for Action Perception Development

In their young lives, infants experience a wide variety of events. Part of the input is sampled more frequently and more thoroughly than other inputs due to infants' early preferences. However, to use this input to acquire an understanding of others' actions requires more than passively gazing around. Compare it to a video camera: despite its ability to register a vast amount of visual scenes, it does not build up an understanding of the contents of the scenes. Seeing and memorizing seen events is thus not sufficient. Infants have and develop a number of capacities that allow action perception abilities to develop.

Categorization

Categorization is a rarely mentioned but likely very relevant skill for action perception. No two instances of an observed or produced act are ever exactly the same, which means that generalization is needed for learning to take place. Generalization is useful and a reasonable step if the encountered instances are sufficiently similar to each other. The question is how similarity is computed, on which dimensions it is computed, and how the infant might decide which dimensions are relevant for the comparison and which are not (see for a phonetic learning account: Pierrehumbert [106]). Then, if encountered instances do not fit the existing categories, novel classes need to be created. Forming novel categories is therefore a computationally complex skill; nevertheless, empirical work demonstrates that infants are capable of categorization. For instance, when 3- to 4-month-olds are familiarized with various pictures of dogs, they prefer to look at cats thereafter [38]. Categories are likely to be acquired through learning about the underlying statistical regularities [75–81]. If that is indeed true, then dense categories (categories in which items correlate on multiple dimensions) are likely to be acquired earlier in life than sparse categories. Hints in that direction come from Kloos and Sloutsky [70], who showed that 4- and 5-year-old children indeed are better able to learn dense categories than sparse categories if no explanation is provided. Sparse categories are better learnt when the rules governing the categorization are explained. Whereas adults seem to represent sparse categories in terms of rules and dense categories in terms of similarities, children represent both in terms of similarity, which gives the impression that children's category retrieval is statistics-based rather than ruled-based [70].

Mental Rotation

Another skill that is frequently omitted in action perception literature is mental rotation. However, there are reasons to think that mental rotation is necessary for action perception [27, 101]. Even if infants do not rely on their own sensorimotor experience when perceiving others' actions, they still might need to mentally rotate the action they observe to allow for generalization across different previously encountered instances of the observed action.

Habituation studies have revealed that already at 3-months of age, the first signs of mental rotation (MR) can be registered. When confronted with dynamic 3D stimuli, male 3-month-old infants can discriminate objects and their mirrored counterparts presented from a novel angle [93]. Mental rotation is regarded as more challenging when it is based on static images than when it is based on dynamic stimuli. Nevertheless, 3.5-month-olds were found able of mentally rotating objects presented by means of static images. Likely, 3.5-month-olds were able to do so because 2D objects were used, which are thought to be easier objects for mental rotation than 3D objects. Here again, MR abilities were only found in the male half of the group and not in the female half of the group [109]. The sex difference seems consistent in the early months [92, 110]. However, around 6- to 9-months, this difference in task performance disappears [39, 97, 117].

Results from MR studies in children, however, paint a completely different picture. Two studies with 4–5-year-olds demonstrated that the tested children were capable of mental rotation [83, 84], but a follow-up study using the same procedure but different stimuli failed to replicate the result [26]. Many children seem to perform at chance level at this age [32, 72, 98]. At least part of the discrepancy between the infant and child findings might stem from differences in task difficulty. In the habituation studies, infants merely need to detect incongruities between the stimuli they are presented with, whereas the tasks used with children demand from the children that they prospectively mentally simulate potential outcomes [40]. It might be that these more challenging tasks can be accomplished only when executive functioning is developed further [40].

Together, these findings posit a challenge for the idea that mental rotation is necessary for action perception development. The MR task for infants in action perception is not to passively respond to incongruent and congruent action stimuli, but rather their task is to actively construe whether the observed action is similar to a previously seen action. Given that this active construction appears to be already hard for children let alone infants, it might be that MR is not utilized in action perception. Potentially, infants learn to generalize across different instantiations of actions due to other invariant properties of the observed actions. More empirical research is needed to elucidate the role of MR for action perception development.

Statistical Learning

Whereas categorization and mental rotation often go unmentioned in the action perception literature, the importance of learning processes is not overlooked. Experience might shape action perception development permitted that the infant has the capacity to learn from these experiences. The associative sequence learning (ASL) model [56] for instance postulates that associative learning is the fundamental learning process that connects experience and action perception development. The most rudimentary and ever-present process underlying associative learning is Hebbian learning. 'What fires together, wires together' is the phrase often used to describe Hebbian learning. More formally, Hebbian learning is the phenomenon that synapses are strengthened when postsynaptic firing occurs shortly after a presynaptic action potential [124]. The strengthening of the synapse goes together with weakening of other, competing, synapses. This temporal correspondence typically takes place when there are multiple input streams that originate from the same event. When a drop of water falls on your hand, visual input arrives via the eyes, auditory input arrives via the ears, and tactile information arrives via touch receptors, and all of this input is received simultaneously and belongs to the same external event. Hence, Hebbian learning is a useful mechanism to strengthen relevant connections and weaken irrelevant connections in the brain. It forms a neuronal explanation for how associative learning can take place. However, Hebbian learning requires a tight temporal coupling between events as the presynaptic action potential should precede postsynaptic firing by no more than 50 ms. If the presynaptic action potential follows rather than precedes postsynaptic firing, synapse strength is weakened [124]. In other words, not all associative learning can be explained by Hebbian learning. In associative learning, the learner starts associating two related stimuli or events through repeated experience with the paired events [113]. The relation between the events can take a variety of forms as the relation may for instance be temporal or spatial in its nature. For example, dropping a ball consistently leads to the ball hitting the floor, which is a temporal relationship with a relatively loose coupling as the delay between dropping the ball and it hitting the floor is variable. An example of a spatial relationship is that a tap is often placed above a sink. According to associationism, contingency between events is a necessary condition for associating two events. Events A and B are said to be contingent if and only if the probability that A occurs is higher when B occurs as well, than the probability of A occurring in the absence of event B. Learning to associate two events is most frequently studied using stimuli with a deterministic relationship that are presented very closely in time. Within associationism, contiguity, closeness in time, is considered a necessary and supportive factor in associative learning [14]. The idea would be that chances that a pairing is learnt decreases with increasing time between the events. However, research has shown that it is not the time in-between the to-be-coupled events that matters but rather the rate of occurrence [47]. The ratio of the time between to-be-paired events and the time between consecutive pairs influences the learning rate. In other words, associative learning is time-invariant but rate dependent [42]. Keep in mind that these conclusions are drawn from experiments in which no other events took place. In real-world situations, many events happen, which all might need to be tracked in order to learn what comes with what. Due to that combinatorily explosion, it is likely that there is upper bound to the delay between the events that can be learnt to belong together. For instance, causal inference strongly drops if the delay between the potential cause, and the presumed effect increases beyond 5 s [14]. This limit can be stretched with knowledge about the to-be-expected delay [14], but it illustrates that spontaneous learning requires contiguity.

Empirical studies from various domains demonstrate that infants are capable of learning about regularities, which includes associative learning. Work from Haith and colleagues has shown that infants at 2 or 3 months can learn simple spatiotemporal regularities and form expectations based on these regularities [52, 51]. This was tested by showing infants sequences of pictures of which the locations of appearance formed a regular order. Beyond simple spatiotemporal regularities, 3-month-olds were also found to be able to predict the location of upcoming visual stimuli based on the contents of the current visual stimulus, indicating that they learned to associate the content of a picture with the location of the next picture, and could use this association as a basis for anticipatory eye movements [135]. In a broader sense, developmental scientists are keen on unraveling infants' potential to learn from statistical regularities in the environment, a skill also known as statistical learning. For instance, 9-month-olds are found capable of learning that some items have a fixed spatial relationship, a problem infants not merely solve based on the heightened probability of co-occurrence [34]. Temporal relationships, and specifically, which items follow each other and which do not, can also be learnt. Saffran and colleagues presented 8-month-olds with a continuous artificial speech stream in which some syllables deterministically followed each other and some syllables followed each other with a probability lower than 100% [114]. The probability that items follow each other is called the transitional probability. The tested infants were able to distinguish artificial words from the continuous speech stream based on these transitional probabilities. Follow-up research highlighted that being exposed to a continuous speech stream containing elements with high transitional probabilities (artificial words) facilitated infants' object-label learning of these artificial words [55]. Moreover, acquiring word-referent mappings is a feat that 12-month-olds accomplish by combining cross-situational statistics, meaning that auditory labels are learnt to belong to specific pictures by means of a process of combining, comparing and eliminating options [121].

Assuming these abilities also hold for action perception would imply that infants might be able to parse streams of observed actions into smaller chunks, and might use the transitional probabilities between these chunks to learn which parts of actions belong together. Indeed, eye-tracking results illustrate that infants are able to learn to segment streams of actions on the basis of transitional probabilities [125]. Furthermore, toddlers can predict which action comes next based on transitional probabilities [91]. Infants might use cross-situational statistics to find out which action steps are and which steps are not strictly necessary in a chain of sub-actions, but more empirical work is needed to demonstrate whether this is indeed the case.

Sensorimotor Development

Apart from statistical learning abilities, sensorimotor development, or motor development in short, is frequently regarded as an important basis for action perception development. Human infants are born with a strikingly limited set of motor skills, of which most are acquired in its basic form within the first years of life. Listing motor milestones has numerous downsides as it obscures the reality of motor development which is characterized by large interindividual differences, protracted skill acquisition, non-deterministic ordering, overlap in acquisition periods and infants going back-and-forth in their performance on tested skills [2]. Nevertheless, for those unfamiliar with infant motor development, it might be of interest to mention a few average onset ages to give the reader a flavor. Reaching toward objects starts immediately when infants are born [59], but this behavior is first labeled 'prereaching' as infants are unsuccessful in reaching the objects. Successful reaching and grasping emerges around 5 months of age [9]. Reaching furthermore improves when infants gain the ability to control their posture when sitting upright [12, 128], a skill that emerges around 6–9 months of age [9, 53]. Around the same age, infants start to locomote by means of crawling [9]. The 'average' infant starts walking at 12 months of age. It is important to realize that the reported ages stem from western, mainly North-American studies. Motor development does not take place in a vacuum but is embedded within the culture the infant is raised in [3, 43].

Imitation

Motor acts can be performed independently, but also in response to others' actions. Copying others' actions, imitation, is an example of such a response. Although on the one hand, the ability to imitate can be viewed as an outcome of action perception development, it may on the other hand also function as a driving force for action perception development. Following Meltzoff's 'Like me' hypothesis, infants have the opportunity to gain a deeper understanding of what others do by copying their behavior [88]. Take the imaginary case in which Ann has never drunk a warm beverage. Now, say Ann observes Peter reaching out for a cup of coffee. Peter carefully lifts the filled cup to his mouth using just a few fingers, meanwhile opens his mouth and then slowly tilts the cup such that a small bit of coffee flows into his mouth. Ann sees him swallow the coffee. By going through the same set of motions, Ann will discover why Peter handled the cup so carefully and took such a small sip, namely because of the heat. This thought experiment illustrates that action understanding might arise from reproducing the actions of others. The ability to overtly or covertly—by means of imagining—reproduce the other's action can form the basis for linking and comparing own experiences to the experiences of others.

Seminal work from Meltzoff and Moore suggests that infants are born with the ability to imitate others, even if the actions are opaque [89]. Opaque actions are actions you cannot see yourself do, or more broadly speaking, the action cannot be reproduced simply by means of comparing and adjusting one's action to the other's action in the same sensory modality. Although the idea of neonatal imitation is very appealing, researchers report difficulties in replicating the initial results [5, 100]. Newborns do consistently imitate one action, namely tongue protrusion, but there is an alternative explanation for that behavior. Newborns reliably stick out their tongue in response to music [67], which indicates that tongue protrusion can be sign of arousal, and, watching someone else protrude his tongue seems arousing to infants as they generally look longer to tongue protrusion than to other actions they observe [66]. If infants are not born with the ability to imitate, then imitative abilities must develop, as adults can imitate opaque actions (but see [21]). Depending on the age of acquisition, imitation can or cannot fulfill a supporting role in the development of action understanding abilities. In a broader sense, whether it is triggered by observing others or not, own experience can lead to a deeper understanding of others' actions.

5.4.3 How Experience Can Shape Action Perception Development

The Formation of Sensorimotor Associations: A Theoretical Account

Being able to tie in own experience when attempting to decipher others' actions bears large potential for action perception development. Associative learning might be a way to build the connection between own and others' actions. The Associative Sequence Learning (ASL) model posits that correlated sensorimotor experience is the prerequisite for infants to develop the ability to overtly or covertly imitate others [56, 58]. For instance, through repeatedly watching their own arm movements, infants might learn to associate the sight of their arm movement with the motor command that caused the arm to move. Once this association has been established, the infant might activate this motor command again when observing someone else making the same movement. Self-observation is not a sufficient explanation for the acquisition of all possible sensorimotor links needed in action perception. That is, opaque actions are perceived through different sensory modalities when self-performed than when performed by another person, and consequently, the percept of the other's action does not match the percept of one's own action. However, correlated visuomotor experiences might arise for opaque actions through mirrors or through being imitated [111]. Parents have a tendency to imitate their baby [104, 131] and by imitating the infant, they provide their infants with opportunities to experience seeing a facial action while performing the same facial action themselves. Linking sensory and motor codes might also take an indirect path, such that if the infant hears and sees a person saying 'ba', and produce the same speech sound later themselves, they can link the observation of 'ba' with the motor code for 'ba' through the similarity of the auditory streams that went with both expressions of 'ba'. A comparable situation may arise for emotions: The visual percept of a smile or an unhappy face might be coupled with the motor code for these actions by a shared joyful or a shared frustrating experience.

The Formation of Sensorimotor Associations: An Empirical Account

The ASL model leads to testable predictions regarding the development of action perception, namely that if the infant is capable of associative learning, then sensorimotor experience will drive the development of action perception development. The data on infant learning discussed above illustrate that infants have the capacity to learn from correlated experiences from early on. The question whether sensorimotor experience forms a basis for action perception development has been the focus of investigation of numerous empirical studies in the infant domain.

The habituation paradigm developed by Woodward classically shows that from about 5 or 6 months of age, infants dishabituate when observing a change in reach target but not when observing a change in reach path [139]. This ability emerges at the same age as the ability to successfully reach and grasp objects. At 3-months of age, infants do not distinguish new path from new target events when tested in the Woodward paradigm [123], and at this age, infants are typically also unable to pick up and manipulate objects. However, when they wear mittens with Velcro attached at the palmar sides, 3-month-olds can pick up objects that have some pieces of Velcro attached to it as well. Sommerville and colleagues [123] did exactly that: They had 3-month-old infants wearing 'sticky' mittens and gave them the opportunity to play with objects that had some Velcro on it. When subsequently tested in the habituation paradigm featuring the same toys as they had just played with, behavior of the 3-month-olds resembled the 5- to 6-month-olds as they now dishabituated to the change in target object. The results suggest that short-term sensorimotor experience may alter the perception of others' actions.

Action prediction studies reveal a similar pattern. Falck-Ytter and colleagues [33] were the first to show that infants make anticipatory eye movements when observing others' actions. Specifically, the tested infants observed an actor reaching out for a toy, picking it up and transporting it to a bucket on the other end of the scene. Twelve-month-olds looked ahead of the action to the bucket, whereas 6-month-olds did not. This corresponds to their action abilities, as 12-month-olds are capable of transport actions whereas 6-month-olds are not. In similar vein, Ambrosini and colleagues found that 10-month-olds anticipated reaching actions to small and large target objects, whereas 6- and 8-months only anticipated reaching to large targets [4]. Visual anticipations of others' actions corresponded to their own motor abilities as 10-month-olds are able to reach for and grasp large and small objects, whereas younger infants can only successfully reach for and grasp large objects. Sensorimotor experience furthermore alters predictions regarding the timing of observed actions. In an eye-tracking study with 14- and 18- to 20-month-old infants, the role of walking and crawling experience on temporal predictions of walking and crawling was investigated [127]. The older age group, experienced in walking and crawling, was accurate in predicting the timing of both actions. The younger age group, with little to no walking experience, was only accurate in predicting crawling.

Based on the ASL model, perceptual experience can alter action perception as well if the perceived action is already part of the perceiver's motor repertoire. That is, upon observing someone performing an action with an auditory effect, the observer might learn to associate the sound with the action [28, 31]. At the same time, the observer might recruit her cortical motor system, mentally simulating the observed action [69]. Once the action-sound association is formed, merely hearing the sound might already activate motor areas in the brain [102]. Empirical data are in line with this hypothesis. Nine-month-olds who had seen their parent shake a rattle subsequently displayed stronger motor activation when hearing the sound produced by the rattle compared to an entirely novel sound and a familiar but action-unrelated sound [103].

That active sensorimotor experience was fundamental for this effect to emerge was illustrated by a study in which infants were trained to perform a novel tool-use action that elicited a sound and observed their parent performing a comparable novel tool-use action that elicited another sound [46]. Infants who demonstrated to be able to produce the trained action after repeatedly training at home showed stronger motor activation when hearing the sound belonging to their hands-on trained action compared to the sound heard when observing their parent perform the other action. In contrast, infants who were despite the training unable to perform the action did not show differential motor activation for the sound associated with the parent's action and the sound associated with their own action experience. In sum, the empirical data on action perception development are largely in line with the hypothesis derived from ASL, namely that sensorimotor experience plays a crucial role.

Beyond Contingent Sensorimotor Experiences

The ASL model provides an example of how action perception abilities may develop in early life. However, other routes are possible and turn out to be utilized as well. For instance, Hunnius and Bekkering [61] demonstrated that already at 6-months of age, infants more frequently show anticipatory eye movements to someone's ear when that person is picking up a phone rather than a cup. It is quite unlikely that knowledge about where phones go is inborn, and at 6-months, infants are not yet motorically capable of bringing objects to their own ear. This implies that purely perceptual, in this case observational experience, forms the building block for action prediction development.

But if observational experience is sufficient for action prediction to emerge, why would sensorimotor experience then play a role as well? Potentially, there are limits to what can be easily learnt from observation. It might be that observational experience can form the basis for target predictions—where the action will end—but not for temporal predictions—when the action will end. Circumstantial evidence comes from the previously mentioned study on temporal predictions of crawling and walking: Quite likely, all tested participants had more visual experience with walking than with crawling actions, but still, the youngest group was better able to predict crawling than walking [127].

Learning from purely observational experience has the advantage over learning from sensorimotor experience that action perception abilities would not be confined to the range of actions one can perform herself. When looking at complicated gymnastic moves, most viewers will not be able to reproduce the observed action. However, when learning from observation is a route to acquiring action perception capacities, then frequently observing these complex movements may allow later prediction of these movements. Especially in development, acquiring action perception abilities through visual experience is beneficial as motor skills take numerous months if not years to develop.

Beyond the classic sensorimotor contingency learning as outlined in the ASL model, experience has the potential to drive action perception development in ways that cross the bounds of the action-specific type of learning explained by the ASL model. Experience, be it purely sensory or sensorimotor in nature, allows—at least theoretically—for the acquisition of rules and regularities that govern others' actions. For instance, reaching trajectories of adults are typically straight lines, and even if external perturbations are applied that naturally evoke a detour in the trajectory, adults tend to adapt their reaches to become straight again [137]. In movement sciences, it is generally agreed upon that motor control strives to minimize costs [138] and hence of all possible ways of performing a movement, we tend to select the most efficient one [94]. Infants and young children are generally not efficient in their own actions [1] but rather seem to first prioritize exploration over exploitation [20]. With increasing action experience, infants might learn which actions or paths are efficient and which are not. Some scholars have postulated that principles of efficiency, mostly implicitly operationalized as shortest path, are understood and applied to action perception already in the first year of life [44]. Evidence supporting the efficiency claim stems primarily from habituation studies featuring non-human agents [45]. Sodian et al. [122] replicated the original effects portraying human agents in the stimuli. Problematic in these studies is, however, that the most efficient action is also observed more frequently in daily life, which makes it hard to dissociate whether the principle of efficiency arises through maturational processes or through experience.

5.5 Conclusion

Infants acquire, within the first few years of life, the ability to dissociate observed actions from each other, segment action streams into smaller parts, form expectations and predict others' actions, and ultimately form an understanding of others' actions. These action perception abilities develop rapidly and are highly relevant to the infant as action perception is foundational for becoming a proficient social partner. Other people provide a wealth of resources for the developing child as they are the gateway to nutrition, comfort, exciting opportunities and guidance in the dazzling world around them.

Action perception gets a head start through early preferences for faces, manipulable objects, visible mouth movements and biological movement in general. The initial tendencies to look at relevant aspects of the visual scene provide infants with important input. Categorization learning and learning in general are employed to transform the input into useful building blocks for the emerging action perception abilities. By seeing others act and by acting themselves, infants can, thanks to their learning capacities, learn to decipher what others are doing.

References

1. Adolph, K. E., Cole, W. G., Komati, M., Garciaguirre, J. S., Badaly, D., Lingeman, J. M., Chan, G. L., et al. (2012). How do you learn to walk? Thousands of steps and dozens of falls per day. *Psychological Science, 23*(11), 1387–1394.
2. Adolph, K. E., Cole, W. G., & Vereijken, B. (2014). Intraindividual variability in the development of motor skills in childhood. In M. Diehl, K. Hooker, & M. Sliwinski (Eds.), *Handbook of intraindividual variability across the life span* (pp. 79–103). London: Routledge.
3. Adolph, K. E., Karasik, L. B., & Tamis, C. S. (2014). Motor skill. In M. H. Bornstein (Ed.), *Handbook of cultural developmental science* (pp. 73–100). England: Psychology Press.
4. Ambrosini, E., Reddy, V., De Looper, A., Costantini, M., Lopez, B., & Sinigaglia, C. (2013). Looking ahead: Anticipatory gaze and motor ability in infancy. *PLoS ONE, 8*(7), e67916.
5. Anisfeld, M. (1991). Neonatal imitation. *Developmental Review, 11*(1), 60–97.
6. Atkinson, A. P., Dittrich, W. H., Gemmell, A. J., & Young, A. W. (2004). Emotion perception from dynamic and static body expressions in point-light and full-light displays. *Perception, 33*(6), 717–746.
7. Baldwin, D. A., Baird, J. A., Saylor, M. M., & Clark, M. A. (2001). Infants parse dynamic action. *Child Development, 72*(3), 708–717.
8. Bardi, L., Regolin, L., & Simion, F. (2011). Biological motion preference in humans at birth: Role of dynamic and configural properties. *Developmental Science, 14*(2), 353–359.
9. Bayley, Nancy. (2006). *Bayley scales of infant and toddler development: administration manual*. San Antonio, TX: Harcourt Assessment.
10. Bekkering, H., Wohlschlager, A., & Gattis, M. (2000). Imitation of gestures in children is goal-directed. *The Quarterly Journal of Experimental Psychology: Section A, 53*(1), 153–164.
11. Bertenthal, B. I., & Davis, P. (1988). *Dynamical pattern analysis predicts recognition and discrimination of biomechanical motions*. Chicago: Paper presented at the annual meeting of the Psychonomic Society.
12. Bertenthal, B., & von Hofsten, C. (1998). Eye, head and trunk control: The foundation for manual development. *Neuroscience and Biobehavioral Reviews, 22*(4), 515–520.
13. Bertenthal, B. I., Proffitt, D. R., Kramer, S. J., & Spetner, N. B. (1987). Infants' encoding of kinetic displays varying in relative coherence. *Developmental Psychology, 23*(2), 171.
14. Buehner, M. J. (2005). Contiguity and covariation in human causal inference. *Learning & Behavior, 33*(2), 230–238.
15. Buresh, J. S., & Woodward, A. L. (2007). Infants track action goals within and across agents. *Cognition, 104*(2), 287–314.
16. Butterworth, G., & Jarrett, N. (1991). What minds have in common is space: Spatial mechanisms serving joint visual attention in infancy. *British Journal of Developmental Psychology, 9*, 55–72.
17. Carpenter, M., Akhtar, N., & Tomasello, M. (1998). Fourteen-through 18-month-old infants differentially imitate intentional and accidental actions. *Infant behavior and Development, 21*, 315–330.
18. Choisdealbha, Á. N., & Reid, V. (2014). The developmental cognitive neuroscience of action: Semantics, motor resonance and social processing. *Experimental Brain Research, 232*(6), 1585–1597.
19. Claxton, L. J., Keen, R., & McCarty, M. E. (2003). Evidence of motor planning in infant reaching behavior. *Psychological Science, 14*(4), 354–356.
20. Comalli, D. M., Keen, R., Abraham, E. S., Foo, V. J., Lee, M. H., & Adolph, K. E. (2016). The development of tool use: Planning for end-state comfort. *Developmental Psychology, 52*(11), 1878.
21. Cook, R., Johnston, A., & Heyes, C. (2013). Facial self-imitation: Objective measurement reveals no improvement without visual feedback. *Psychological Science, 24*(1), 93–98.
22. Cooper, F. S., Delattre, P. C., Liberman, A. M., Borst, J. M., & Gerstman, L. J. (1952). Some experiments on the perception of synthetic speech sounds. *The Journal of the Acoustical Society of America, 24*(6), 597–606.

23. Corkum, V., & Moore, C. (1995). Development of joint visual attention in infants. In C. Moore & P. Dunham (Eds.), *Joint attention: Its origin and role in development*. Hillsdale, NJ: Erlbaum.
24. Cutting, J. E., & Kozlowski, L. T. (1977). Recognizing friends by their walk: Gait perception without familiarity cues. *Bulletin of the Psychonomic Society, 9*(5), 353–356.
25. Daum, M. M., & Gredebäck, G. (2011). The development of grasping comprehension in infancy: Covert shifts of attention caused by referential actions. *Experimental Brain Research, 208*(2), 297–307.
26. Dean, A. L., & Harvey, W. O. (1979). An information-processing analysis of a Piagetian imagery task. *Developmental Psychology, 15*(474–476), 9.
27. Dittrich, W. H. (1993). Action categories and the perception of biological motion. *Perception, 22*(1), 15–22.
28. Drost, U. C., Rieger, M., Brass, M., Gunter, T. C., & Prinz, W. (2005). Action-effect coupling in pianists. *Psychological Research, 69*(4), 233–241.
29. Ekman, P. (1993). Facial expression and emotion. *American Psychologist, 48*(4), 384.
30. Elbers, L. (1982). Operating principles in repetitive babbling: A cognitive continuity approach. *Cognition, 12,* 45–63.
31. Elsner, B., & Hommel, B. (2001). Effect anticipation and action control. *Journal of Experimental Psychology: Human Perception and Performance, 27*(1), 229–240.
32. Estes, D. (1998). Young children's awareness of their mental activity: The case of mental rotation. *Child Development, 69,* 1345–1360.
33. Falck-Ytter, T., Gredebäck, G., & von Hofsten, C. (2006). Infants predict other people's action goals. *Nature Neuroscience, 9*(7), 878–879.
34. Fiser, J., & Aslin, R. N. (2002). Statistical learning of new visual feature combinations by infants. *Proceedings of the National Academy of Sciences, 99*(24), 15822–15826.
35. Fitts, P. M. (1954). The information capacity of the human motor system in controlling the amplitude of movement. *Journal of Experimental Psychology, 47,* 381–391.
36. Flanagan, J. R., & Johansson, R. S. (2003). Action plans used in action observation. *Nature, 424*(6950), 769–771.
37. Fox, R., & McDaniel, C. (1982). The perception of biological motion by human infants. *Science, 218*(4571), 486–487.
38. French, R. M., Mareschal, D., Mermillod, M., & Quinn, P. C. (2004). The role of bottom-up processing in perceptual categorization by 3-to 4-month-old infants: Simulations and data. *Journal of Experimental Psychology: General, 133*(3), 382.
39. Frick, A., & Möhring, W. (2013). Mental object rotation and motor development in 8-and 10-month-old infants. *Journal of Experimental Child Psychology, 115*(4), 708–720.
40. Frick, A., Möhring, W., & Newcombe, N. S. (2014). Development of mental transformation abilities. *Trends in Cognitive Sciences, 18*(10), 536–542.
41. Friederici, A. D. (2002). Towards a neural basis of auditory sentence processing. *Trends in cognitive sciences, 6*(2), 78–84.
42. Gallistel, C. R., & Gibbon, J. (2000). Time, rate, and conditioning. *Psychological Review, 107*(2), 289.
43. Geber, M., & Dean, R. F. A. (1957). The state of development of newborn African children. *The Lancet, 269*(6981), 1216–1219.
44. Gergely, G., & Csibra, G. (2003). Teleological reasoning in infancy: The naıve theory of rational action. *Trends in Cognitive Sciences, 7*(7), 287–292.
45. Gergely, G., Nádasdy, Z., Csibra, G., & Bíró, S. (1995). Taking the intentional stance at 12 months of age. *Cognition, 56*(2), 165–193.
46. Gerson, S. A., Bekkering, H., & Hunnius, S. (2015). Short-term motor training, but not observational training, alters neurocognitive mechanisms of action processing in infancy. *Journal of Cognitive Neuroscience, 27*(6), 1207–1214.
47. Gibbon, J., Farrell, L., Locurto, C. M., Duncan, H. J., & Terrace, H. S. (1980). Partial reinforcement in autoshaping with pigeons. *Animal Learning & Behavior, 8*(1), 45–59.

48. Goren, C. C., Sarty, M., & Wu, P. Y. (1975). Visual following and pattern discrimination of face-like stimuli by newborn infants. *Pediatrics, 56*(4), 544–549.
49. Gottwald, J. M., De Bortoli Vizioli, A., Lindskog, M., Nyström, P., Ekberg, T. L., von Hofsten, C., et al. (2017). Infants prospectively control reaching based on the difficulty of future actions: To what extent can infants' multiple-step actions be explained by Fitts' law? *Developmental Psychology, 53*(1), 4–12.
50. Haith, M. M., Bergman, T., & Moore, M. J. (1977). Eye contact and face scanning in early infancy. *Science, 198*(4319), 853–855.
51. Haith, M. M., Wentworth, N., & Canfield, R. L. (1993). The formation of expectations in early infancy. *Advances in Infancy Research, 8,* 251–297.
52. Haith, M. M., Hazan, C., & Goodman, G. S. (1988). Expectation and anticipation of dynamic visual events by 3.5-month-old babies. *Child Development*, 467–479.
53. Harbourne, R. T., & Stergiou, N. (2003). Nonlinear analysis of the development of sitting postural control. *Developmental Psychobiology: The Journal of the International Society for Developmental Psychobiology, 42*(4), 368–377.
54. Hardison, D. M. (2003). Acquisition of second-language speech: Effects of visual cues, context, and talker variability. *Applied Psycholinguistics, 24*(4), 495–522.
55. Hay, J. F., Pelucchi, B., Estes, K. G., & Saffran, J. R. (2011). Linking sounds to meanings: Infant statistical learning in a natural language. *Cognitive Psychology, 63*(2), 93–106.
56. Heyes, C. (2001). Causes and consequences of imitation. *Trends in Cognitive Sciences, 5*(6), 253–261.
57. Heyes, C. (2016). Homo imitans? Seven reasons why imitation couldn't possibly be associative. *Philosophical Transactions of the Royal Society B: Biological Sciences, 371*(1686), 20150069.
58. Heyes, C. M., & Ray, E. D. (2000). What is the significance of imitation in animals? In *Advances in the study of behavior* (Vol. 29, pp. 215–245). New York: Academic.
59. Von Hofsten, C. (1982). Eye–hand coordination in the newborn. *Developmental Psychology, 18*(3), 450.
60. Hunnius, S., & Geuze, R. H. (2004). Developmental changes in visual scanning of dynamic faces and abstract stimuli in infants: A longitudinal study. *Infancy, 6*(2), 231–255.
61. Hunnius, S., & Bekkering, H. (2010). The early development of object knowledge: A study of infants' visual anticipations during action observation. *Developmental Psychology, 46*(2), 446–454.
62. Iacoboni, M., Molnar-Szakacs, I., Gallese, V., Buccino, G., Mazziotta, J. C., & Rizzolatti, G. (2005). Grasping the intentions of others with one's own mirror neuron system. *PLoS Biology, 3*(3), e79.
63. Johnson, A. H., & Barrett, J. (2003). The role of control in attributing intentional agency to inanimate objects. *Journal of Cognition and Culture, 3*(3), 208–217.
64. Johnson, M. H., Dziurawiec, S., Ellis, H., & Morton, J. (1991). Newborns' preferential tracking of face-like stimuli and its subsequent decline. *Cognition, 40*(1–2), 1–19.
65. Johnson, S. C., Ok, S. J., & Luo, Y. (2007). The attribution of attention: 9-month-olds' interpretation of gaze as goal-directed action. *Developmental Science, 10*(5), 530–537.
66. Jones, S. S. (1996). Imitation or exploration? Young infants' matching of adults' oral gestures. *Child Development, 67*(5), 1952–1969.
67. Jones, S. S. (2006). Exploration or imitation? The effect of music on 4-week-old infants' tongue protrusions. *Infant Behavior and Development, 29*(1), 126–130.
68. Kanakogi, Y., & Itakura, S. (2011). Developmental correspondence between action prediction and motor ability in early infancy. *Nature Communications, 2,* 341.
69. Keysers, C., Kohler, E., Umiltà, M. A., Nanetti, L., Fogassi, L., & Gallese, V. (2003). Audiovisual mirror neurons and action recognition. *Experimental Brain Research, 153*(4), 628–636.
70. Kloos, H., & Sloutsky, V. M. (2008). What's behind different kinds of kinds: Effects of statistical density on learning and representation of categories. *Journal of Experimental Psychology: General, 137*(1), 52–72.

71. Koch, B., & Stapel, J. (2017). The role of head and hand movements for infants' predictions of others' actions. *Psychological Research*, 1–12.
72. Krüger, M., & Krist, H. (2009). Imagery and motor processes—When are they connected? The mental rotation of body parts in development. *Journal of Cognition and Development, 10,* 239–261.
73. Kutas, M., & Federmeier, K. D. (2011). Thirty years and counting: finding meaning in the N400 component of the event-related brain potential (ERP). *Annual Review of Psychology, 62,* 621–647.
74. Land, M. F., & Hayhoe, M. (2001). In what ways do eye movements contribute to everyday activities? *Vision Research, 41*(25–26), 3559–3565.
75. Langton, S. R., Watt, R. J., & Bruce, V. (2000). Do the eyes have it? Cues to the direction of social attention. *Trends in Cognitive Sciences, 4*(2), 50–59.
76. Lewkowicz, D. J., & Hansen-Tift, A. M. (2012). Infants deploy selective attention to the mouth of a talking face when learning speech. *Proceedings of the National Academy of Sciences, 109*(5), 1431–1436.
77. Liberman, A. M., Cooper, F. S., Shankweiler, D. P., & Studdert-Kennedy, M. (1967). Perception of the speech code. *Psychological Review, 74*(6), 431–461.
78. Lloyd-Fox, S., Blasi, A., Everdell, N., Elwell, C. E., & Johnson, M. H. (2011). Selective cortical mapping of biological motion processing in young infants. *Journal of Cognitive Neuroscience, 23*(9), 2521–2532.
79. Loula, F., Prasad, S., Harber, K., & Shiffrar, M. (2005). Recognizing people from their movement. *Journal of Experimental Psychology: Human Perception and Performance, 31*(1), 210.
80. Mareschal, D., & French, R. (2000). Mechanisms of categorization in infancy. *Infancy, 1*(1), 59–76.
81. Mareschal, D., French, R. M., & Quinn, P. C. (2000). A connectionist account of asymmetric category learning in early infancy. *Developmental Psychology, 36*(5), 635.
82. Mareschal, D., & French, R. M. (1997). A connectionist account of interference effects in early infant memory and categorization. In M.G. Shafto & P. Langley (Eds.), *Proceedings of the 19th Annual Conference of the Cognitive Science Society* (pp. 484–489). Mahwah, NJ: Erlbaum.
83. Marmor, G. S. (1975). Development of kinetic images: When does the child first represent movement in mental images? *Cognitive Psychology, 7,* 548–559.
84. Marmor, G. S. (1977). Mental rotation and number conservation: Are they related? *Developmental Psychology, 13,* 320–325.
85. Marshall, P. J., Young, T., & Meltzoff, A. N. (2011). Neural correlates of action observation and execution in 14-month-old infants: An event-related EEG desynchronization study. *Developmental Science, 14*(3), 474–480.
86. Maurer, D., & Young, R. E. (1983). Newborn's following of natural and distorted arrangements of facial features. *Infant Behavior and Development, 6*(1), 127–131.
87. Meltzoff, A. N. (1995). Understanding the intentions of others: Re-enactment of intended acts by 18-month-old children. *Developmental Psychology, 31*(5), 838.
88. Meltzoff, A. N. (2007). 'Like me': A foundation for social cognition. *Developmental Science, 10*(1), 126–134.
89. Meltzoff, A. N., & Moore, M. K. (1977). Imitation of facial and manual gestures by human neonates. *Science, 198*(4312), 75–78.
90. Merin, N., Young, G. S., Ozonoff, S., & Rogers, S. J. (2007). Visual fixation patterns during reciprocal social interaction distinguish a subgroup of 6-month-old infants at-risk for autism from comparison infants. *Journal of Autism and Developmental Disorders, 37*(1), 108–121.
91. Monroy, C. D., Gerson, S. A., & Hunnius, S. (2017). Toddlers' action prediction: Statistical learning of continuous action sequences. *Journal of Experimental Child Psychology, 157,* 14–28.
92. Moore, D. S., & Johnson, S. P. (2008). Mental rotation in human infants: A sex difference. *Psychological Science, 19*(11), 1063–1066.

93. Moore, D. S., & Johnson, S. P. (2011). Mental rotation of dynamic, three-dimensional stimuli by 3-month-old infants. *Infancy, 16*(4), 435–445.
94. Morel, P., Ulbrich, P., & Gail, A. (2017). What makes a reach movement effortful? Physical effort discounting supports common minimization principles in decision making and motor control. *PLoS Biology, 15*(6), e2001323.
95. Morton, J., & Johnson, M. H. (1991). CONSPEC and CONLERN: a two-process theory of infant face recognition. *Psychological Review, 98*(2), 164.
96. Muthukumaraswamy, S. D., Johnson, B. W., & McNair, N. A. (2004). Mu rhythm modulation during observation of an object-directed grasp. *Cognitive Brain Research, 19*(2), 195–201.
97. Möhring, W., & Frick, A. (2013). Touching up mental rotation: Effects of manual experience on 6-month-old infants' mental object rotation. *Child Development, 84*(5), 1554–1565.
98. Noda, M. (2010). Manipulative strategies prepare for mental rotation in young children. *European Journal Developmental Psychology, 7,* 746–762.
99. Oller, D. K., & Eilers, R. E. (1988). The role of audition in infant babbling. *Child Development, 59*(2), 441–449.
100. Oostenbroek, J., Suddendorf, T., Nielsen, M., Redshaw, J., Kennedy-Costantini, S., Davis, J., Clark, S., et al. (2016). Comprehensive longitudinal study challenges the existence of neonatal imitation in humans. *Current Biology, 26(10),* 1334–1338.
101. Parsons, L. M. (1994). Temporal and kinematic properties of motor behavior reflected in mentally simulated action. *Journal of Experimental Psychology: Human Perception and Performance, 20,* 709.
102. Paulus, M., van Dam, W., Hunnius, S., Lindemann, O., & Bekkering, H. (2011). Action-effect binding by observational learning. *Psychonomic Bulletin & Review, 18*(5), 1022–1028.
103. Paulus, M., Hunnius, S., & Bekkering, H. (2012). Neurocognitive mechanisms underlying social learning in infancy: Infants' neural processing of the effects of others' actions. *Social Cognitive and Affective Neuroscience, 8*(7), 774–779.
104. Pawlby, S. J. (1977). Imitative interaction. In H. Schaffer (Ed.), *Studies in mother–infant interaction* (pp. 203–224). New York: Academic.
105. Phillips, A. T., & Wellman, H. M. (2005). Infants' understanding of object-directed action. *Cognition, 98*(2), 137–155.
106. Pierrehumbert, J. B. (2003). Phonetic diversity, statistical learning, and acquisition of phonology. *Language and Speech, 46*(2–3), 115–154.
107. Posner, M. I. (1980). Orienting of attention. *Quarterly Journal of Experimental Psychology, 32*(1), 3–25.
108. Proffit, D. R., & Bertenthal, B. I. (1990). Converging operations revisited: Assessing what infants perceive using discrimination measures. *Perception and Psychophysics, 47,* 1–11.
109. Quinn, P. C., & Liben, L. S. (2008). A sex difference in mental rotation in young infants. *Psychological Science, 19*(11), 1067–1070.
110. Quinn, P. C., & Liben, L. S. (2014). A sex difference in mental rotation in infants: Convergent evidence. *Infancy, 19*(1), 103–116.
111. Ray, E., & Heyes, C. (2011). Imitation in infancy: the wealth of the stimulus. *Developmental Science, 14*(1), 92–105.
112. Reid, V. M., Hoehl, S., Grigutsch, M., Groendahl, A., Parise, E., & Striano, T. (2009). The neural correlates of infant and adult goal prediction: evidence for semantic processing systems. *Developmental Psychology, 45*(3), 620–629.
113. Rescorla, R. A., & Wagner, A. R. (1972). A theory of Pavlovian conditioning: Variations in the effectiveness of reinforcement and nonreinforcement. *Classical Conditioning II: Current Research and Theory, 2,* 64–99.
114. Saffran, J. R., Aslin, R. N., & Newport, E. L. (1996). Statistical learning by 8-month-old infants. *Science, 274*(5294), 1926–1928.
115. Saylor, M. M., Baldwin, D. A., Baird, J. A., & LaBounty, J. (2007). Infants' on-line segmentation of dynamic human action. *Journal of Cognition and Development, 8*(1), 113–128.

116. Scaife, M., & Bruner, J. S. (1975). The capacity for joint visual attention in the infant. *Nature, 253,* 265–266.
117. Schwarzer, G., Freitag, C., Buckel, R., & Lofruthe, A. (2013). Crawling is associated with mental rotation ability by 9-month-old infants. *Infancy, 18*(3), 432–441.
118. Shimada, S., & Hiraki, K. (2006). Infant's brain responses to live and televised action. *Neuroimage, 32*(2), 930–939.
119. Simion, F., Regolin, L., & Bulf, H. (2008). A predisposition for biological motion in the newborn baby. *Proceedings of the National Academy of Sciences, 105*(2), 809–813.
120. Sitnikova, T., Holcomb, P. J., Kiyonaga, K. A., & Kuperberg, G. R. (2008). Two neurocognitive mechanisms of semantic integration during the comprehension of visual real-world events. *Journal of Cognitive Neuroscience, 20*(11), 2037–2057.
121. Smith, L., & Yu, C. (2008). Infants rapidly learn word-referent mappings via cross-situational statistics. *Cognition, 106*(3), 1558–1568.
122. Sodian, B., Schoeppner, B., & Metz, U. (2004). Do infants apply the principle of rational action to human agents? *Infant Behavior and Development, 27*(1), 31–41.
123. Sommerville, J. A., Woodward, A. L., & Needham, A. (2005). Action experience alters 3-month-old infants' perception of others' actions. *Cognition, 96*(1), B1–B11.
124. Song, S., Miller, K. D., & Abbott, L. F. (2000). Competitive Hebbian learning through spike-timing-dependent synaptic plasticity. *Nature Neuroscience, 3*(9), 919.
125. Stahl, A. E., Romberg, A. R., Roseberry, S., Golinkoff, R. M., & Hirsh-Pasek, K. (2014). Infants segment continuous events using transitional probabilities. *Child Development, 85*(5), 1821–1826.
126. Stapel, J. C., Hunnius, S., & Bekkering, H. (2015). Fifteen-month-old infants use velocity information to predict others' action targets. *Frontiers in Psychology, 6,* 1092.
127. Stapel, J. C., Hunnius, S., Meyer, M., & Bekkering, H. (2016). Motor system contribution to action prediction: Temporal accuracy depends on motor experience. *Cognition, 148,* 71–78.
128. Thelen, E., & Spencer, J. P. (1998). Postural control during reaching in young infants: A dynamic systems approach. *Neuroscience and Biobehavioral Reviews, 22*(4), 507–514.
129. Thomson, J. A. (1983). Is continuous visual monitoring necessary in visually guided locomotion? *Journal of Experimental Psychology: Human Perception and Performance, 9*(3), 427.
130. Thut, G., Nietzel, A., Brandt, S. A., & Pascual-Leone, A. (2006). α-Band electroencephalographic activity over occipital cortex indexes visuospatial attention bias and predicts visual target detection. *Journal of Neuroscience, 26*(37), 9494–9502.
131. Uzgiris, I. C., Benson, J. B., Kruper, J. C., & Vasek, M. E. (1989). Contextual influences on imitative interactions between mothers and infants. In J. Lockman & N. Hazen (Eds.), *Action in social context: Perspectives on early development* (pp. 103–127). New York: Plenum Press.
132. Valenza, E., Simion, F., Cassia, V. M., & Umiltà, C. (1996). Face preference at birth. *Journal of Experimental Psychology: Human Perception and Performance, 22*(4), 892–903.
133. Vallortigara, G., & Regolin, L. (2006). Gravity bias in the interpretation of biological motion by inexperienced chicks. *Current Biology, 16*(8), R279–R280.
134. Vallortigara, G., Regolin, L., & Marconato, F. (2005). Visually inexperienced chicks exhibit spontaneous preference for biological motion patterns. *PLoS Biology, 3*(7), e208.
135. Wentworth, N., Haith, M. M., & Hood, R. (2002). Spatiotemporal regularity and interevent contingencies as information for infants' visual expectations. *Infancy, 3*(3), 303–321.
136. Wilmer, J. B., Germine, L., Chabris, C. F., Chatterjee, G., Williams, M., Loken, E., Nakayama, K., et al. (2010). Human face recognition ability is specific and highly heritable. *Proceedings of the National Academy ofSsciences, 107*(11), 5238–5241.
137. Wolpert, D. M. (1997). Computational approaches to motor control. *Trends in Cognitive Sciences, 1*(6), 209–216.
138. Wolpert, D. M., & Ghahramani, Z. (2000). Computational principles of movement neuroscience. *Nature Neuroscience, 3*(11s), 1212.
139. Woodward, A. L. (1998). Infants selectively encode the goal object of an actor's reach. *Cognition, 69*(1), 1–34.

140. Woodward, A. L. (2003). Infants' developing understanding of the link between looker and object. *Developmental Science, 6*(3), 297–311.
141. Woodward, A. L. (2009). Infants' grasp of others' intentions. *Current Directions in Psychological Science, 18*(1), 53–57.
142. Woodward, A. L., & Sommerville, J. A. (2000). Twelve-month-old infants interpret action in context. *Psychological Science, 11*(1), 73–77.
143. Young, A. W. (1998). *Face and mind*. Oxford: Oxford University Press.
144. Yu, C., & Smith, L. B. (2013). Joint attention without gaze following: Human infants and their parents coordinate visual attention to objects through eye-hand coordination. *PLoS ONE, 8*, e79659.
145. Yu, C., & Smith, L. B. (2017). Hand–eye coordination predicts joint attention. *Child Development, 88*(6), 2060–2078.

Chapter 6
The Importance of the Affective Component of Movement in Action Understanding

Giuseppe Di Cesare

Abstract Social interactions require the ability to evaluate the attitudes of others according to the way in which actions are performed. For example, a hand gesture can be kind or vigorous or the tone of voice can be pleasant or rude providing information about the attitude of the agent. Daniel Stern called these aspects of social communication vitality forms. Vitality forms continuously pervade the life of individuals and play a fundamental role in social relations. Despite the importance of vitality forms, very little is known on their neural basis. The aim of the present chapter is to provide an overview of the neural substrates underpinning the encoding of these aspects of social communication. This chapter is organized in four sections. Section 6.1 describes the structural and functional domains of the insular cortex. Section 6.2 provides evidence that the dorso-central insula plays a central role in the perception and expression of action vitality forms. Section 6.3 demonstrates that the same insular sector is also involved in the perception of words conveying gentle and rude vitality forms. Finally, Sect. 6.4 discusses the important role of vitality forms in social interactions and proposes some future perspectives.

6.1 Introduction

When observing actions performed by others, we are able to understand the action-goals as well as their intentions. These abilities are related to the existence of a basic brain mechanism known as "*mirror mechanism*" that transforms sensory representation of others' behavior into one's own motor representation of that behavior [12]. This mechanism is based on the activity of a distinct class of neurons that discharge both when individuals perform a goal directed action and when individuals observe another person performing the same action. Originally, mirror neurons were discovered in the ventral premotor cortex of the macaque monkey (area F5; di Pellegrino et al. [8]). Subsequently, the mirror mechanism has been found in humans in the

G. Di Cesare (✉)
Istituto Italiano di Tecnologia, Genoa, Italy
e-mail: giuseppe.dicesare@iit.it

N. Noceti et al. (eds.), *Modelling Human Motion*,
https://doi.org/10.1007/978-3-030-46732-6_6

parietal and premotor cortices as well as in anterior cingulate cortex [2] and in the anterior insula [16].

In addition to goal (what) and motor intention (why), there is another fundamental aspect of the action: the form (how). The aim of the present chapter is to focus on the action form highlighting its fundamental role in social communication. Indeed, during interpersonal relations, actions can be performed in different ways. For example, a hand shake can be gentle or vigorous, and a caress can be delicate or rushed communicating the positive or negative attitude of the agent. Similarly, words can be pronounced with a kind or unkind tone also conveying the agent's attitude. These different forms of communication have been named "vitality forms" by Stern [15]. Vitality forms continuously pervade the life of individuals characterizing their behaviors. The execution of vitality forms allows the agent to express his own mood/attitude, while the perception of vitality forms allows the receiver to understand the mood/attitude of others. For example, observing a person interacting with you, you may instantly understand if that person is glad or not and the same thing goes for words. Indeed, answering the phone, it is possible to understand how the other person feels by listening to the tone of voice. An interesting question is to investigate the neural correlates of these forms of communication. Results obtained in a functional magnetic resonance imaging study (fMRI) showed that the observation of actions performed with rude and gentle vitality forms produced the activation of a small part of the brain named dorso-central insula [3]. Most importantly, this brain area is activated not only during the perception of gentle or rude actions but also during their execution [6]. Thus, the activation of the same area for both the observation and execution of vitality forms strongly suggests the existence of a mirror mechanism for action vitality forms in the dorso-central insula. Differently from the mirror mechanism located in the parietal and frontal areas specific for the action goal understanding, the mirror mechanism located in the insula might allow one to express own mood/attitude and to understand those of others. It is important to note that, the same mechanism is also involved in the perception (listening) and expression of action verbs pronounced gently or rudely (speech vitality forms; Di Cesare et al. [4, 7]).

All these findings highlight that the insular cortex is the key node involved in the processing of vitality forms and suggest its plausible role in modulating the affective aspect of actions and words. The ability to express and recognize vitality forms allow people to be socially connected. Indeed, during interpersonal relations, the expression of vitality forms allows the agent to communicate by gestures or words his own affective state, while the perception of vitality forms allows the receiver to understand the positive or negative attitude of the agent and prepare an adequate motor response.

In this regard, vitality forms are a valuable feature of social communication useful to promote human–human and human–robot interactions.

The present chapter will provide an overview of action and speech vitality forms highlighting the neural substrates underpinning the encoding of these aspects of social communication. In particular, this chapter is subdivided into four sections devoted to the following topics:

1. *Structural and functional domains of the insula*. In this first section, it will be illustrated and described the anatomical structure of the insula indicating the location of the dorso-central insula, which is involved in the processing of vitality forms.
2. *The encoding of action vitality forms*. The second section will present fMRI data showing that the dorso-central insula is active during the observation and the execution of action vitality forms.
3. *The encoding of auditory vitality forms*. The third section will describe fMRI data showing that listening to action verbs pronounced with gentle and rude vitality forms activates the dorso-central insula.
4. *The role of vitality forms in social interactions*. Finally, the fourth section will show behavioral data highlighting that, during social interactions, the vitality form expressed by the agent influences the subsequent motor response of the receiver.

6.2 Structure and Function of the Insula

In humans, the anatomical structure of the insula has been described for the first time by Johann Christian Reil (1809). It is a small part of the brain, located in both the left and right hemispheres in the depth of the Sylvian fissure (Fig. 6.1a). Anatomically, the insula is made up of anterior and posterior parts separated by the central insular sulcus (CIS). The anterior insula includes the anterior, middle, and posterior short gyri (asg, msg, psg), while the posterior insula includes the anterior and the posterior long gyri (alg, plg) (Fig. 6.1b). Being anatomically connected with the amygdala, thalamic nuclei, and with many other cortical areas, the insula is involved in several different functions, such as attention, pain, gustation, and the processing of emotions. On the basis of a meta-analysis carried out on 1768 functional neuroimaging studies, Kurth et al. [10] described the functional organization of the insular cortex. In particular, the authors identified four distinct functional domains in the insula: the sensory-motor (SM), the olfactory-gustatory (OG), the socio-emotional (SE), and the cognitive domain (CG) (Fig. 6.1c). In this chapter, it will be discussed the role of dorso-central insula (DCI), which is composed by the middle and posterior short gyri (msg and psg; Fig. 6.1b), in the encoding of action and speech vitality forms.

6.3 The Encoding of Action Vitality Forms

During social interactions, important information about others' behavior is carried out by the form of the action. Action vitality form describes "how" an action is performed, representing an important aspect that an observer may capture viewing an action performed by others. Differently from the action goal (what) and intention

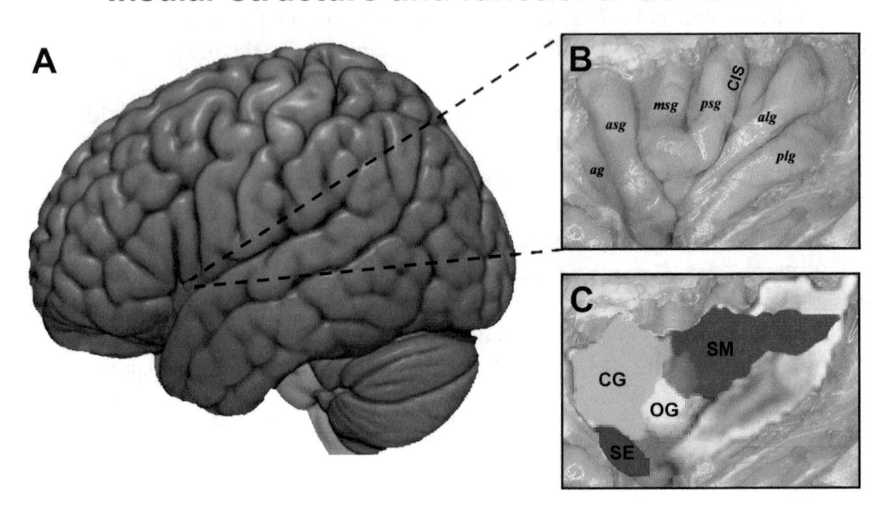

Fig. 6.1 Localization of the insula in the human brain (**a**). Anatomical structure of the insula (**b**): accessory gyrus (ag); anterior short gyrus (asg); middle short gyrus (msg); posterior short gyrus (psg); central insular sulcus (CIS); anterior long gyrus (alg); posterior long gyrus (plg). Functional domains identified in the right insular cortex: sensory-motor (red), olfactory-gustatory (yellow), socio-emotional (blue), and cognitive domain (green) (**c**). Figure adapted from Kurth et al. [10]

(why), vitality form (how) reflects the internal psychological state of the agent, providing also an appraisal of the affective quality underlying the relation between the agent and the action recipient [15]. In the first, fMRI study was investigated the neural correlates involved in the recognition of vitality forms. To this purpose, 19 healthy right-handed participants were presented with video clips lasting 3 s showing interactions between two actors that performed 4 actions without object (stroke the other actor's backhand, shake hands, clap hands, stop gesture; Fig. 6.2a) and 4 actions with object (pass a bottle, hand a cup, pass a ball, give a packet of crackers; Fig. 6.2b). Most importantly, each action was performed with a gentle or rude vitality form (Fig. 6.2c). During the fMRI experiment, participants were requested to pay attention either to the action goal (*what* task) or to the action vitality form (*how* task).

Results showed that the contrast between the two tasks (*what vs. how*) revealed activations for the *what task*, in the posterior parietal lobe and premotor cortex bilaterally, and in the caudal part of the inferior frontal gyrus of the left hemisphere (Fig. 6.3a). The opposite contrast (*how vs. what*) revealed a specific activation for the *how* task in the dorso-central insula of the right hemisphere (Fig. 6.3b). These data indicate that, paying attention to the action goal (*what* task) produces the activation of the parieto-frontal circuit classically involved in the action goal understanding [11]. In contrast, paying attention to the action vitality form (*how* task) produces the activation of the dorso-central insula. The main finding of this first fMRI study

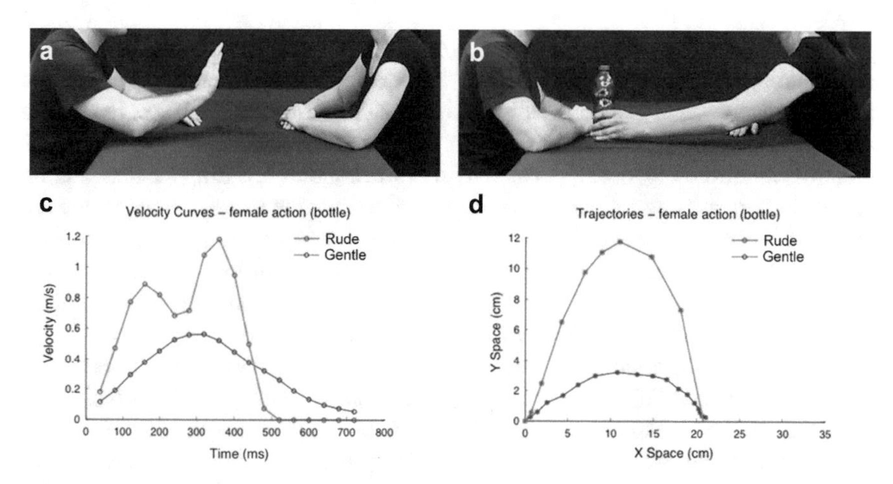

Fig. 6.2 Example of video clips observed by participants during the experiment. Frame representing an actor executing a stop gesture (**a**); frame representing an action with an object (passing a bottle; **b**). Velocity profiles (**c**), and trajectories (**d**) associated with one of the actions (passing a bottle) performed by the female actress with two vitality forms (rude: red line; gentle: blue line). As shown by graphs, the rude action was characterized by a hither velocity and a wider trajectory (*Y* space) than that observed for the gentle one. Figure adapted from Di Cesare et al. [3]

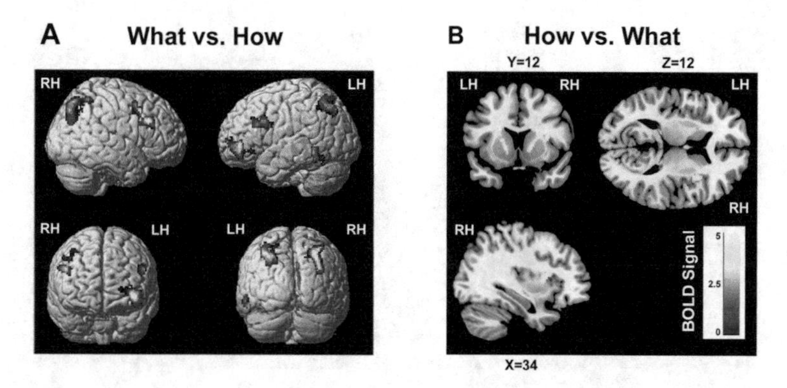

Fig. 6.3 Brain activations resulting from the direct contrasts *what task versus how task* (**a**) and *how task versus what task* (**b**). These activations are rendered into a standard MNI brain template (PFWE < 0.05 at cluster level). *LH* Left hemisphere; *RH*, right hemisphere. Figure adapted from Di Cesare et al. [3]

was the demonstration that, during action observation, the insula is the brain region involved in the processing of gentle and rude action vitality forms.

During social interactions, people not only observe vitality forms but also perform them. An interesting question is to investigate whether the dorso-central insula underlies both observation and execution of vitality forms. This issue was assessed in a subsequent fMRI study carried out on 15 healthy right-handed participants.

In this experiment, participants were requested to perform three different tasks: observation (OBS), imagination (IMA), and execution (EXE). In the observation task, participants observed video clips showing an actor passing an object in a gentle and rude way (vitality form condition; Fig. 6.4a1–b1) or placing a small ball in a box (control condition; Fig. 6.4c1). In the vitality form condition, during the imagination and execution tasks participants were requested to imagine to pass an object toward the actor facing them (Fig. 6.4a2–b2) or to move the object in a rude (Fig. 6.4a3) or gentle way (Fig. 6.4b3). In contrast, in the control condition, participants were requested to imagine to place a small ball in the box (imagination task; Fig. 6.4c2) or to place it without any explicit vitality form (execution task; Fig. 6.4c3).

The results of the conjunction analysis showed that in all three tasks (observation, imagination, execution) for each condition (rude, gentle, Ctrl), there was a bilateral activation of the premotor and parietal cortices plus a strong activation of the left somatosensory cortex, motor cortex and the dorsal part of the cerebellum (Fig. 6.5). In addition, in *rude* and *gentle* conditions, there was also the activation of the middle temporal area, the inferior frontal gyrus and the posterior parietal cortex. Most

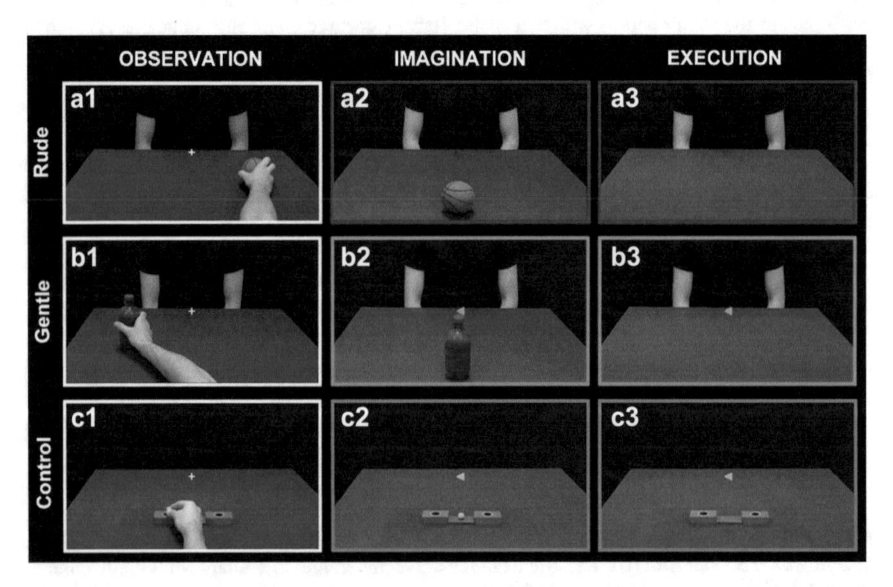

Fig. 6.4 Experimental design. Left column: *Observation task*. Participants observed the right hand of an actor moving an object in rightward (**a1**) or leftward (**b1**) directions. The observed action could be performed with a gentle or rude vitality form and the task request was to pay attention on the action vitality form. As a control participants observed the actor's hand placing a small ball in the right or left box (**c1**). Middle column: *Imagination task*. According with the edge screen color (red or blue), participants were requested to imagine themselves to pass an object toward another actor with a rude (red color; **a2**) or gentle (blue color; **b2**) vitality form. As a control the participants imagined to place a small ball in the right or left box (**c2**). Right column: *Execution task*. The participants moved a packet of crackers with a rude (red color; **a3**) or a gentle (blue color; **b3**) vitality form toward the actor facing them. As a control the participants had to place a small ball in the box (**c3**). Figure adapted from Di Cesare et al. [6]

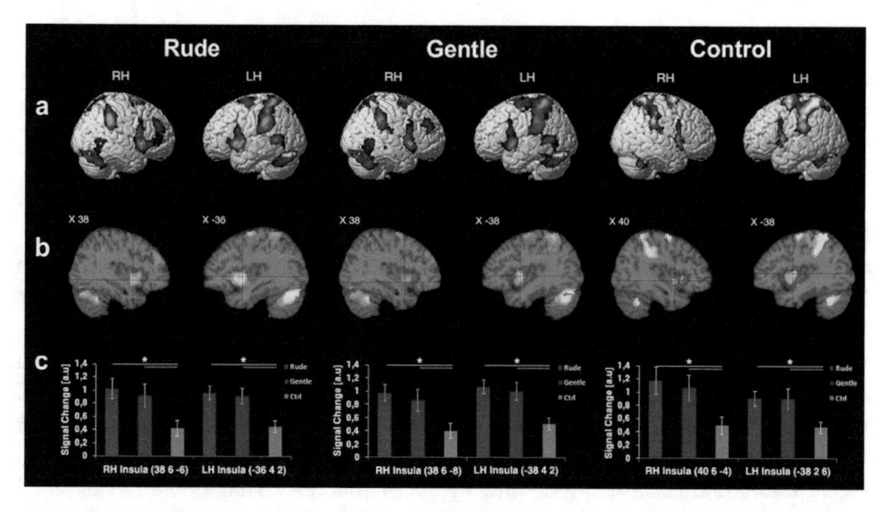

Fig. 6.5 Overlapping of areas activated in all three tasks (OBS, IMA, EXE). Lateral views of the brain activations obtained in the right and left hemispheres (**a**). Parasagittal sections showing the insular activations in the two hemispheres during the three tasks (**b**). These activations are rendered on a standard MNI brain template (PFWE < 0.05 at cluster level). BOLD signal were extracted from six regions of interest (ROIs) created on the dorso-central insula. All ROIs were defined centering the sphere (radium 10 mm) around the maxima of the functional maps resulting from a conjunction analysis of OBS, IMA, and EXE tasks. The horizontal lines indicate the comparisons between gentle, rude, and control conditions. Asterisks indicate significant differences ($p < 0.05$, Bonferroni correction). Figure adapted from Di Cesare et al. [6]

importantly, this analysis revealed a selective activation of the dorso-central part of the insula when the action was observed, imagined to perform and performed with a gentle or rude vitality form (Fig. 6.5b, c).

The finding that the dorso-central insula is involved in both vitality form perception and expression suggests that neurons of this insular sector might be endowed with the mirror mechanism transforming visual representation of the perceived vitality forms in their motor representation. This view is in line with other fMRI findings demonstrating that the anterior sector of the insula is active during both the expression and recognition of disgust in others [16]. A similar matching mechanism is likely involved in feeling of pain and in recognizing it in other [13]. Thus, the anterior and dorso-central sectors of the insula, although underlying different functions appear to be both endowed with the mirror mechanism.

6.4 The Encoding of Auditory Vitality Forms

During social interactions, words may be pronounced in gentle or rude way conveying different vitality forms. Listening to different speech vitality forms allows the receiver to understand the positive or negative attitude of the speaker. For example, answering the phone, it is possible to understand how the other person feels by hearing the tone of voice. As described for the action, the speech vitality forms allow people to communicate their internal state and to understand those of others by modulating the tone of voice [5, 7]. The ability to express and to understand the auditory vitality forms is already present in infants [14]. During mother–child interactions, the mother pronounces words by using a childish language. In particular, during the verbal communication with their children, mothers voluntary slow down the pronunciation of the verbal material adapting their language to the perceptive and expressive capacities of their children [1].

An interesting question is to understand whether the dorso-central insula, involved in the encoding of action vitality forms, is also involved in the encoding of speech vitality forms. In order to address this issue, an fMRI study was carried out on 16 healthy right-handed participants [4]. In particular, participants were presented with audio stimuli consisting of four Italian action verbs [Italian verbs: "dammi" (give), "prendi" (take), "tocca" (touch), "strappa" (tear)] pronounced by a male actor and a female actress. Most importantly, all the action verbs were pronounced using two different vitality forms: rude and gentle (vitality condition; Fig. 6.6a1–b1). For each action verb, two controls were presented: a robotic voice (robot condition) pronouncing the same action verbs as the actors; a scrambled version of the action verbs pronounced with gentle and rude vitality forms (scrambled VF condition). With regard to the robot condition, the robotic voice pronounced the same action verbs maintaining the meaning but not conveying any vitality form (Fig. 6.6a2–b2). In contrast, concerning the scrambled condition, the scrambled stimuli maintained the

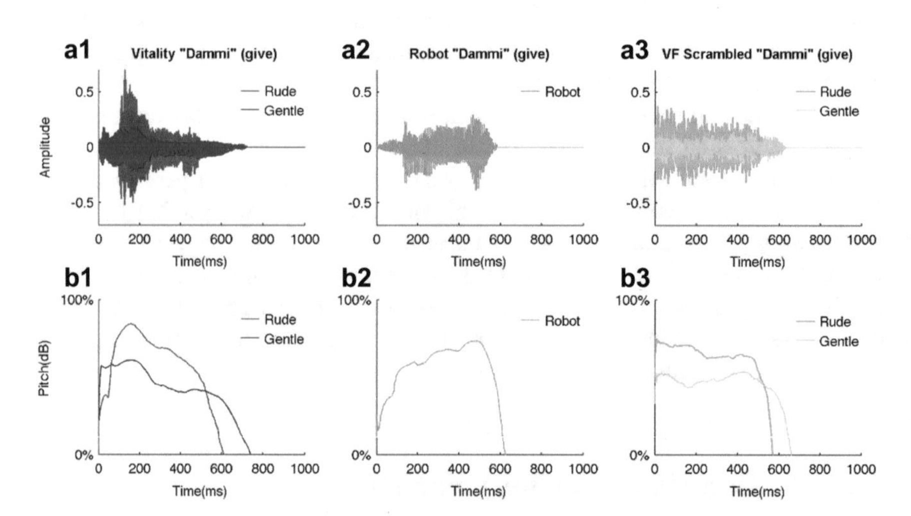

Fig. 6.6 Physical characteristics relative to the action verb "*dammi*" (give). Graphs **a** show the audio wave amplitude for all three categories [**a1** vitality: rude (red color), gentle (blue color); **a2** robot (gray color); **a3** scrambled VF: rude (green color), gentle (cyan color)]. Graphs **b** show the sound intensity of each stimulus category. Figure adapted from Di Cesare et al. [4]

physical properties of the stimuli (pitch, amplitude) but did not convey any meaning (Fig. 6.6a3–b3).

The results indicated that hearing vitality forms action verbs produced activations of the superior temporal gyrus, left inferior parietal lobule, left premotor, left prefrontal cortex, and posterior part of the inferior frontal gyrus plus a bilateral activation of the insula (Fig. 6.7a, left side). A very similar activation pattern was observed for the robot condition except for the insula activation (Fig. 6.7a, center). In contrast, listening to scrambled stimuli produced only the activation of the auditory temporal areas (Fig. 6.7a, right side). Most importantly, the direct contrasts *vitality forms versus robot* and *vitality forms versus scrambled vf* revealed a significant activation of the left central part of the insula (Fig. 6.7b).

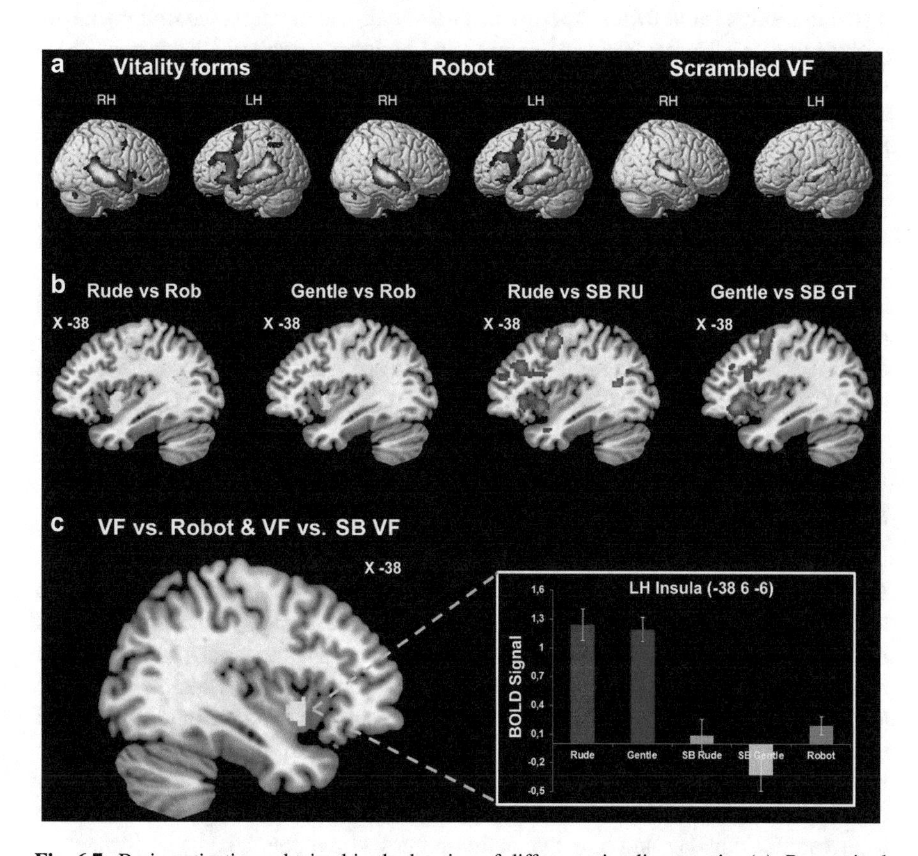

Fig. 6.7 Brain activations obtained in the hearing of different stimuli categories (**a**). Parasagittal sections showing the activations resulting from the contrast *rude versus robot*, *gentle versus robot*, *rude versus scrambled rude*; *gentle versus scrambled gentle* (**b**). Conjunction analysis evidences voxels activated in both contrasts *vitality forms versus robot* and *vitality forms versus scrambled vf* (**c**; left side). BOLD signal recorded in the insular area highlighted from the conjunction analysis (**c**; right side). To avoid circular analysis, statistical comparison was not carried out between conditions [9]. Figure adapted from Di Cesare et al. [4]

The finding that the dorso-central insula is activated during the hearing of vitality forms cannot be merely accounted for the meaning of action verbs. Indeed, although the robotic voice conveyed the same verbal message, the insula was activated only when participants listened to action verbs conveying vitality forms. Additionally, listening to scrambled stimuli, although the physical properties (intensity, frequency) were the same between scrambled and vitality forms conditions, it did not produce the activation of the insula excluding the possibility that the insular activity could be due to the physical properties of the auditory stimuli. It is plausible that listening to action verbs pronounced with different vitality forms evokes in the participants an internal simulation activating in them the same areas involved in the pronunciation of those action verbs except for the primary motor cortex. This hypothesis has been tested in a subsequent fMRI experiment showing that in the dorso-central insula are present voxels selective for both listening and imaging speech vitality forms. These findings strongly suggest the existence of the same neural substrate located in the insula which is involved in the perception and expression of speech vitality forms [7].

The activity of the insula in response to auditory stimuli endowed with vitality forms is in agreement with previous fMRI studies described above on action vitality forms [3, 6]. These data corroborate the idea that the central sector of the insular cortex is the key region for vitality forms processing. During social interactions, this area is triggered not only by action vitality forms (observation, imagination, execution) but also by speech vitality forms (listening, imagination) indicating that the dorso-central insula plays a crucial role in the processing of vitality forms regardless of the modality with which they are conveyed.

6.5 The Role of Vitality Forms in Social Interactions

In everyday life, people socially interact expressing their positive or negative attitudes by performing actions or pronouncing words. The expression of vitality forms allows individuals to communicate their own internal state while the perception of vitality forms allow them to understand those of others. For example, if actions or words are performed/pronounced gently or rudely, the receiver can understand if the agent is angry or calm. It is still unknown whether, during social interactions, gentle and rude vitality forms expressed by the agent may influence positively or negatively the motor behavior of the receiver. In this regard, a kinematic study was carried out to investigate whether and how two action requests (give me; take it) performed by an actor or an actress with different vitality forms (rude and gentle) may affect the kinematics of a subsequent motor response performed by participants [5]. Fourteen right-handed participants took part in the study. For each participant, a reflective marker was placed on the nails of right thumb, the index finger (grasping markers) and on the wrist (reaching marker). The two grasping markers allowed to record the grasping phase of the action characterized by an initial phase of fingers opening up to a maximum (maximal finger aperture), followed by a phase of the finger closing on the object. Differently, the reaching marker allowed to analyze the kinematics of the reaching phase. During the experiment, participants were presented with video clips showing an actor/actress performing a giving request (asking for a bottle; task 1, Fig. 6.8a) or a taking request (handing a bottle; task 2, Fig. 6.8b). Most importantly, each request was presented as visual action (V: visual modality) or auditory action verb (A: auditory modality) or both (AV: audio–visual) (Fig. 6.8). All the requests were expressed with rude and gentle vitality forms. After the actor's request (V, A, AV), participants performed a subsequent action (reach-to-grasp the bottle with the goal to give or to take it).

The results indicated that, for both tasks (giving action, taking action), the perception of vitality forms modulated the kinematic parameters (velocity and trajectory) of the subsequent action performed by participants. In particular, concerning the reaching phase (Fig. 6.9a, b), vitality forms modulated the temporal (acceleration and velocity) and spatial parameters (trajectory) of the reach component, showing a wider trajectory and higher velocity in response to the rude requests compared to the gentle ones. Additionally, concerning the grasping phase (Fig. 6.9c, d), results showed a wider maximal finger aperture in response to rude vitality form than the gentle one. Taken together, these data indicate that vitality forms expressed by the actors influenced both the reach and grasp components of the motor acts performed by participants.

It is important to note that the effect of vitality forms expressed by the actor/actress on the motor response of the receiver also occurred when participants simply heard the verbal requests pronounced gently or rudely. This suggests that the influence of vitality forms on the participants' motor responses cannot be merely ascribed to an imitation mechanism of the observed actions. During the perception of vitality forms, the physical parameters characterizing actions execution (velocity, trajectory) or words pronunciation (pitch, intensity) are encoded in the dorso-central insula. The

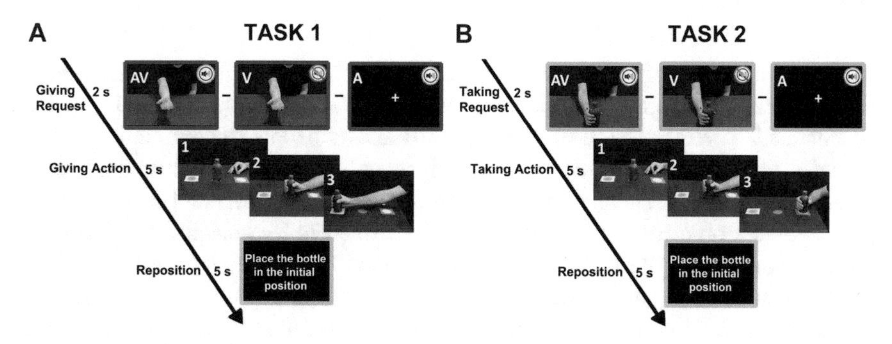

Fig. 6.8 Experimental paradigm. Participants were presented with audio–visual (AV), visual (V), and auditory (A) (**a**) stimuli. In the task 1, after the request, participants were requested to give the bottle (**a**). In the task 2, after the request, participants were requested to take the bottle (**b**). Panels with numbers display the phases of the participants' movement during the experimental trial: 1, starting position; 2, grasping the bottle; 3, taking (or giving) the bottle. Time line reports the timing of different trial phases. Figure adapted from Di Cesare et al. [5]

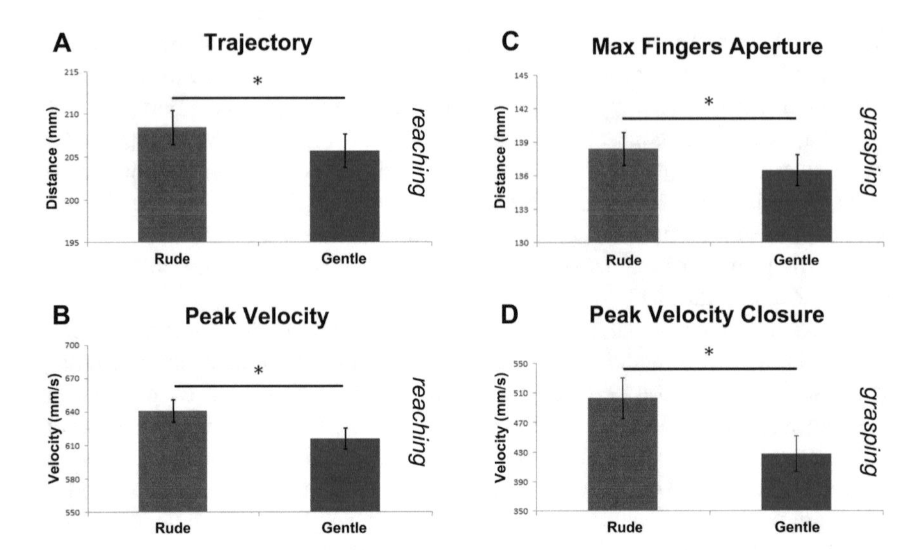

Fig. 6.9 Graphs display the mean values of kinematic parameters recorded in participants in response to a rude or gentle vitality forms during the reaching (**a**, **b**) and the grasping phases (**c**, **d**). Vertical bars represent the standard error of the mean (SEM). The asterisk (*) indicates the statistical significance ($p < 0.05$). Figure adapted from Di Cesare et al. [5]

role of the insula would be to transform the visual and acoustic information of the perceived vitality forms into a motor domain allowing the receiver to understand the positive or negative attitude of the agent and prepare the adequate motor response.

6.6 Conclusions

An important aspect of the action that characterizes human interactions is the vitality form. Vitality form represents the way in which actions and words are performed or pronounced. The expression of vitality forms allows people to communicate their attitudes while the perception of vitality forms allows them to understand those of others. This mechanism is important to relate to and understand others from a psychological point of view. The findings described in this chapter highlight the fundamental role of vitality forms in social communication and lay the foundations for future studies on human–human and human–robot interactions. The concept of vitality form could be used in the future operating systems of robots, which would allow them, on the one hand, to detect the positive or negative attitudes of humans, and, on the other hand, to assume the correct role in different contexts such as an authoritative role in the security context. From this perspective, vitality forms could become a future fundamental source of social communication to promote not only human–human interactions but also human–robot interactions.

Acknowledgements Giuseppe Di Cesare is supported by a Starting Grant from the European Research Council (ERC) under the European Union's Horizon 2020 research and innovation program. G.A. No. 804388, wHiSPER.

References

1. Anderson, W., & Jaffe, J. (1972). The definition and timing of vocalic syllables in speech signals. *Scientific Reports, 12.*
2. Caruana, F., Avanzini, P., Gozzo, F., Francione, S., Cardinale, F., & Rizzolatti, G. (2015). Mirth and laughter elicited by electrical stimulation of the human anterior cingulate cortex. *Cortex, 71,* 323–331.
3. Di Cesare, G., Di Dio, C., Rochat, M. J., Sinigaglia, C., Bruschweiler-Stern, N., Stern, D. N., et al. (2013). The neural correlates of "vitality form" recognition: An fMRI study. *Social Cognitive and Affective Neuroscience, 9*(7), 951–960.
4. Di Cesare, G., Fasano, F., Errante, A., Marchi, M., & Rizzolatti, G. (2016). Understanding the internal states of others by listening to action verbs. *Neuropsychologia, 89,* 172–179.
5. Di Cesare, G., De Stefani, E., Gentilucci, M., & De Marco, D. (2017). Vitality forms expressed by others modulate our own motor response: A kinematic study. *Frontiers in Human Neuroscience, 11,* 565.
6. Di Cesare, G., Di Dio, C., Marchi, M., & Rizzolatti, G. (2015). Expressing our internal states and understanding those of others. *Procedure of the National Academic Science of the United States of America, 112*(33).
7. Di Cesare, G., Marchi, M., Errante, A., Fasano, F., & Rizzolatti, G. (2017). Mirroring the social aspects of speech and actions: The role of the insula. *Cerebral Cortex*, 1–10.
8. di Pellegrino, G., Fadiga, L., Fogassi, L., Gallese, V., & Rizzolatti, G. (1992). Understanding motor events: A neurophysiological study. *Experimental Brain Research, 91*(1), 176–180.
9. Kriegeskorte, N., Simmons, W. K., Bellgowan, P. S., & Baker, C. I. (2009). Circular analysis in systems neuroscience: The dangers of double dipping. *Nature Neuroscience, 12*(5), 535–540. https://doi.org/10.1038/nn.2303.
10. Kurth, F., Zilles, K., Fox, P. T., Laird, A. R., & Eickhoff, S. B. (2010). A link between the systems: Functional differentiation and integration within the human insula revealed by meta-analysis. *Brain Structure and Function, 214,* 519–534.
11. Rizzolatti, G., Cattaneo, L., Fabbri-Destro, M., & Rozzi, S. (2014). Cortical mechanisms underlying the organization of goal-directed actions and mirror neuron-based action understanding. *Physiological Reviews, 94,* 655–706.
12. Rizzolatti, G., & Sinigaglia, C. (2016). The mirror mechanism: A basic principle of brain function. *Nature Reviews Neuroscience, 12,* 757–765.
13. Singer, T., Seymour, B., O'Doherty, J., Kaube, H., Dolan, R. J., & Frith, C. D. (2004). Empathy for pain involves the affective but not sensory components of pain. *Science, 303*(5661), 1157–1162.
14. Stern, D. N. (1985). *The interpersonal world of the infant.* New York: Basic Books.
15. Stern, D. N. (2010). *Forms of vitality exploring dynamic experience in psychology, arts, psychotherapy, and development.* UK: Oxford University Press.
16. Wicker, B., Keysers, C., Plailly, J., Royet, J. P., Gallese, V., & Rizzolatti, G. (2003). Both of us disgusted in my insula: The common neural basis of seeing and feeling disgust. *Neuron, 40,* 655–664.

Part II
Computational Models for Motion Understanding

Chapter 7
Optical Flow Estimation in the Deep Learning Age

Junhwa Hur and Stefan Roth

Abstract Akin to many subareas of computer vision, the recent advances in deep learning have also significantly influenced the literature on optical flow. Previously, the literature had been dominated by classical energy-based models, which formulate optical flow estimation as an energy minimization problem. However, as the practical benefits of Convolutional Neural Networks (CNNs) over conventional methods have become apparent in numerous areas of computer vision and beyond, they have also seen increased adoption in the context of motion estimation to the point where the current state of the art in terms of accuracy is set by CNN approaches. We first review this transition as well as the developments from early work to the current state of CNNs for optical flow estimation. Alongside, we discuss some of their technical details and compare them to recapitulate which technical contribution led to the most significant accuracy improvements. Then we provide an overview of the various optical flow approaches introduced in the deep learning age, including those based on alternative learning paradigms (e.g., unsupervised and semi-supervised methods) as well as the extension to the multi-frame case, which is able to yield further accuracy improvements.

7.1 Emergence and Advances of Deep Learning-Based Optical Flow Estimation

The recent advances in deep learning have significantly influenced the literature on optical flow estimation and fueled a transition from classical energy-based formulations, which were mostly hand defined, to end-to-end trained models. We first review how this transition proceeded by recapitulating early work that started to utilize deep learning, typically as one of several components. Then, we summa-

J. Hur (✉) · S. Roth
Department of Computer Science, Technische Universität Darmstadt,
Hochschulstr. 10, 64289 Darmstadt, Germany
e-mail: junhwa.hur@visinf.tu-darmstadt.de

S. Roth
e-mail: stefan.roth@visinf.tu-darmstadt.de

N. Noceti et al. (eds.), *Modelling Human Motion*,
https://doi.org/10.1007/978-3-030-46732-6_7

rize several canonical end-to-end approaches that have successfully adopted CNNs for optical flow estimation and have highly influenced the mainstream of research, including other subareas of vision in which optical flow serves as an input.

7.1.1 From Classical Energy-Based Approaches to CNNs

For more than three decades, research on optical flow estimation has been heavily influenced by the variational approach of Horn and Schunck [20]. Their basic energy minimization formulation consists of a data term, which encourages brightness constancy between temporally corresponding pixels, and a spatial smoothness term, which regularizes neighboring pixels to have similar motion in order to overcome the aperture problem. The spatially continuous optical flow field $\mathbf{u} = (u_x, u_y)$ is obtained by minimizing

$$E(\mathbf{u}) = \int \Big((I_x u_x + I_y u_y + I_t)^2 + \alpha^2 \big(\|\nabla u_x\|^2 + \|\nabla u_y\|^2 \big) \Big) \, \mathrm{d}x \, \mathrm{d}y, \tag{7.1}$$

where I_x, I_y, I_t are the partial derivatives of the image intensity I with respect to x, y, and t (Fig. 7.1a). To minimize Eq. (7.1) in practice, spatial discretization is necessary. In such a spatially discrete form, the Horn and Schunck model [20] can also be rewritten in the framework of standard pairwise Markov random fields (MRFs) [7, 31] through a combination of a unary data term $D(\cdot)$ and a pairwise smoothness term $S(\cdot, \cdot)$,

$$E(\mathbf{u}) = \sum_{\mathbf{p} \in \mathcal{I}} D(\mathbf{u_p}) + \sum_{\mathbf{p},\mathbf{q} \in \mathcal{N}} S(\mathbf{u_p}, \mathbf{u_q}), \tag{7.2}$$

where $\mathcal{I}$ is the set of image pixels and the set $\mathcal{N}$ denotes spatially neighboring pixels. Starting from this basic formulation, much research has focused on designing better energy models that more accurately describe the flow estimation problem (see [11, 49] for reviews of such methods).

Concurrently with pursing better energy models, the establishment of public benchmark datasets for optical flow, such as the Middlebury [4], MPI Sintel [8], and KITTI Optical Flow benchmarks [14, 37], has kept revealing the challenges and limitations of existing methods. These include large displacements, severe illumination changes, and occlusions. Besides allowing for the fair comparison between existing methods on the same provided data, these public benchmarks have moreover stimulated research on more faithful energy models that address some of the specific challenges mentioned above.

Meanwhile, the relatively recent success of applying Convolutional Neural Networks (CNNs) with backpropagation on a large-scale image classification task [29] paved the way for applying CNNs to various other computer vision problems, including optical flow as well. Early work that applied CNNs to optical flow used them as an advanced feature extractor [2, 3, 12, 16], as sketched in Fig. 7.1b. The main

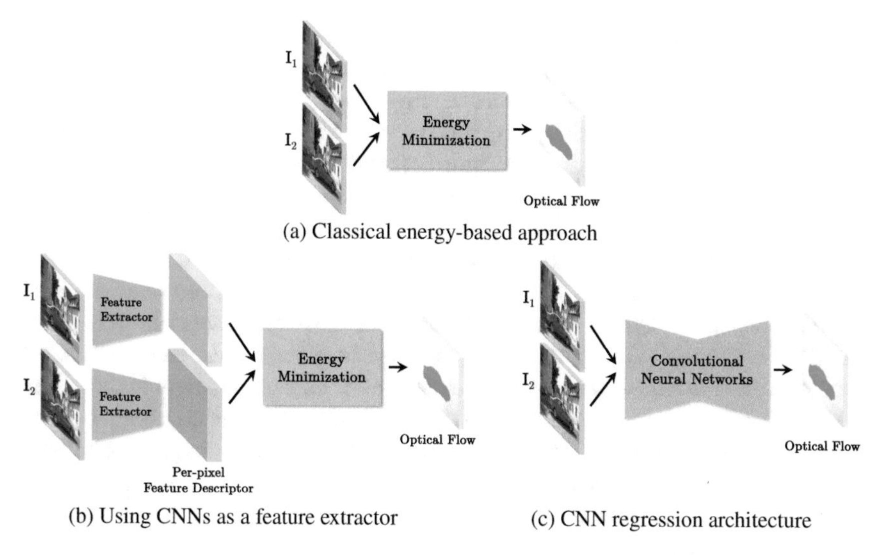

Fig. 7.1 Transition from **a** classical energy-based approaches to **b** CNN-based approaches that use CNNs as a feature extractor or to **c** end-to-end trainable CNN regression architectures

idea behind this is to substitute the data term (e.g., in Eqs. (7.1) and (7.2)) in classical energy-based formulations with a CNN-based feature matching term. Instead of using image intensities, image gradients, or other hand-crafted features as before, CNNs enable learning feature extractors such that each pixel can be represented with a high-dimensional feature vector that combines a suitable amount of distinctiveness and invariance, for example to appearance changes. The putative similarity between regions is given by the feature distance. The remaining pipeline, including using the smoothness term as well as the optimization strategies, remain the same. As we will review in more detail below, several methods [2, 3, 12, 16] demonstrated an accuracy benefit of such CNN-based feature extractors.

At the same time, another line of research investigated regression-based CNN architectures that can directly estimate optical flow from a pair of input images and can be trained end-to-end, as sketched in Fig. 7.1c. Unlike methods that combine CNN feature extractors with classical regularizers and energy minimization, such regression frameworks employ CNNs for the entire pipeline by virtue of their ability to act as a function approximator, which effectively learns the relationship between the input images and the desired flow output given the labeled training dataset. FlowNet [10] is the first work that demonstrated an end-to-end CNN regression approach for estimating optical flow based on an encoder-decoder architecture. Owing to the difficulty of obtaining dense ground truth optical flow in real-world images, Dosovitskiy et al. [10] generated a synthetic dataset from CAD models of chairs, which move in front of a static background. Pairs of images with ground truth optical flow serve to train the network. FlowNet [10] demonstrated that a CNN-based regression architecture is able to predict optical flow directly, yet the accuracy remained behind that of state-of-the-art energy-based methods at the time [44, 52]. Unlike in other areas of computer vision, this left it initially unclear whether end-to-end CNN architectures can compete with classical energy-based methods in terms of accuracy.

However, later research cleared up this question by developing better end-to-end architectures that eventually outperformed classical energy-based methods, reaching new accuracy levels on public benchmarks [8, 14, 37]. These advances mainly stem from discovering new architecture designs, for example, by stacking multiple networks to refine previous estimates [25] or constructing a CNN pyramid to estimate flow in a coarse-to-fine fashion [23, 41, 48], as had been done in classical methods before. Unlike energy-based models, CNN regressors run in real time on GPUs combined with much better accuracy. In other words, end-to-end CNN regressors have established themselves by now as dominant paradigm in the current literature on optical flow estimation. Yet, they have not remained without limitations, hence much research continues to be carried out. For example, recent work aims to overcome the reliance on large amounts of labeled data as well as accuracy drops on unseen domains and datasets, for example by pursuing unsupervised or semi-supervised learning paradigms.

In the following, we will give a detailed overview of the two major CNN paradigms in optical flow estimation and survey other recent trends.

7.1.2 CNNs as Feature Extractor

Not restricted to the problem domain of optical flow estimation but rather correspondence estimation more generally, several early works [18, 46, 60, 61] employed CNNs for matching descriptors or patches. In most cases, the underlying network uses a so-called Siamese architecture that extracts a learned feature descriptor separately for each of two input image patches, followed by a shallow joint network that computes a matching score between the two feature representations. The name Siamese alludes to the fact that the two feature extractor sub-networks are identical including their weights. Inspired by these successes, significant amounts of earlier work that adopted deep learning for optical flow estimation focused on utilizing CNNs as a feature extractor on top of conventional energy-based formulations such as MRFs. Their main idea is to utilize CNNs as a powerful tool for extracting discriminative features and then use well-proven conventional energy-based frameworks for regularization.

Gadot and Wolf [12] proposed a method called **PatchBatch**, which was among the first flow approaches to adopt CNNs for feature extraction. PatchBatch [12] is based on a Siamese CNN feature extractor that is fed 51×51 input patches and outputs a 512-dimensional feature vector using a shallow 5-layer CNN. Then, PatchBatch [12] adopts Generalized PatchMatch [5] as an Approximate Nearest Neighbor (ANN) algorithm for correspondence search, i.e., matching the extracted features between two images. The method constructs its training set by collecting positive corresponding patch examples given ground-truth flow and negative non-matching examples by randomly shifting the image patch in the vicinity of where the ground-truth flow directs. The intuition of collecting negative examples in such a way is to train CNNs to be able to separate non-trivial cases and extract more discriminative features. The shallow CNNs are trained using a variant of the DrLIM [17] loss, which minimizes the squared L_2 distance between positive patch pairs and maximizes the squared L_2 distance between negative pairs above a certain margin.

In a similar line of work, Bailer et al. [3] proposed to use the thresholded hinge embedding loss for training the feature extractor network. The hinge embedding loss based on the L_2 loss function has been commonly used to minimize the feature distance between two matching patches and to maximize the feature distance above m between non-matching patches:

$$l_{\text{hinge}}(\mathbf{P}_1, \mathbf{P}_2) = \begin{cases} L_2(\mathbf{P}_1, \mathbf{P}_2), & (\mathbf{P}_1, \mathbf{P}_2) \in M^+ \\ \max\big(0, m - L_2(\mathbf{P}_1, \mathbf{P}_2)\big), & (\mathbf{P}_1, \mathbf{P}_2) \in M^- \end{cases} \tag{7.3}$$

$$L_2(\mathbf{P}_1, \mathbf{P}_2) = \big\| F(\mathbf{P}_1) - F(\mathbf{P}_2) \big\|_2, \tag{7.4}$$

where $F(\mathbf{P}_1)$ and $F(\mathbf{P}_2)$ are the extracted descriptors from CNNs applied to $\mathbf{P}_1$ in the first image and $\mathbf{P}_2$ in the second image, respectively, $L_2(\mathbf{P}_1, \mathbf{P}_2)$ calculates the L_2 loss between the two descriptors, and M^+ and M^- are collected sets of positive and negative samples, respectively.

However, minimizing the L_2 loss of some challenging positive examples (e.g., with appearance difference or illumination changes) can move the decision boundary into an undesired direction and lead to misclassification near the decision boundary. Thus, Bailer et al. [3] proposed to put a threshold t on the hinge embedding loss in order to prevent the network from minimizing the L_2 distance too aggressively:

$$l_{t\text{-hinge}}(\mathbf{P}_1, \mathbf{P}_2) = \begin{cases} \max\big(0, L_2(\mathbf{P}_1, \mathbf{P}_2) - t\big), & (\mathbf{P}_1, \mathbf{P}_2) \in M^+ \\ \max\big(0, m - (L_2(\mathbf{P}_1, \mathbf{P}_2) - t)\big), & (\mathbf{P}_1, \mathbf{P}_2) \in M^-. \end{cases} \tag{7.5}$$

Compared to standard losses, such as the hinge embedding loss in Eq. (7.3) or the DrLIM loss [17], this has led to more accurate flow estimates.

Meanwhile, Güney and Geiger [16] demonstrated successfully combining a CNN feature matching module with a discrete MAP estimation approach based on a pair-wise Markov random field (MRF) (MRFs) model. The proposed CNN module outputs per-pixel descriptors, from which a cost volume is constructed by calculating feature distances between sample matches. This is input to a discrete MAP estimation approach [38] to infer the optical flow. To keep training efficient, Güney and Geiger [16] followed a piece-wise setting that first trains the CNN module alone and only then trains the joint CNN-MRF module together. Bai et al. [2] followed a similar setup overall, but utilized semi-global block matching (SGM) [19] to regress the output optical flow from the cost volume, which is constructed by calculating a distance between features from CNNs.

Taken together, these approaches have successfully demonstrated that the benefits of the representational power of CNNs can be combined with well-proven classical energy-based models. Specifically, they demonstrated more accurate estimates on inliers and more precise estimates on object boundaries than previous baselines with hand-constructed features.

7.1.3 End-to-End Regression Architectures for Optical Flow Estimation

Concurrently with the development of feature extraction-based networks, active research also started on developing end-to-end CNN architectures for optical flow estimation based on regression. Unlike methods that use CNNs only for feature extraction as addressed above, such regression methods exploit CNNs for the entire pipeline and directly output optical flow from a pair of input images. By substituting classical regularizers and avoiding energy minimization, these CNN-based methods combine the advantages of end-to-end trainability and runtime efficiency.

Dosovitskiy et al. proposed the first end-to-end CNN architecture for estimating optical flow, called FlowNet [10], which has two main architectural lines, **FlowNetS** and **FlowNetC**. The two models are fundamentally based on an hourglass-shaped neural network architecture that consists of an encoder and a decoder, and differs only in the encoder part. In FlowNetS, a pair of input images is simply concatenated and then input to the hourglass-shaped network that directly outputs optical flow. On the other hand, FlowNetC has a shared encoder for both images, which extracts a feature map for each input image, and a cost volume is constructed by measuring patch-level similarity between the two feature maps with a correlation operation. The result is fed into the subsequent network layers.

To train the networks in a supervised way, a training dataset with a large number of image pairs and their ground truth flow are required, but at the time only datasets with few hundreds of images or even fewer were available [4, 14, 37]; the challenge of obtaining dense optical flow ground truth in real-world images remains until today. In order to overcome the shortage of suitable training data, Dosovitskiy et al. [10] established a synthetic dataset, called FlyingChairs, by layering natural images with rendered CAD models of chairs; their parameterized affine motion is designed to follow the motion statistics of existing real-world datasets. However, due to the intrinsic differences between synthetic and real-world images, unfortunately FlowNet trained on the synthetic dataset alone did not generalize well to real images. In fact, even after fine-tuning on real-world images, the accuracy initially remained behind that of classical energy-based models at the time. This left the question whether such a generic CNN regression architecture can actually outperform classical energy-based methods, or why it did not (yet). Importantly, however, FlowNet [10] demonstrated the possibility of employing an end-to-end regression architecture for optical flow estimation. Moreover, FlowNet established several standard practices for training optical flow networks such as learning rate schedules, basic network architectures,

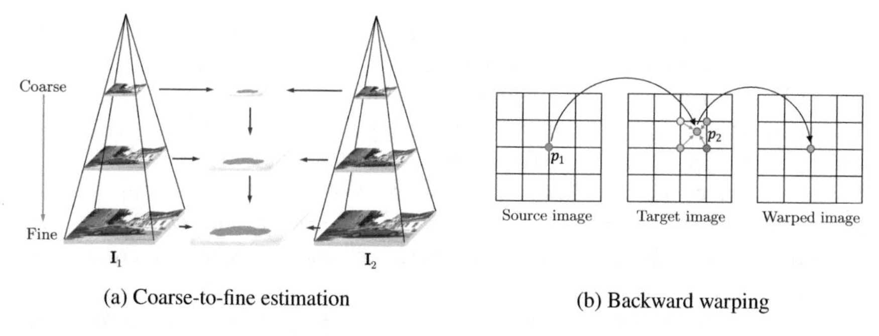

(a) Coarse-to-fine estimation

(b) Backward warping

Fig. 7.2 **a** The classical coarse-to-fine concept proceeds by estimating optical flow using a multi-scale image pyramid, starting from the coarsest level to the finest level. By gradually estimating and refining optical flow through the pyramid levels, this approach can handle large displacements better and improve accuracy. **b** Backward warping is commonly used in optical flow estimation. For each pixel p_1 in the source image, the warped image obtains the intensity from (sub)pixel location p_2, which is obtained from the estimated flow. Bilinear interpolation is often used to obtain the pixel intensity at the non-integer coordinate

data augmentation schemes, and the necessity of pre-training on synthetic datasets, which have substantially impacted follow-up research.

Ranjan and Black proposed **SPyNet** [41], which incorporates the classical "coarse-to-fine" concept (please refer to Fig. 7.2a for an illustration) into a CNN model and updates the residual flow over multiple pyramid levels. SPyNet consists of 5 pyramid levels, and each pyramid level consists of a shallow CNN that estimates flow between a source image and a target image, which is warped by the current flow estimate (see Fig. 7.2b). This estimate is updated so that the network can residually refine optical flow through a spatial pyramid and possibly handle large displacements. Compared to FlowNet, SPyNet significantly reduces the number of model parameters by 96% by using a pyramid-shaped architecture, while achieving comparable and sometimes even better results than FlowNet. Although SPyNet [41] is still outperformed by classical energy-based methods, it demonstrates a promising way of designing flow architectures by integrating classical principles into deep learning.

Meanwhile, Ilg et al. [25] proposed **FlowNet2**, which significantly improves the flow accuracy over their previous FlowNet architecture and started to outperform classical energy-based approaches. The main limitations of FlowNet are blurry outputs from the CNN decoder and lower accuracy compared to classical approaches. To overcome these limitations, Ilg et al. proposed the key idea that by stacking multiple FlowNet-style networks, one can sequentially refine the output from the previous network modules. Despite of the conceptual simplicity, stacking multiple networks is very powerful and significantly improves the flow accuracy by more than 50% over FlowNet. Additionally, Ilg et al. revealed several important practices for training their networks, including the necessity of pre-training and fine-tuning on synthetic datasets, the effectiveness of using a correlation layer, and the guidance of proper learning rate schedules, followed by in-depth empirical analyses. In practice, Ilg et al. [25] suggest to pre-train their networks on a less challenging synthetic dataset first (i.e., the FlyingChairs dataset [10]) and then further train on a more challenging synthetic dataset with 3D motion and photometric effects (i.e., the FlyingThings3D dataset [35]). Their empirical study revealed a more than 20% accuracy difference depending on the usage of the proper pre-training dataset (see Table 7.1 in [25]). The underlying conjecture is that making the network first learn the general concept of motion estimation with a simpler dataset is more important than learning to handle various challenging examples from the start. Also, the proposed learning rate schedules for pre-training and fine-tuning have become a standard and guidance for follow-up research.

After the successful demonstration of FlowNet2 [25] that end-to-end regression architectures can outperform energy-based approaches, further investigations on finding better network architectures have continued. Sun et al. proposed an advanced architecture called **PWC-Net** [48] by exploiting well-known design principles from classical approaches. PWC-Net relies on three main design principles: (i) pyramid, (ii) warping, and (iii) cost volume. Similar to SPyNet [41], PWC-Net estimates optical flow in a coarse-to-fine way with several pyramid levels, but PWC-Net constructs a feature pyramid by using CNNs, while SPyNet constructs an image pyramid by simply downsampling images. Next, PWC-Net constructs a cost volume with a feature map from the source image and the warped feature map from the target image based on the current flow. Then, the subsequent CNN modules act as a decoder that outputs optical flow from the cost volume. In terms of both accuracy and practicality, PWC-Net [48] set a new state of the art with its light-weight architecture allowing for shorter training times, faster inference, and more importantly, clearly improved accuracy. Comparing to FlowNet2 [25], PWC-Net is 17 times smaller in model size and twice as fast during inference while being more accurate. Similar to SPyNet, the computational efficiency stems from using coarse-to-fine estimation, but PWC-Net crucially demonstrates that constructing and warping feature maps instead of using downsampled warped images yields much better accuracy.

As a concurrent work and similar to PWC-Net [48], **LiteFlowNet** [23] also demonstrated utilizing a multi-level pyramid architecture that estimates flow in a coarse-to-fine manner, proposing another light-weight regression architecture for optical flow. The major technical differences to PWC-Net are that LiteFlowNet residually updates optical flow estimates over the pyramid levels and proposes a flow regularization module. The proposed flow regularization module creates per-pixel local filters using CNNs and applies the filters to each pixel so that customized filters refine flow fields by considering neighboring estimates. The regularization module is given the optical flow, feature maps, and occlusion probability maps as inputs to take motion boundary information and occluded areas into account in creating per-pixel local filters. The experimental results demonstrate clear benefits, especially from using the regularization module that smoothes the flow fields while effectively sharpening motion boundaries, which reduces the error by more than 13% on the training domain.

Afterwards, Hur and Roth [24] proposed an iterative estimation scheme with weight sharing entitled *iterative residual refinement* (**IRR**), which can be applied to several backbone architectures and improves the accuracy further. Its main idea is to take the output from a previous pass through the network as input and iteratively refine it by only using a single network block with shared weights; this allows the network to residually refine the previous estimate. The IRR scheme can be used on top of various flow architectures, for example FlowNet [10] and PWC-Net [48]. For FlowNet [10], the whole hourglass shape network is iteratively re-used to keep refining its previous estimate and, in contrast to FlowNet2 [25], increases the accuracy without adding any parameters. For PWC-Net [48], a repetitive but separate flow decoder module at each pyramid level is replaced with only one common decoder for all levels, and then iteratively refines the estimation through the pyramid levels. Applying the scheme on top of PWC-Net [48] is more interesting as it makes an already lean model even more compact by removing repetitive modules that perform the same functionality. Yet, the accuracy is improved, especially on unseen datasets (i.e. allowing better generalization). Furthermore, Hur and Roth [24] also demonstrated an extension to joint occlusion and bi-directional flow estimation that leads to further flow accuracy improvements of up to 17.7% while reducing the number of parameters by 26.4% in case of PWC-Net; this model is termed **IRR-PWC** [24].

Yin et al. [57] proposed a general probabilistic framework termed **HD**^{3} for dense pixel correspondence estimation, exploiting the concept of the so-called match density, which enables the joint estimation of optical flow and its uncertainty. Mainly following the architectural design of PWC-Net (i.e., using a multi-scale pyramid, warping, and a cost volume), the method estimates the full match density in a hierarchical and computationally efficient manner. The estimated spatially discretized match density can then be converted into optical flow vectors while providing an uncertainty assessment at the same time. This output representation of estimating the match density is rather different from all previous works above, which directly regress optical flow with CNNs. On established benchmarks datasets, their experimental results demonstrate clear advantages, achieving state-of-the-art accuracy regarding both optical flow and uncertainly measures.

While the cost volume has been commonly used in backbone architectures [10, 23, 25, 48, 57], its representation is mainly based on a heuristic design. Instead of representing the matching costs between all pixels (x, y) with their possible 2D displacements (u, v) into a 4D tensor (x, y, u, v), the conventional design is based on a 3D cost volume—a 2D array (x, y) augmented with a uv channel, which is computationally efficient but often yields limited accuracy and overfitting. To overcome this limitation, Yang and Ramanan [55] proposed Volumetric Correspondence Networks (**VCN**), which are based on true 4D volumetric processing: constructing a proper 4D cost volume and processing with 4D convolution kernels. For reducing the computational cost and memory of 4D processing, Yang and Ramanan [55] used separable 4D convolutions, which approximate the 4D convolution operation with two 2D convolutions, reducing the complexity by N^2 (please refer the original paper for technical details). Through proper 4D volumetric processing with computationally cheaper operations, the method further pushes both accuracy and practicality on

Table 7.1 Overview of the main technical design principles of end-to-end optical flow architectures

Methods	FlowNetS [10]	FlowNetC [10]	SPyNet [41]	FlowNet2 [25]	PWC-Net [48]	LiteFlowNet [23]	HD^3 [57]	VCN [55]
Pyramid	–	3-level feature	5-level image	3-level feature	6-level feature	6-level feature	5-level feature	6-level feature
Warping	–	–	Image	Image	Feature	Feature	Feature	Feature
Cost volume	–	3D	–	3D	3D	3D	3D	4D
Network stacking	–	–	–	5	–	–	–	–
Flow inference	Direct	Direct	Residual	Direct	Direct	Residual	Residual	Hypothesis selection
Parameters (M)	38.67	39.17	1.20	162.49	8.75	5.37	39.6	6.20

widely used public benchmarks, improving generalization and demonstrating faster training convergence—requiring 7 times fewer training iterations than its competitors.

Table 7.1 summarizes the main differences in technical design of the various end-to-end optical flow architectures discussed above. Starting from FlowNetS [10], the methods are listed in chronological order. We omit the IRR scheme as it can be applied on top of several backbone architectures. Table 7.2 compares the quantitative results of each method on the MPI Sintel [8] and KITTI benchmarks [14, 37]. Each method is pre-trained on synthetic datasets first and then fine-tuned on each benchmark. Looking at the two tables, we can gain some first insights into which design choices lead to the observed accuracy improvements. First, having a pyramid structure by adopting a "coarse-to-fine" strategy makes networks more compact and improves the flow estimation accuracy (e.g., from FlowNet [10] to SPyNet [41],

Table 7.2 Quantitative comparison on public benchmarks: MPI Sintel [8] and KITTI [14, 37]

Methods	MPI Sintel [a]		KITTI [b]	
	Clean	Final	2012	2015
FlowNetS [10]	6.158	7.218	37.05%	–
FlowNetC [10]	6.081	7.883	–	–
SPyNet [41]	6.640	8.360	12.31%	35.07%
FlowNet2 [25]	3.959	6.016	4.82%	10.41%
PWC-Net [48]	4.386	5.042	4.22%	9.60%
LiteFlowNet [23]	3.449	5.381	3.27%	9.38%
IRR-PWC [24]	3.844	4.579	3.21%	7.65%
HD^3 [57]	4.788	4.666	2.26%	6.55%
VCN [55]	*2.808*	*4.404*	–	*6.30%*

[a]Evaluation metric: end point error (EPE)
[b]Evaluation metric: outlier rate (i.e. less than 3 pixel or 5% error is considered an inlier)

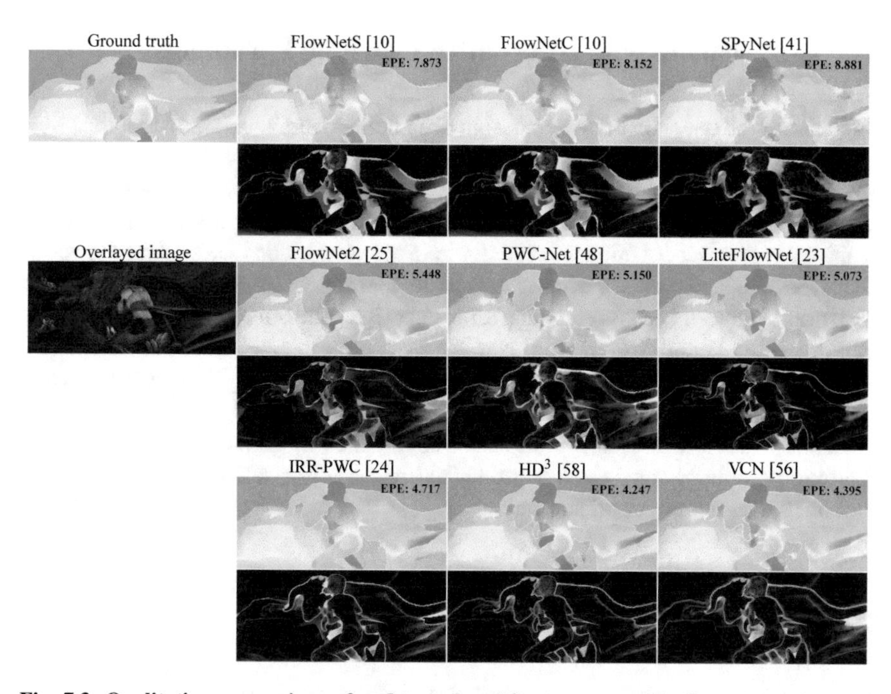

Fig. 7.3 **Qualitative comparison of end-to-end architectures**: example from Sintel final test [8]. The first column shows the ground-truth flow and the overlayed input images. In the further columns, we show the color-coded flow visualization of each method, overlayed with the end point error (EPE) and their error maps (the brighter a pixel, the higher its error)

PWC-Net [48], and LiteFlowNet [23]). Second, stacking networks can also improve the flow accuracy while linearly increasing the number of parameters (e.g., from FlowNet [10] to FlowNet2 [25]). Third, constructing a cost volume by calculating a patch-wise correlation between two feature maps has become a standard approach and is more beneficial than not using it (e.g., FlowNetS vs. FlowNetC, according to a study from [25]). Fourth, even if based on similar conceptual designs, subtle design differences or additional modules can further lead to accuracy improvements (e.g., LiteFlowNet [23] vs. PWC-Net [48]). Fifth, the iterative residual refinement scheme IRR [24] can further boost the accuracy of existing backbone architectures (e.g., from PWC-Net [48] to IRR-PWC [24]). Lastly, investigating better fundamental designs such as the output representation (e.g., the match density [57]) or the cost volume representation (e.g., 4D cost volume [55]) can lead to further improvement, sometimes quite significantly so.

Figure 7.3 shows a qualitative comparison of each method on an example from the Sintel Final Test set [8]. The optical flow visualizations and the error maps demonstrate how significantly end-to-end methods have been improved over the past few years, especially near motion boundaries and in non-textured areas.

7.2 Approaches with Alternative Learning Paradigms

Aside from the question of how to design deep network architectures for optical flow estimation, another problem dimension has grown into prominence recently—how to train such CNNs for optical flow especially in the context of the limited quantities of ground-truth data available in practice. Most (early) CNN approaches are based on standard supervised learning and directly train the network on labeled data. However, real-world labeled data is available only in comparatively small quantities and often constrained to certain settings, which turns out to have the limitation that the accuracy can drop significantly on unseen data. To overcome this, a number of alternative approaches based on unsupervised or semi-supervised learning have been proposed to lighten the necessity of and reliance on large amounts of labeled data. In this section, we review and categorize CNN approaches in terms of their underlying learning paradigm: supervised learning, unsupervised or self-supervised learning, and finally semi-supervised learning.

7.2.1 Challenges of Supervised Learning

Based on the end-to-end trainability of CNNs, the most straightforward way to train CNNs for optical flow estimation is in a supervised fashion using a labeled dataset. In the supervised learning setting—but not only there—the dataset plays an important role, and details such as the size and design of the dataset, the type of loss function, and training schedules become critical factors in achieving high accuracy.

Approaches that are based on CNNs as feature extractor [2, 3, 12, 16], as already discussed above, collect positive matching samples and negative non-matching samples as a training set and train the CNNs by applying a loss function at the final output of the network. Different types of loss functions has been investigated to obtain discriminative features that are invariant to common appearance and illumination changes (please refer to Sect. 7.1.2 for further details). When training CNNs in general, having a large labeled dataset is crucial to avoid overfitting on the training dataset and enable the network to generalize to unseen data. As the networks tend to be comparatively lean and do not have to (and in fact cannot) learn something about plausible motions, but rather only classify when patches match in terms of their appearance, the issue of overfitting is less prominent than in end-to-end regression approaches.

For training end-to-end optical flow architectures in a supervised fashion, on the other hand, we need to have a training dataset with many temporally consecutive image pairs with dense ground-truth flow, representing the range of possible optical flow fields. The entire flow map with per-pixel labels is used to train the network by minimizing the per-pixel Euclidean distance between the ground truth flow and the output from the network. However, collecting such a dataset with real-world images has been challenging due to the difficulty of measuring the true motion for all pixels [4]. Establishing synthetic datasets instead is a viable alternative (e.g., the FlyingChairs [10], Sintel [8], and FlyingThings3D [35] datasets), as it is much easier to generate a large amount of synthesized images with accurate ground-truth flow.

Yet, using a synthetic dataset for training flow networks still does not completely solve the issue of dataset suitability. The generalization to an unseen setting remains a challenge. According to the empirical studies of [25] and [48], the flow accuracy significantly depends on the dataset used for training and on how close the test-time domain is to the training domain. Consequently, overfitting on the training dataset domain is a problem. As a solution, FlowNet2 [25] is accompanied with a training dataset schedule that leads to a better local parameter optimum so that the trained networks can perform reasonably on unseen data: pre-training on synthetic datasets before fine-tuning on the target domain dataset in the end (please refer to Sect. 7.1.3 for further details). Both FlowNet2 [25] and PWC-Net [48] empirically demonstrated that training networks with this schedule allows for better generalization to an unseen target domain. In fact, pre-training on a synthetic dataset followed by fine-tuning on the target domain yields much better accuracy than directly training on the target domain, even on the target domain itself.

All regression architectures mentioned above have multi-scale intermediate optical flow outputs along the decoder (e.g., FlowNet [10] and FlowNet2 [25]) or at each pyramid level (e.g., PWC-Net [48], SPyNet [41], and LiteFlowNet [23]). For all intermediate outputs, an L_2 loss between the output and the downscaled ground truth is applied per pixel so that the network learns to estimate optical flow in a coarse-to-fine manner and achieves better accuracy at the final output resolution. The final training loss becomes the weighted sum of all intermediate losses.

7.2.2 *Unsupervised or Self-supervised Learning*

While synthetic datasets enable training CNNs with a large amount of labeled data, the networks only trained on synthetic datasets perform relatively poorly on real-world datasets due to the domain mismatch between the training domain and the target domain. As just discussed, supervised approaches thus require fine-tuning on the target domain for better accuracy. However, this can be problematic if there is no ground truth optical flow available for the target domain. To resolve this issue, unsupervised learning approaches have been proposed to directly train CNNs on the target domain without having access to any ground truth flow. Such methods are also called self-supervised, as the supervisory signal comes from the input images themselves. In this section, we will overview existing unsupervised or self-supervised learning methods and discuss how they have progressed to achieve results that are competitive with many supervised methods.

Ahmadi and Patras [1] pioneered unsupervised learning-based optical flow using CNNs. Inspired by the classical Horn and Schunck [20] method, Ahmadi and Patras used the classical optical flow constraint equation as a loss function for training the network. By minimizing this unsupervised loss function, the network learns to predict optical flow fields that satisfy the optical flow constraint equation on the input images, i.e., the brightness constancy assumption. [1] further combines this with classical coarse-to-fine estimation so that the flow field improves through multi-scale estimation. By demonstrating that the flow accuracy is close to the best supervised method at the time, i.e. FlowNet [10], Ahmadi and Patras [1] suggest that unsupervised learning of networks for optical flow estimation is possible and can overcome some of the limitations of supervised learning approaches.

Concurrently, Yu et al. [58] and Ren et al. [43] proposed to use a proxy unsupervised loss that is inspired by a standard MRF formulation. Following classical concepts, the proposed unsupervised proxy loss consists of a data term and a smoothness term as in Eq. (7.2). The data term directly minimizes the intensity difference between the first image and the warped second image from estimated optical flow, and the smoothness term penalizes flow differences between neighboring pixels. Both methods demonstrate that directly training on a target domain (e.g., the KITTI datasets [14]) in an unsupervised manner performs competitive to or sometimes even outperforms the same network that is trained on a different domain (e.g., the FlyingChairs dataset [10]) in a supervised manner. This observation suggests that unsupervised learning approaches can be a viable alternative to supervised learning, if labeled data for training is not available in the target domain.

In a follow-up work, Zhu and Newsam [63] showed that the backbone network can be improved by using a dense connectivity. They built on DenseNet [22], which uses dense connections with skip connections between all convolutional layers to improve the accuracy over the previous state of the art for image classification. Inspired by DenseNet, Zhu et al. [63] adopted the such dense connections in an hourglass-shaped architecture by using dense blocks before every downsampling and upsampling step; each dense block has four convolutional layers with dense skip connections between each other. [63] improves the flow accuracy by more than 10% on public benchmark datasets over [58] on average, which uses FlowNet [10] as a backbone network, indicating the importance of choosing the right backbone network in the unsupervised learning setting as well.

Zhu et al. [62] also proposed a different direction of unsupervised learning, combining an unsupervised proxy loss and a guided supervision loss using proxy ground truth obtained from an off-the-shelf classical energy-based method. As in [43, 58], the unsupervised proxy loss makes the network learn to estimate optical flow to satisfy the brightness constancy assumption while the guided loss helps the network perform close to off-the-shelf classical energy-based method. In the circumstance that learning with the unsupervised proxy loss is outperformed by the classical energy-based method, the guided loss can help and even achieve better accuracy than either of the two losses alone.

Unsupervised or self-supervised learning of optical flow relies on minimizing a proxy loss rather than estimating optical flow close to some ground truth. Thus, designing a faithful proxy loss is critical to its success. Meister et al. [36] proposed a proxy loss function that additionally considers occlusions, demonstrates better accuracy than previous unsupervised methods, and outperforms the supervised backbone network (i.e., FlowNet [10]). Further, bi-directional flow is estimated from the same network by only switching the order of input images and occlusions are detected using a bi-directional consistency check. The proxy loss is applied only to non-occluded regions as the brightness constancy assumption does not hold for occluded pixels. In addition, Meister et al. [36] suggested to use a higher-order smoothness term and a ternary census loss [47, 59] to obtain a data term that is robust to brightness changes. This advanced proxy loss significantly improves the accuracy by halving the error compared to previous unsupervised learning approaches. Meister et al.[36] resulting in better accuracy than supervised approaches pre-trained on synthetic data alone (assuming the same backbone), which suggests that directly training on the target domain in an unsupervised manner can be a good alternative to supervised pre-training with synthetic data.

Wang et al. [51] also introduced an advanced proxy loss that takes occlusion into account and is applied only to non-occluded regions. Similar to [36], Wang et al. [51] estimate bi-directional optical flow and then obtain an occlusion mask for the forward motion by directly calculating disocclusion from the backward flow. They exploit the fact that occlusion from the forward motion is the inverse of disocclusion from the backward motion. Disocclusions can be obtained by forward-warping the given flow and detecting the holes to which no pixels have been mapped. In addition to occlusion handling, their approach contains other innovations such as a modified architecture and pre-processing. According to their ablation study, the accuracy is improved overall by 25% on public benchmark datasets compared to the unsupervised approach of Yu et al. [58]. In addition, the method demonstrates good occlusion estimation results, close to those of classical energy-based approaches.

Janai et al. [26] extended unsupervised learning of optical flow to a multi-frame setting, taking in three consecutive frames and jointly estimating an occlusion map. Based on the PWC-Net [48] architecture, they estimate bi-directional flow from the reference frame and occlusion maps for both directions as well. After the cost volume of PWC-Net, Janai et al. use three different decoders: (i) a future frame decoder that estimates flow from the reference frame to the future frame, (ii) a past flow decoder, and (iii) an occlusion decoder. A basic unsupervised loss consisting of photometric and smoothness terms is applied only on non-occluded regions for estimating flow, and a constant velocity constraint is also used, which encourages the magnitude of forward flow and backward flow to be similar but going in opposite directions. Their experimental results demonstrate the benefits of using multiple frames, outperforming all two-frame based methods. Furthermore, the accuracy of occlusion estimation is competitive with classical energy-based methods.

Liu et al. [32, 33] demonstrated another direction for unsupervised (or self-supervised) learning by using a data distillation framework with student-teacher networks. Their two methods, DDFlow [32] and its extension SelFlow [33], distill reliable predictions from a teacher network, which is trained in an unsupervised manner [36], and use them as pseudo ground truth for training the student network, which is used at inference time. The accuracy of this framework depends on how to best distill the knowledge for the student network. For better accuracy especially in occluded regions, the two methods focus on how to provide more reliable labels for occluded pixels to the student network. DDFlow [32] proposes to randomly crop the predicted flow map from the teacher network as well as the input images. Then in the cropped images, some of the non-occluded pixels near the image boundaries become out-of-bounds pixels (i.e., occluded pixels), and its reliably predicted optical flow from the non-occluded pixels in the teacher network can work as reliable pseudo ground truth for occluded pixels in the student network. In the experiments, DDFlow [32] showed data distillation to significantly improve the accuracy on average up to 34.7% on public benchmark datasets, achieving the best accuracy among existing unsupervised learning-based approaches.

SelFlow [33] suggests a better data distillation strategy by exploiting superpixel knowledge and hallucinating occlusions in non-occluded regions. Given the prediction from the teacher network, SelFlow [33] superpixelizes the target frame and perturbs random superpixels by injecting random noise as if non-occluded pixels in the target images were occluded by randomly looking superpixels. Then likewise, those non-occluded pixels with reliable predictions from the teacher network become occluded pixels when training the student network, guiding to estimate reliable optical flow in occluded areas. In addition, SelFlow [33] further demonstrates multi-frame extensions using 3 frames as input for improving the accuracy by exploiting temporal coherence. Evaluating on public benchmark datasets, SelFlow [33] further improves the accuracy over DDFlow [32], demonstrating the importance of having a better data distillation strategy and suggesting a promising direction for self-supervised learning.

7.2.3 Semi-supervised Learning

Complementary to supervised and unsupervised learning methods, semi-supervised learning approaches have been also proposed recently. Lai et al. [30] utilized Generative Adversarial Networks (GANs) [15] and proposed an adversarial loss that captures the structural pattern of the flow warp error, allowing to train a network in a semi-supervised way. First, a generator network produces optical flow from the two given input images. Next, the flow warp error map is obtained by calculating the image intensity difference between the first image and the warped second image using the flow output. Then, a discriminator network tries to distinguish whether the warp error map is created by the generator or is the ground truth. The generator aims to fool the discriminator network by producing optical flow whose warp error patterns look close to those of the ground truth. Meanwhile, the discriminator keeps trying to correctly distinguish whether the flow warp error pattern is from the generated flow or the ground truth flow, challenging the generator. To train the networks, a combination of labeled and unlabeled data has been used, equally distributed in each mini-batch. For labeled data in each mini-batch, the standard L_2 loss is applied to the output of the generator to ensure closeness of the flow estimate to the ground truth. The adversarial loss is applied to the output of the discriminator to both labeled and unlabeled data. The experiments demonstrate benefits over purely supervised and purely unsupervised methods: the results are more accurate than when training with a synthetic dataset only in a supervised way and they also outperform training with unlabeled real data in the target domain only in an unsupervised way.

Yang and Soatto [56] proposed another semi-supervised approach by learning a conditional prior for predicting optical flow. They posit that current learning-based approaches to optical flow do not rely on any explicit regularizer (which refers to any prior, model, or assumption that adds any restrictions to the solution space), which results in a risk of overfitting on the training domain, relating to the domain mismatch problem regarding the testing domain. To address the issue, they propose a network that contains prior information of possible optical flows that an input image can give rise to and then use the network as a regularizer for training a standard off-the-shelf optical flow network. They first train the conditional prior network in a supervised manner to learn prior knowledge on the possible optical flows of an input image, and then train FlowNet [10] in an unsupervised manner with a regularization loss from the trained conditional prior network. The experiments demonstrate that the conditional prior network enables the same network trained on the same dataset (i) to outperform typical unsupervised training and (ii) to give results that are competitive with the usual supervised training, yet showing better generalization across different dataset domains. This observation suggests that semi-supervised learning can benefit domain generalization without labeled data by leveraging the available ground truth from another domain.

7.3 Multi-frame Optical Flow Estimation

In the literature of classical optical flow methods, utilizing multiple frames has a long history (e.g., [39]). When additional temporally consecutive frames are available, different kinds of assumptions and strategies can be exploited. One basic and straightforward way is to utilize the temporal coherence assumption that optical flow smoothly changes over time [6, 27, 28, 50, 53]. This property is sometimes also referred to as constant velocity or acceleration assumption. Another way is to parameterize and model the trajectories of motion, which allows to exploit higher-level motion information instead of simply enforcing temporal smoothness on optical flow [9, 13, 45] in 2D. Recently, there has been initial work on adopting these proven ideas in the context of deep learning to improve the flow accuracy.

Ren et al. [42] proposed a multi-frame optical flow network by extending the two-frame, state-of-the-art PWC-Net [48]. Given three temporally consecutive frames, I_{t-1}, I_t, and I_{t+1}, the proposed method fuses the two optical flows from I_{t-1} to I_t and from I_t to I_{t+1} to exploit the temporal coherence between the three frames. Each optical flow is obtained using PWC-Net. In order to fuse the two optical flows, the method also estimates the flow from I_t to I_{t-1} to backwardwarp the flow from I_{t-1} to I_t to match the spatial coordinates of corresponding pixels. When fusing the two flows, Ren et al. use an extra network that inputs the flows with their brightness error and outputs the refined final flow. The underlying idea of inputting the brightness error together is to guide regions to refine to where optical flow may be inaccurate.

In their experiments, Ren et al. [42] demonstrated that utilizing two adjacent optical flows and fusing them improves the flow accuracy especially in occluded areas and out-of-bound areas.

Maurer and Bruhn [34] also proposed a multi-frame optical flow method that exploits the temporal coherence but in a different direction by learning to predict forward flow from the backward flow in an online manner. Similarly given three temporally consecutive frames, I_{t-1}, I_t, and I_{t+1}, the proposed method first estimates the forward flow (i.e., from I_t to I_{t+1}) and the backward flow (i.e., from I_t to I_{t-1}) using an off-the-shelf energy-based approach [21]. Next, the method finds inliers for each flow by estimating the opposite directions of each flow (i.e., from I_{t+1} to I_t and from I_{t-1} to I_t) and performing a consistency check. Given the inlier flow for both directions as ground truth data, the method then trains shallow 3-layer CNNs that predict the forward flow (i.e., from I_t to I_{t+1}) from the input backward flow (i.e., from I_t to I_{t-1}). The idea to predict the forward flow from the backward flow is to exploit the valuable motion information from the previous time step including in occluded regions, which the current step is not able to properly handle but that are visible in the previous time step. This training is done in an online manner so that the network can be trained adaptively to input samples while exploiting temporal coherence. Finally, the method fuses the predicted forward flow and the estimated forward flow to obtain a refined forward flow. On major benchmark datasets, the method demonstrates the advantages of exploiting temporal coherence by improving the accuracy especially in occluded regions by up to 27% overall over a baseline model that does not use temporal coherence.

Finally, Neoral et al. [40] proposed an extended version of PWC-Net [48] in the multi-frame setting, jointly estimating optical flow and occlusion. Given a temporal sequence of frames, Neoral et al. proposed to improve the flow and occlusion accuracy by leveraging each other in a recursive manner in the temporal domain. First, they propose a sequential estimation of optical flow and occlusion: estimating occlusion first and then estimating optical flow, feeding the estimated occlusion as one of inputs into the flow decoder. They found that providing the estimated occlusion as an additional input improves the flow accuracy by more than 25%. Second, they input the estimated flow from the previous time step into the occlusion and flow decoders as well, which yields additional accuracy improvements for both tasks, especially improving the flow accuracy by more than 12% on public benchmark datasets. Similar to other multi-frame based methods above, the flow accuracy improvement is especially prominent in occluded areas and also near motion boundaries.

7.4 Conclusion

The recent advances in deep learning have significantly influenced the transition from classical energy-based formulations to CNN-based approaches for optical flow estimation. We reviewed this transition here. Two main families of CNN approaches to optical flow have been pursued: (i) using CNNs as a feature extractor on top of conventional energy-based formulations and (ii) end-to-end trainable, regression-based CNN architectures. While methods proposed in the initial stages of this transition were outperformed by classical energy-based formulations at the time, steady research progress, e.g. discovering better backbone architectures, synthetic training datasets, and learning strategies eventually led CNN-based methods to yield the most accurate results today and to dominate the current literature. To overcome the (domain) overfitting tendency of supervised learning, unsupervised or self-supervised methods, as well as semi-supervised learning methods have been recently investigated as alternatives. Finally, multi-frame CNN approaches, exploiting temporal smoothness or coherency, have demonstrated the potential of improving the flow estimation accuracy even further.

Despite the significant progress, a number of limitations of current approaches remain including, e.g., (i) the domain overfitting tendency, i.e. trained models do not generalize well to unseen domains yet, and (ii) the necessity of complex training schemes, which require pre-training on synthetic datasets first before fine-tuning on the target domain and can make training models complicated in practice. These and other challenges leave significant room for future work on deep learning methods for optical flow.

References

1. Ahmadi, A., & Patras, I. (2016). Unsupervised convolutional neural networks for motion estimation. In *Proceedings of the IEEE International Conference on Image Processing* (pp. 1629–1633).
2. Bai, M., Luo, W., Kundu, K., & Urtasun, R. (2016). Exploiting semantic information and deep matching for optical flow. In B. Leibe, J. Matas, N. Sebe, & M. Welling (Eds.), *Proceedings of the 14th European Conference on Computer Vision. Lecture Notes in Computer Science* (Vol. 9908, pp. 154–170). Springer.
3. Bailer, C., Varanasi, K., & Stricker, D. (2017). CNN-based patch matching for optical flow with thresholded hinge embedding loss. In *Proceedings of the IEEE Computer Society Conference on Computer Vision and Pattern Recognition* (pp. 3250–3259). Honolulu, Hawaii.
4. Baker, S., Scharstein, D., Lewis, J. P., Roth, S., Black, M. J., & Szeliski, R. (2011). A database and evaluation methodology for optical flow. *International Journal of Computer Vision, 92*(1), 1–31.
5. Barnes, C., Shechtman, E., Goldman, D. B., & Finkelstein, A. (2010). The generalized PatchMatch correspondence algorithm. In K. Daniilidis, P. Maragos, & N. Paragios (Eds.), *Proceedings of the 11th European Conference on Computer Vision. Lecture Notes in Computer Science* (Vol. 6313, pp. 29–43). Springer.
6. Black, M. J., & Anandan, P. (1991). Robust dynamic motion estimation over time. In *Proceedings of the IEEE Computer Society Conference on Computer Vision and Pattern Recognition* (pp. 296–302). Lahaina, Maui, Hawaii.
7. Boykov, Y., Veksler, O., & Zabih, R. (1998). Markov random fields with efficient approximations. In *Proceedings of the IEEE Computer Society Conference on Computer Vision and Pattern Recognition* (pp. 648–655). Santa Barabara, California.
8. Butler, D. J., Wulff, J., Stanley, G. B., & Black, M. J. (2012). A naturalistic open source movie for optical flow evaluation. In A. Fitzgibbon, S. Lazebnik, P. Perona, Y. Sato, & C. Schmid (Eds.), *Proceedings of the 12th European Conference on Computer Vision. Lecture Notes in Computer Science* (Vol. 7577, pp. 611–625). Springer.

9. Chaudhury, K., & Mehrotra, R. (1995). A trajectory-based computational model for optical flow estimation. *IEEE Transactions on Robotics and Automation, 11*(5), 733–741.
10. Dosovitskiy, A., Fischer, P., Ilg, E., Häusser, P., Hazırbaş, C., Golkov, V., et al. (2015). FlowNet: Learning optical flow with convolutional networks. In *Proceedings of the Fifteenth IEEE International Conference on Computer Vision* (pp. 2758–2766). Santiago, Chile.
11. Fortun, D., Bouthemy, P., & Kervrann, C. (2015). Optical flow modeling and computation: A survey. *Computer Vision and Image Understanding, 134*(1), 1–21.
12. Gadot, D., & Wolf, L. (2016). PatchBatch: A batch augmented loss for optical flow. In *Proceedings of the IEEE Computer Society Conference on Computer Vision and Pattern Recognition* (pp. 4236–4245). Las Vegas, Nevada.
13. Garg, R., Roussos, A., & Agapito, L. (2013). A variational approach to video registration with subspace constraints. *International Journal of Computer Vision, 104*(3), 286–314.
14. Geiger, A., Lenz, P., & Urtasun, R. (2012). Are we ready for autonomous driving? The KITTI vision benchmark suite. In *Proceedings of the IEEE Computer Society Conference on Computer Vision and Pattern Recognition* (pp. 3354–3361). Providence, Rhode Island.
15. Goodfellow, I. J., Pouget-Abadie, J., Mirza, M., Xu, B., Warde-Farley, D., Ozair, S., et al. (2014). Generative adversarial nets. *Advances in Neural Information Processing Systems*, 2672–2680.
16. Güney, F., & Geiger, A. (2016). Deep discrete flow. In S. H. Lai, V. Lepetit, K. Nishino, & Y. Sato (Eds.), *Proceedings of the Thirteenth Asian Conference on Computer Vision. Lecture Notes in Computer Science* (Vol. 10115, pp. 207–224). Springer.
17. Hadsell, R., Chopra, S., & LeCun, Y. (2006). Dimensionality reduction by learning an invariant mapping. In *Proceedings of the IEEE Computer Society Conference on Computer Vision and Pattern Recognition* (pp. 1735–1742). New York.
18. Han, X., Leung, T., Jia, Y., Sukthankar, R., & Berg, A. C. (2015). MatchNet: Unifying feature and metric learning for patch-based matching. In *Proceedings of the IEEE Computer Society Conference on Computer Vision and Pattern Recognition* (pp. 3279–3286). Boston, Massachusetts.
19. Hirschmüller, H. (2008). Stereo processing by semiglobal matching and mutual information. *IEEE Transactions on Pattern Analysis and Machine Intelligence, 30*(2), 328–341.
20. Horn, B. K. P., & Schunck, B. G. (1981). Determining optical flow. *Artificial Intelligence, 17*(1–3), 185–203.
21. Hu, Y., Song, R., & Li, Y. (2016). Efficient coarse-to-fine PatchMatch for large displacement optical flow. In *Proceedings of the IEEE Computer Society Conference on Computer Vision and Pattern Recognition* (pp. 5704–5712). Las Vegas, Nevada.
22. Huang, G., Liu, Z., vd Maaten, L., & Weinberger, K. Q. (2017). Densely connected convolutional networks. In *Proceedings of the IEEE Computer Society Conference on Computer Vision and Pattern Recognition* (pp. 2261–2269).
23. Hui, T. W., Tang, X., & Loy, C. C. (2018). LiteFlowNet: A lightweight convolutional neural network for optical flow estimation. In *Proceedings of the IEEE Computer Society Conference on Computer Vision and Pattern Recognition* (pp. 8981–8989). Salt Lake City, Utah.
24. Hur, J., & Roth, S. (2019). Iterative residual refinement for joint optical flow and occlusion estimation. In *Proceedings of the IEEE Computer Society Conference on Computer Vision and Pattern Recognition* (pp. 5754–5763).
25. Ilg, E., Mayer, N., Saikia, T., Keuper, M., Dosovitskiy, A., & Brox, T. (2017). FlowNet 2.0: Evolution of optical flow estimation with deep networks. In *Proceedings of the IEEE Computer Society Conference on Computer Vision and Pattern Recognition* (pp. 1647–1655). Honolulu, Hawaii.
26. Janai, J., Güney, F., Ranjan, A., Black, M. J., & Geiger, A. (2018). Unsupervised learning of multi-frame optical flow with occlusions. In V. Ferrari, M. Hebert, C. Sminchisescu, & Y. Weiss (Eds.), *Proceedings of the 15th European Conference on Computer Vision. Lecture Notes in Computer Science* (pp. 713–731). Springer.
27. Janai, J., Guney, F., Wulff, J., Black, M. J., & Geiger, A. (2017). Slow flow: Exploiting high-speed cameras for accurate and diverse optical flow reference data. In *Proceedings of the IEEE Computer Society Conference on Computer Vision and Pattern Recognition* (pp. 1406–1416).

28. Kennedy, R., & Taylor, C. J. (2015). Optical flow with geometric occlusion estimation and fusion of multiple frames. In X. C. Tai, E. Bae, T. Chan, & M. Lysaker (Eds.), *Proceedings of the 10th International Conference on Energy Minimization Methods in Computer Vision and Pattern Recognition. Lecture Notes in Computer Science* (pp. 364–377).
29. Krizhevsky, A., Sutskever, I., & Hinton, G. E. (2012). ImageNet classification with deep convolutional neural networks. *Advances in Neural Information Processing Systems, 25*, 1097–1105.
30. Lai, W. S., Huang, J. B., & Yang, M. H. (2017). Semi-supervised learning for optical flow with generative adversarial networks. *Advances in Neural Information Processing Systems*, 354–364.
31. Li, S. Z. (1994). Markov random field models in computer vision. In J. O. Eklundh (Ed.), *Proceedings of the Third European Conference on Computer Vision. Lecture Notes in Computer Science* (pp. 361–370). Springer.
32. Liu, P., King, I., Lyu, M. R., & Xu, J. (2019). DDFlow: Learning optical flow with unlabeled data distillation. In *Proceedings of the Thirty-Third AAAI Conference on Artificial Intelligence* (pp. 8770–8777). Honolulu, Hawaii.
33. Liu, P., Lyu, M., King, I., & Xu, J. (2019). SelFlow: Self-supervised learning of optical flow. In *Proceedings of the IEEE Computer Society Conference on Computer Vision and Pattern Recognition* (pp. 4571–4580). Long Beach, California.
34. Maurer, D., & Bruhn, A. (2018). ProFlow: Learning to predict optical flow. In *Proceedings of the British Machine Vision Conference*. Newcastle, UK.
35. Mayer, N., Ilg, E., Häusser, P., Fischer, P., Cremers, D., Dosovitskiy, A., et al. (2016). Large dataset to train convolutional networks for disparity, optical flow, and scene flow estimation. In *Proceedings of the IEEE Computer Society Conference on Computer Vision and Pattern Recognition* (pp. 4040–4048). Las Vegas, Nevada.
36. Meister, S., Hur, J., & Roth, S. (2018). UnFlow: Unsupervised learning of optical flow with a bidirectional census loss. In *Proceedings of the Thirty-Second AAAI Conference on Artificial Intelligence*. New Orleans, Louisiana.
37. Menze, M., & Geiger, A. (2015). Object scene flow for autonomous vehicles. In *Proceedings of the IEEE Computer Society Conference on Computer Vision and Pattern Recognition* (pp. 3061–3070). Boston, Massachusetts.
38. Menze, M., Heipke, C., & Geiger, A. (2015). Discrete optimization for optical flow. In *Proceedings of the 37th German Conference on Pattern Recognition, Lecture Notes in Computer Science* (pp. 16–28). Springer.
39. Nagel, H. H. (1990). Extending the 'oriented smoothness constraint' into the temporal domain and the estimation of derivatives of optical flow. In O. D. Faugeras (Ed.), *Proceedings of the First European Conference on Computer Vision. Lecture Notes in Computer Science* (Vol. 427, pp. 139–148). Springer.
40. Neoral, M., Šochman, J., & Matas, J. (2018). Continual occlusions and optical flow estimation. In C. Jawahar, H. Li, G. Mori, & K. Schindler (Eds.), *Proceedings of the Fourtheenth Asian Conference on Computer Vision. Lecture Notes in Computer Science*. Springer.
41. Ranjan, A., & Black, M. J. (2017). Optical flow estimation using a spatial pyramid network. In *Proceedings of the IEEE Computer Society Conference on Computer Vision and Pattern Recognition* (pp. 2720–2729). Honolulu, Hawaii.
42. Ren, Z., Gallo, O., Sun, D., Yang, M. H., Sudderth, E. B., & Kautz, J. (2019). A fusion approach for multi-frame optical flow estimation. In *IEEE Winter Conference on Applications of Computer Vision* (pp. 2077–2086). Waikoloa Village, HI.
43. Ren, Z., Yan, J., Ni, B., Liu, B., Yang, X., & Zha, H. (2017). Unsupervised deep learning for optical flow estimation. In *Proceedings of the Thirty-First AAAI Conference on Artificial Intelligence* (pp. 1495–1501). San Francisco, California.
44. Revaud, J., Weinzaepfel, P., Harchaoui, Z., & Schmid, C. (2015). EpicFlow: Edge-preserving interpolation of correspondences for optical flow. In *Proceedings of the Fifteenth IEEE International Conference on Computer Vision* (pp. 1164–1172). Santiago, Chile.
45. Ricco, S., & Tomasi, C. (2012). Dense Lagrangian motion estimation with occlusions. In *Proceedings of the IEEE Computer Society Conference on Computer Vision and Pattern Recognition* (pp. 1800–1807). Providence, Rhode Island.

46. Simo-Serra, E., Trulls, E., Ferraz, L., Kokkinos, I., Fua, P., & Moreno-Noguer, F. (2015). Discriminative learning of deep convolutional feature point descriptors. In *Proceedings of the Fifteenth IEEE International Conference on Computer Vision* (pp. 118–126). Santiago, Chile.
47. Stein, F. (2004). Efficient computation of optical flow using the census transform. In C. Rasmussen, H. Bülthoff, B. Schölkopf, & M. Giese (Eds.), *Pattern Recognition, Proceedings of the 26th DAGM-Symposium. Lecture Notes in Computer Science* (Vol. 3175, pp. 79–86). Springer.
48. Sun, D., Yang, X., Liu, M. Y., & Kautz, J. (2018). PWC-Net: CNNs for optical flow using pyramid, warping, and cost volume. In *Proceedings of the IEEE Computer Society Conference on Computer Vision and Pattern Recognition* (pp. 8934–8943). Salt Lake City, Utah.
49. Tu, Z., Xie, W., Zhang, D., Poppe, R., Veltkamp, R. C., Li, B., et al. (2019). A survey of variational and CNN-based optical flow techniques. *Signal Processing: Image Communication, 72*, 9–24.
50. Volz, S., Bruhn, A., Valgaerts, L., & Zimmer, H. (2011). Modeling temporal coherence for optical flow. In *Proceedings of the Thirteenth IEEE International Conference on Computer Vision* (pp. 1116–1123). Barcelona, Spain.
51. Wang, Y., Yang, Y., Yang, Z., Zhao, L., Wang, P., & Xu, W. (2018). Occlusion aware unsupervised learning of optical flow. In *Proceedings of the IEEE Computer Society Conference on Computer Vision and Pattern Recognition* (pp. 4884–4893).
52. Weinzaepfel, P., Revaud, J., Harchaoui, Z., & Schmid, C. (2013). DeepFlow: Large displacement optical flow with deep matching. In *Proceedings of the Fourteenth IEEE International Conference on Computer Vision* (pp. 1385–1392). Sydney, Australia.
53. Werlberger, M., Trobin, W., Pock, T., Wedel, A., Cremers, D., & Bischof, H. (2009). Anisotropic huber-L1 optical flow. In *Proceedings of the British Machine Vision Conference*. London, UK.
54. Xu, J., Ranftl, R., & Koltun, V. (2017). Accurate optical flow via direct cost volume processing. In *Proceedings of the IEEE Computer Society Conference on Computer Vision and Pattern Recognition* (pp. 1289–1297). Honolulu, Hawaii.
55. Yang, G., & Ramanan, D. (2019). Volumetric correspondence networks for optical flow. *Advances in Neural Information Processing Systems*, 793–803.
56. Yang, Y., & Soatto, S. (2018). Conditional prior networks for optical flow. In V. Ferrari, M. Hebert, C. Sminchisescu, & Y. Weiss (Eds.), *Proceedings of the 15th European Conference on Computer Vision. Lecture Notes in Computer Science* (pp. 271–287). Springer.
57. Yin, Z., Darrell, T., & Yu, F. (2019). Hierarchical discrete distribution decomposition for match density estimation. In *Proceedings of the IEEE Computer Society Conference on Computer Vision and Pattern Recognition* (pp. 6044–6053). Long Beach, California.
58. Yu, J. J., Harley, A. W., & Derpanis, K. G. (2016). Back to basics: Unsupervised learning of optical flow via brightness constancy and motion smoothness. In *Proceedings of the 14th European Conference on Computer Vision Workshops. Lecture Notes in Computer Science* (pp. 3–10). Springer.
59. Zabih, R., & Woodfill, J. (1994). Non-parametric local transforms for computing visual correspondence. In J. O. Eklundh (Ed.), *Proceedings of the Third European Conference on Computer Vision. Lecture Notes in Computer Science* (Vol. 801, pp. 151–158). Springer.
60. Zagoruyko, S., & Komodakis, N. (2015). Learning to compare image patches via convolutional neural networks. In *Proceedings of the IEEE Computer Society Conference on Computer Vision and Pattern Recognition* (pp. 4353–4361). Boston, Massachusetts.
61. Žbontar, J., & LeCun, Y. (2015). Computing the stereo matching cost with a convolutional neural network. In *Proceedings of the IEEE Computer Society Conference on Computer Vision and Pattern Recognition* (pp. 1592–1599). Boston, Massachusetts.
62. Zhu, Y., Lan, Z., Newsam, S., & Hauptmann, A. G. (2017). Guided optical flow learning. In *Proceedings of the IEEE Computer Society Conference on Computer Vision and Pattern Recognition Workshops (CVPRW)*.
63. Zhu, Y., & Newsam, S. (2017). DenseNet for dense flow. In *Proceedings of the IEEE International Conference on Image Processing* (pp. 790–794).

Chapter 8
Spatio-Temporal Action Instance Segmentation and Localisation

Suman Saha, Gurkirt Singh, Michael Sapienza, Philip H. S. Torr, and Fabio Cuzzolin

Abstract Current state-of-the-art human action recognition is focused on the classification of temporally trimmed videos in which only one action occurs per frame. In this work we address the problem of action localisation and instance segmentation in which multiple concurrent actions of the same class may be segmented out of an image sequence. We cast the action tube extraction as an energy maximisation problem in which configurations of region proposals in each frame are assigned a cost and the best action tubes are selected via two passes of dynamic programming. One pass associates region proposals in space and time for each action category, and another pass is used to solve for the tube's temporal extent and to enforce a smooth label sequence through the video. In addition, by taking advantage of recent work on action foreground-background segmentation, we are able to associate each tube with class-specific segmentations. We demonstrate the performance of our algorithm on the challenging LIRIS-HARL dataset and achieve a new state-of-the-art result which is 14.3 times better than previous methods.

8.1 Introduction

The exiting competing approaches [8, 18, 21, 25] address the problem of action detection in a setting where videos contain single action category and most of them

S. Saha (✉) · G. Singh
Computer Vision Lab (CVL), ETH Zurich, Sternwartstrasse 7, 8092 Zurich, Switzerland
e-mail: suman.saha@vision.ee.ethz.ch

M. Sapienza
Think Tank Team, Samsung Research America, Mountain View, CA, United States
e-mail: mikesapi@robots.ox.ac.uk

P. H. S. Torr
Department of Engineering Science, University of Oxford, Parks Road, Oxford OX1 3PJ, United Kingdom
e-mail: philip.torr@eng.ox.ac.uk

F. Cuzzolin
Oxford Brookes University, Wheatley campus, E block, room E208, Oxford, United Kingdom
e-mail: fabio.cuzzolin@brookes.ac.uk

N. Noceti et al. (eds.), *Modelling Human Motion*,
https://doi.org/10.1007/978-3-030-46732-6_8

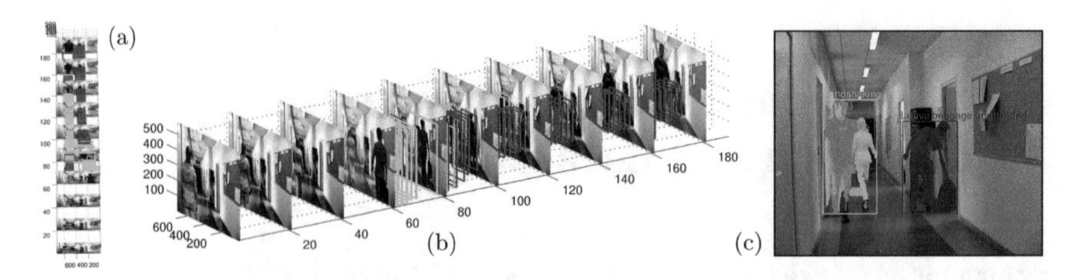

Fig. 8.1 A video sequence taken from the LIRIS-HARL dataset plotted in space-and time. **a** A top down view of the video plotted with the detected action tubes of class "*handshaking*" in green, and "*person leaves baggage unattended*" in red. Each action is located to be within a space-time tube. **b** A side view of the same space-time detections. Note that no action is detected at the beginning of the video when there is human motion present in the video. **c** Action instance segmentation results for two actions occurring simultaneously in a single frame

are temporally trimmed. In contrast, this chapter addresses the problems of both spatio-temporal action instance segmentation and action detection. Here, we consider real-world scenarios where videos often contain co-occurring action instances belong to different action categories. Consider the example shown in Fig. 8.1, where our proposed model performs action instance segmentation and detection of two co-occurring actions "*leaving bag unattended*" and "*handshaking*" which have different spatial and temporal extents within the given video sequence. The video is taken from the LIRIS-HARL dataset [13]. In this chapter, we propose a deep learning based framework for both action instance segmentation and detection, and evaluate the proposed model on the LIRIS-HARL dataset which is more challenging than the standard benchmarks: UCF-101-24 [23] and J-HMDB-21 [13] due to its multi-label and highly temporally untrimmed videos. To demonstrate the generality of the segmentation results on other standard benchmarks, we present some additional qualitative action instance segmentation results on the standard UCF-101-24 dataset (Sect. 8.4.4).

Outline. This chapter is organized as follows. First we present an overview of the approach in Sect. 8.2. We then introduce the detailed methodology in Sect. 8.3. Finally, Sects. 8.4 and 8.5 present the experimental validation and discussion respectively.

Related publication. The work presented in this chapter has appeared in arXiv [20].

8.2 Overview of the Approach

An overview of the algorithm is depicted in Fig. 8.2. At test time, we start by performing binary human motion segmentation **(a)** for each input video frame by leveraging the human action segmentation [17], followed by a frame-level region proposal generation **(b)** (Sect. 8.3.1.1). Proposal bounding boxes are then used to crop patches from both RGB and optical flow frames **(c)**. We refer readers to Section A.1 of [19] for details on optical flow frame computation. Crop image patches are resized to a fixed dimension and fed as inputs to an appearance- and a motion-based detection net-

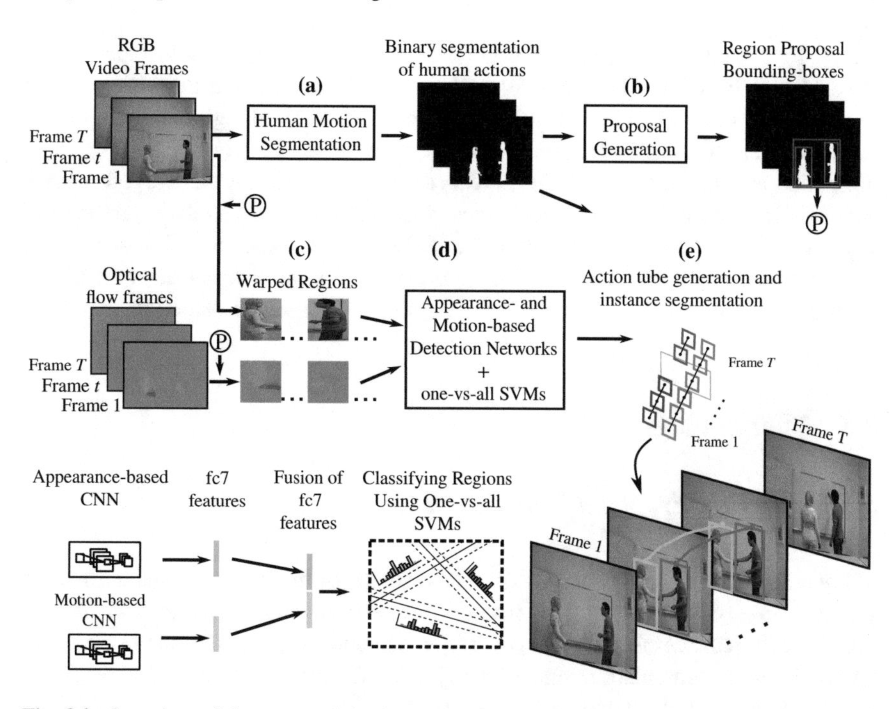

Fig. 8.2 Overview of the proposed spatio-temporal action instance segmentation and detection pipeline. At test time, **a** RGB video frames are fed as inputs to a human motion segmentation algorithm to generate binary segmentation of human actions; at this point these human silhouettes do not carry any class- and instance-aware labels, and they only have binary labels for foreground (and the pixels don't belonging to human silhouettes are labelled as background class). **b** Our region proposal generation algorithm accepts the binary segmented video frames as inputs and computes region proposal bounding boxes using all possible combinations of 2D connected components ($2^N - 1$) present in the binary map. **c** Once the region proposals are computed, warped regions are extracted from both RGB and optical flow frames and fed as inputs to the respective appearance- and motion-based detection networks. **d** The detection networks compute fc7 appearance and motion features for each warped region, features are then fused and subsequently used by a set of one-vs-all SVMs to generate action classification scores for each region. **e** Finally, frame-level detection windows are temporally linked as per their class-specific scores and spatial overlaps to build class-specific action tubes. Further, each pixel within the detection windows is assigned to an class- and instance-aware label by by utilising both the bounding-box detections associated with each class-specific action tubes and the binary segmentation maps (or human silhouettes) generated in (**a**)

work **(d)** (Sect. 8.3.2) to compute CNN fc7 features. Subsequently, these appearance- and motion-based fc7 features are fused, and later, these fused features are classified by a set of one-versus-all SVMs. Each fused feature vector is a high-level image representation of its corresponding warped region and encodes both static appearance (e.g. boundaries, corners, object shapes) and motion pattern of human actions (if there is any). Finally, the top k frame-level detections (regions with high classification scores) are temporally linked in time to build class-specific action tubes **(e)** and then, these tubes are trimmed (as in [21]) to solve for temporal action localisation. Pixels belonging to each action tube are assigned class- and instance-aware action labels by taking advantage of both tube's class score and the binary action segmentation maps computed in **(a)**. At train time, first action region hypotheses are generated for RGB video frames using Selective Search [24] (Sect. 8.3.1.2), then, pretrained appearance and motion CNNs **(d)** are fine-tuned on the warped regions extracted from both RGB and flow frames. Subsequently, fine-tuned appearance and motion CNNs are used to compute fc7 features from both RGB and flow training frames, features are then fused and pass as inputs to a set of one-versus-all SVMs for training. A detailed descriptions of these above steps are presented in Sect. 8.3.

8.3 Methodology

8.3.1 Region Proposal Generation

We denote each 2D region proposal '**r**' as a subset of the image pixels, associated with a minimum bounding box '**b**' around it. In the following sub sections we present our two different region proposal generation schemes: (1) the first one is based on human motion segmentation algorithm [17], and (2) the second one uses Selective Search algorithm [24] to generate 2D action proposals.

8.3.1.1 Proposals Based on Motion Segmentation

The human motion segmentation [17] algorithm generates binary segmentation of human actions (Fig. 8.2a). It extracts human motion from video using long term trajectories [3]. In order to detect static human body parts which don't carry any motion but are still significant in the context of the whole action, it attaches scores to these regions using a human shape prior from a deformable part-based (DPM) model [6]. By striking balance between the human motion and static human-body appearance information, it generates binary silhouettes of human actions in space and time. At test time our region proposal algorithm accepts the binary segmented images produced by [17], and generates region proposal hypotheses using all possible combinations of 2D connected components ($2^N - 1$) present in the binary map (Fig. 8.2b), where N is the number of 2D connected components present in each video frame (Sect. A.3 of [19]). In the following subsection, we briefly introduce the human motion segmentation pipeline.

Human Motion Segmentation. The human motion segmentation algorithm takes as input a sequence of RGB video frames (which contain human action) and outputs binary-labelled space-time video segments where pixels belong to an human action are labelled as foreground and remaining are as background. Firstly, in order to localise and rank "actionness" [4], a human motion saliency feature is computed by exploiting the foreground motion and human appearance information. Foreground motion is estimated by forming a camera model using long term trajectories [3]

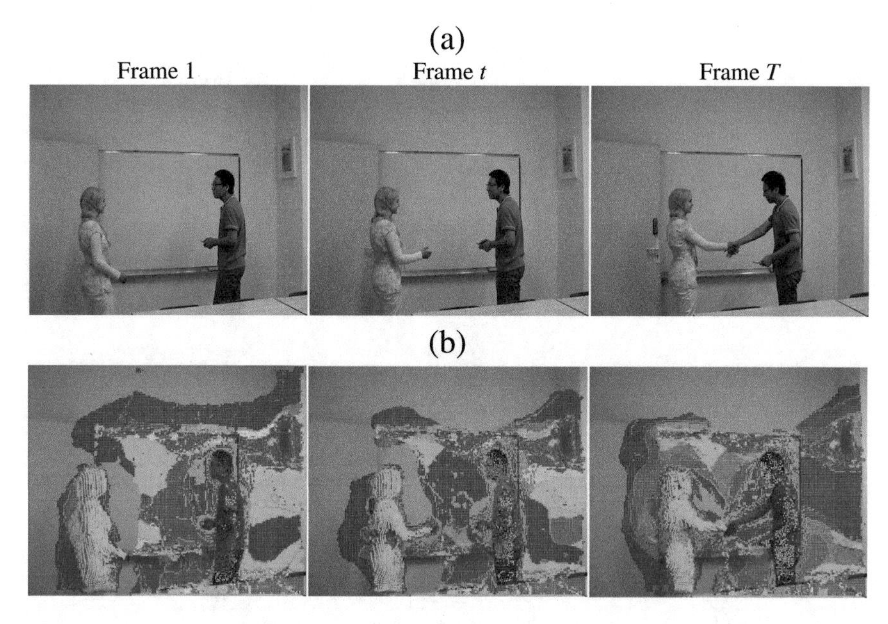

Fig. 8.3 **a** Three sample input video frames showing a "handshaking" action from a test video clip of LIRIS HARL dataset [26]. **b** The corresponding motion saliency response generated using long term trajectories [3] are shown for these three frames. Notice, the motion saliency is relatively higher for the person at the left, who first enters into the room and then approaches towards the person in the right for "handshaking". Also note that, motion saliency is computed on the entire video clip, for the sake of visualization, we pick three sample frames

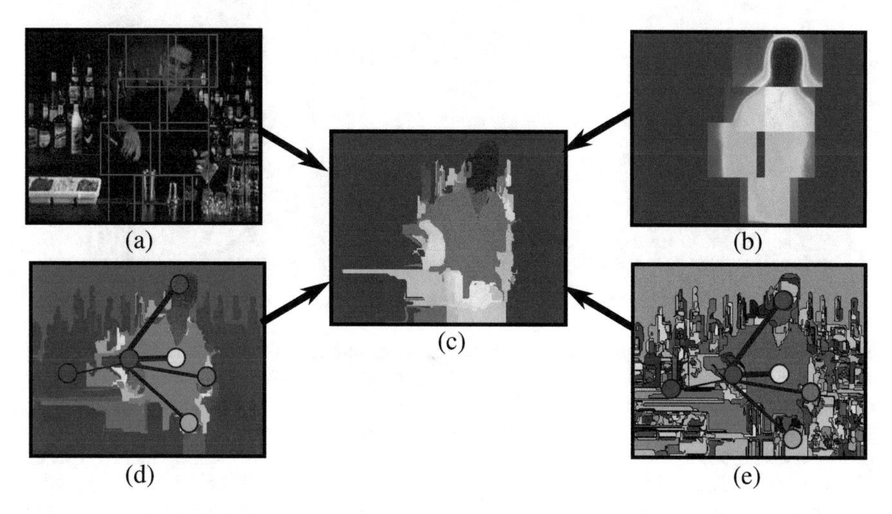

Fig. 8.4 **a** DPM based person detection. **b** Corresponding DPM part mask. **c** Supervoxel response for the DPM mask. **d** and **e** Pairwise connections of motion saliency map and segmentation respectively. This figure is taken from [17] with author's permission

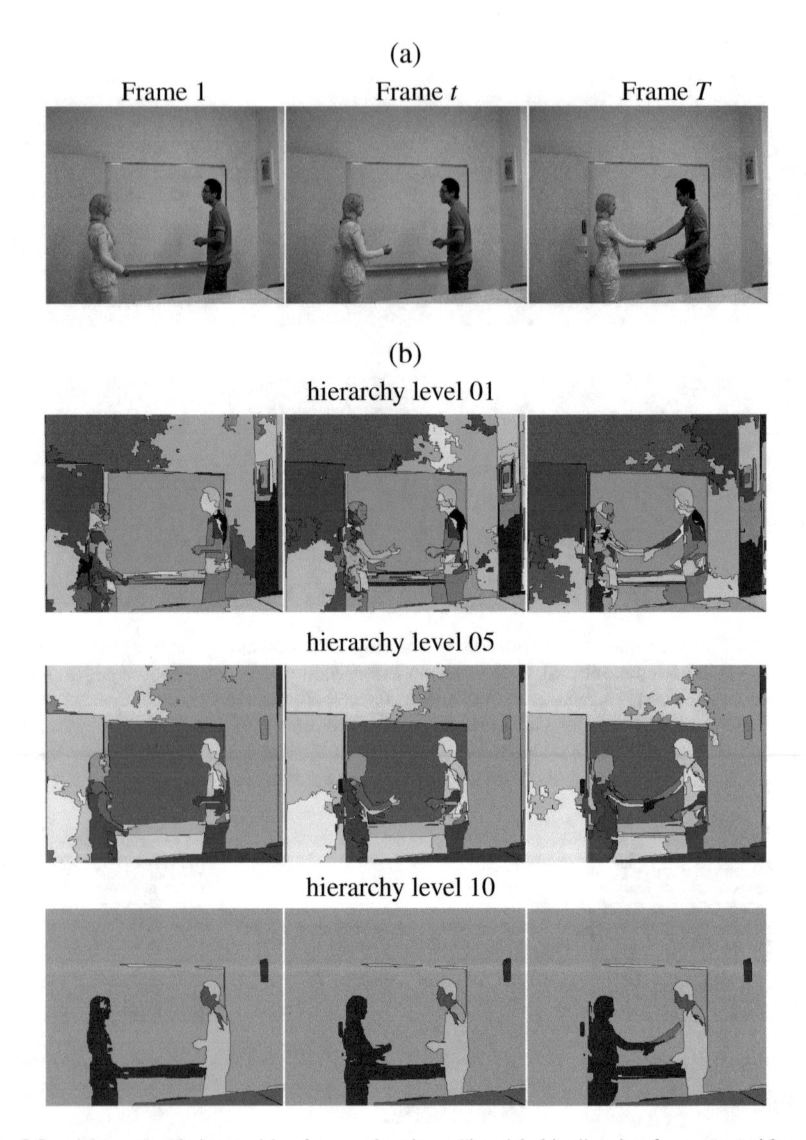

Fig. 8.5 **a** Three sample input video frames showing a "handshaking" action from a test video clip of LIRIS HARL dataset [26]. **b** The hierarchical graph based video segmentation results (at three different levels of hierarchy) are shown for these three frames. The three rows show segmentation results for hierarchy level 1, 5 and 10 respectively where 1 is the lowest level with supervoxels having smaller spatial extents and 10 is the highest level with supervoxels having relatively larger spatial extents. Notice, the supervoxels belong to higher levels of segmentation hierarchy tend to preserve the semantic information and are less prone to leaks. Also note that, video segmentation is computed on the entire video clip, for the sake of visualization, we pick three sample frames

(Fig. 8.3) and human appearance based saliency map is generated using a DPM person detector [6] (Fig. 8.4a–c) trained on PASCAL VOC 2007 [5]. Secondly, to segment human actions, a hierarchical graph-based video segmentation algorithm [28] is used to extract supervoxels at different level of pixel granularity (i.e. different levels of segmentation hierarchy) (Fig. 8.5). The foreground motion and human appearance based saliency features are then encoded in the hierarchy of supervoxels using a hierarchical Markov Random Field (MRF) model. This encoding gives the unary potential components. To avoid a brittle graph due to a large number of supervoxels [12], the MRF graph is built with a smaller subset of supervoxels which are highly likely to contain human actions. Thus, a candidate edge is built between two neighbouring supervoxels based on their optical flow directions and overlaps with a person detection. In the MRF graph structure, supervoxels are nodes and an edge between two supervoxels are built if: (a) they are temporal neighbours i.e. neighbours in the direction of optical flow, or (b) spatial neighbours, i.e. both the supervoxels have high overlaps with a DPM person detection where the person detection has a confidence greater than a threshold. The temporal supervoxel neighbours and the appearance-aware spatial neighbours (Fig. 8.4d, e) give the pairwise potential components. To avoid leaks and encourage better semantic information, supervoxels (constrained by appearance and motion cues) from higher levels in the hierarchy (Fig. 8.5) are supported by the higher-order potential. Finally, the energy of the MRF is minimised using the α-expansion algorithm [1, 15] and GMM estimation is used to automatically learn the model parameters. The final outputs of the human motion segmentation are the human foreground background binary maps as depicted in Fig. 8.6.

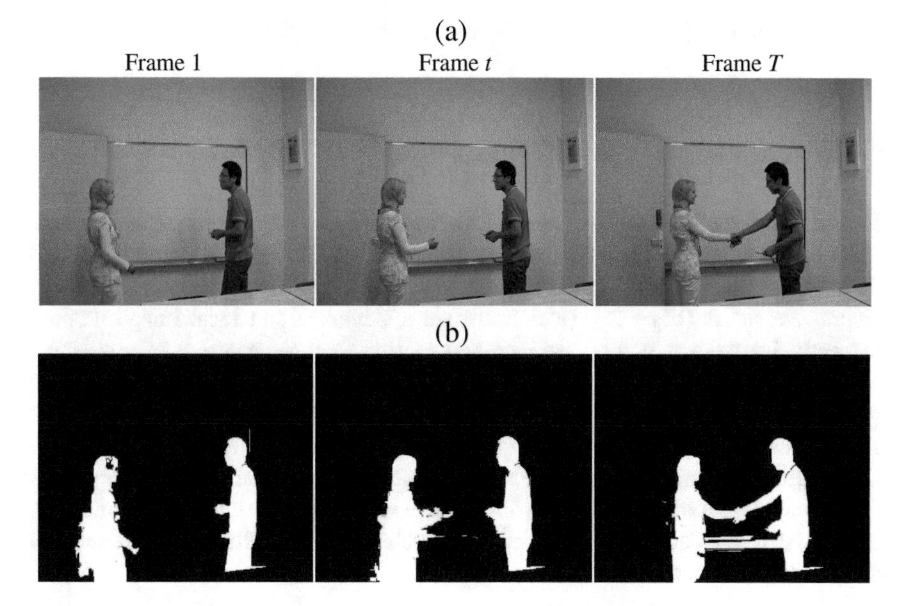

Fig. 8.6 **a** Three sample input video frames showing a "handshaking" action from a test video clip of LIRIS HARL dataset [26]. **b** The human action foreground-background segmentation results are shown for these three frames

8.3.1.2 Proposal Based on Selective Search

We use two competing approaches to generate region proposals for action detection. The first is based upon Selective Search [24], and the second approach is presented in Sect. 8.3.1.1. Whilst using the Selective Search based method for both training and testing, we only use the motion segmentation based method for testing since it does not provide good negative proposals to use during training. Having a sufficient number of negative examples is crucial to train an effective classifier. At test time, the human motion segmentation (Sect. 8.3.1.1) allows us to extract pixel-level action instance segmentation which is superior to what we may obtain by using Selective Search. We validate our action detection pipeline using both algorithms - the results are discussed in Sect. 8.4.

Measuring "Actionness" of Selective Search Proposals. The selective-search region-merging similarity score is based on a combination of colour (histogram intersection), and size properties, encouraging smaller regions to merge early, and avoid holes in the hierarchical grouping. Selective Search (SS) generates on average 2,000 region proposals per frame, most of which do not contain human activities. In order to rank the proposals with an "actionness" score and prune irrelevant regions, we compute dense optical flow between each pair of consecutive frames using the state-of-the-art algorithm in [2]. Unlike Gkioxari and Malik [8], we use a relatively smaller motion threshold value to prune SS boxes, (Sect. A.4 of [19]) to avoid neglecting human activities which exhibit minor body movements exhibited in the LIRIS HARL [26] such as "typing on keyboard", "telephone conversation" and "discussion" activities. In addition to pruning region proposals, the 3-channel optical flow values (i.e., flow-x, flow-y and the flow magnitude) are used to construct 'motion images' from which CNN motion features are extracted [8].

8.3.2 Appearance- and Motion-Based Detection Networks

In the second stage of the pipeline, we use the "actionness" ranked region proposals (Sect. 8.3.1) to select image patches from both the RGB (original video frames) and flow images. The image patches are then fed to a pair of fine-tuned Convolutional Neural Networks (Fig. 8.2d) (which encode appearance and local image motion, respectively) from which appearance and motion feature vectors were extracted. As a result the first network learns static appearance information (both lower-level features such as boundary lines, corners, edges and high level features such as object shapes), while the other encodes action dynamics at frame level. The output of the Convolutional Neural Network may be seen as a highly nonlinear transformation $\Phi(.)$ from local image patches to a high-dimensional vector space in which discrimination may be performed accurately even by a linear classifier. We follow the AlexNet [16] and [29]'s network architectures.

8.3.2.1 Pretraining

We adopt a CNN training strategy similar to [7]. Indeed, for domain-specific tasks on relatively small scale datasets, such as LIRIS HARL [26], it is important to initialise the CNN weights using a model *pre-trained on a larger-scale dataset*, in order to avoid over-fitting [8]. Therefore, to encode object "context" we initialise the appearance-based CNN's weights using a model pre-trained on the PASCAL VOC 2012s object detection dataset. To encode typical motion patterns over a temporal window, the optical motion-based CNN is initialised using a model pre-trained on the UCF101 dataset (split 1) [23]. Both appearance- and motion-based pre-trained models are publicly available online at https://github.com/gkioxari/ActionTubes.

8.3.2.2 Fine Tuning

We use deep learning software tool Caffe [14] to fine-tune pretrained domain-specific appearance- and motion-based CNNs on LIRIS HARL training set. For training CNNs, the Selective Search region proposals (Sect. 8.3.1.2) with an IoU overlap score greater than 0.5 with respect to the ground truth bounding box were considered as positive examples, the rest as negative examples. The image patches specified by the pruned region proposals were randomly cropped and horizontally flipped by the Caffe's *WindowDataLayer* [14] with a crop dimension of 227×227 and a flip probability of 0.5 (Fig. 8.2c). Random cropping and flipping were done for both RGB and flow images. The pre-processed image patches along with the associated ground truth action class labels are then passed as inputs to the appearance and motion CNNs to fine-tune (i.e. updating only the weights of the fully connected layers, in this case, fc6 and fc7 layers, and keeping the weights of the other layers untouched during training) for action classification (Fig. 8.2d). A mini batch of 128 image patches (32 positive and 96 negative examples) are processed by the CNNs at each training forward-pass. Note that the number of batches varies frame-to-fame as per the number of ranked proposals per frame. It makes sense to include fewer positive examples (action regions) as these are relatively rare when compared to background patches (negative examples).

8.3.2.3 Feature Extraction from CNN Layers

We extract the appearance- and motion-based features from the *fc7* layer of the the two networks. Thus, we get two feature vectors (each of dimension 4096): appearance feature '$\mathbf{x}_a = \Phi_a(\mathbf{r})$' and motion feature '$\mathbf{x}_f = \Phi_f(\mathbf{r})$'. We perform L2 normalisation on the obtained feature vectors, to then, scale and merge appearance and motion features (Fig. 8.2d) in an approach similar to that proposed by [8]. This yields a single feature vector $\mathbf{x}$ for each image patch $\mathbf{r}$. Such frame-level region feature vectors are used to train an SVM classifier (Sect. 8.3.3).

8.3.3 *Training Region Proposal Classifiers*

Once discriminative CNN fc7 feature vectors $\mathbf{x} \in \mathbb{R}^n$ are extracted for region proposals (Sect. 8.3.1.2), they can be used to train a set of binary classifiers (Fig. 8.2d) to attach a vector of scores $\mathbf{s}_c$ to each region proposal '**r**', where each element in the score vector $\mathbf{s}_c$ is a confidence measure of each action class $c \in \{1, 2, \ldots, C\}$ to be present within that region. Due to the notable success of linear SVM classifiers when combined with CNN features [7], we trained a set of one-versus-rest linear SVMs to classify region proposals.

8.3.3.1 Class Specific Positive and Negative Examples

In the original RCNN-based one-versus-rest SVM training approach [7], only the ground truth bounding boxes are considered as positive training examples. In contrast, due to extremely high inter- and intra-class variations in LIRIS HARL dataset [26], we use those bounding boxes as positive training examples which have an IoU overlap with the ground truth greater than 75%. In addition, we also consider the ground truth bounding boxes as positives. We believe, our this training data sampling scheme is more intuitive for complex datasets to train SVMs with more positive examples rather than only ground truths. We have achieved almost 5% gain over SVMs classification accuracy with this training strategy. In a similar way, we consider as negative examples only those features vectors whose associated region proposal have an overlap smaller than 30% with respect to the ground truth bounding boxes (possibly several) present in the frame.

8.3.3.2 Training with Hard Negative Mining

We train the set of class specific linear SVMs using hard negative mining [6] to speed up the training process. Namely, in each iteration of the SVM training step we consider only those negative features which fall within the margin of the decision boundary. We use the publicly available toolbox *Liblinear*[1] for SVM training and use $L2$ regularizer and $L1$ hinge-loss with the following parameter values to train the SVMs: positive loss weight $W_{LP} = 2$; SVM regularisation constant $C = 10^{-3}$; bias multiplier $B = 10$.

8.3.4 *Testing Region Proposal Classifiers*

With our actionness-ranked region proposals $\mathbf{r}_i$ (Sect. 8.3.1) we can extract a cropped image patch and pass it to the CNNs for feature extraction in a similar fashion as described in Sect. 8.3.2.3. A prediction takes the form:

[1] http://www.csie.ntu.edu.tw/~cjlin/liblinear/.

$$\mathbf{s}_c(\mathbf{b}) = \mathbf{w}_c^T \Phi(\mathbf{r}) + b_c^{\text{svm}}, \tag{8.1}$$

where, $\Phi(\mathbf{r}) = \{\Phi_a(\mathbf{r}); \Phi_f(\mathbf{r})\}$ is combination of appearance and motion features of $\mathbf{r}$, $\mathbf{w}_c^T$ and b_c^{svm} are the hyperplane parameter and the bias term of the learned SVM model of class c. The confidence measure $\mathbf{s}_c(\mathbf{b})$ that the action 'c' has happened within the bounding-box region '**b**' is based on the appearance and motion features. Here **b** denotes the associated bounding box for a region proposal **r**.

After SVM prediction,each region proposal '**r**' has been assigned a set of class-specific scores $\mathbf{s}_c$, where c denotes the action category label, $c \in \{1, \ldots, C\}$. Once a region proposal has been assigned classification scores $\mathbf{s}_c$, we call it as a detection bounding-box and denote it as **b**. Due to the typically large number of region proposals generated by the Selective Search algorithms (Sect. 8.3.1.2), we further apply non-maximum suppression to prune the regions.

8.3.5 Action Tube Generation and Classification

Once we extract the frame-level detection boxes $\mathbf{b}_t$ (Sect. 8.3.4) for an entire video, we would like to identify sequences of detections most likely to form action tubes. Thus, to extract final detection tubes, linking of these detection boxes in time is essential to generate tubes. We use our two-pass dynamic programming approach as in [21] to formulate the action tube generation problem as a labelling problem where: (i) we link detections $\mathbf{b}_t$ into temporally connected action paths for each action, and (ii) we perform a piece-wise constant temporal labelling on the action paths. A detailed formulation of the tube generation problem can be found in the Appendix A.5 [19].

8.4 Experimental Results

We evaluate two region proposal methods with our pipeline, one based on human motion segmentation (HMS) (Sect. 8.3.1.1) and another one based on selective search (SS) (Sect. 8.3.1.2). We will use "HMS" and "SS" abbreviations in tables and plot to show the performance of our pipeline based on each region proposal technique. Our results are also compared to the current state-of-the-art: VPULABUAM-13 [22] and IACAS-51 [11].

8.4.1 Instance Classification Performance—No Localisation (NL)

This evaluation strategy ignores the localisation information (i.e. the bounding boxes) and only focuses on whether an action is present in a video or not. If a video con-

Table 8.1 Quantitative measures precision and recall on LIRIS HARL dataset

Method	Recall	Precision	F1-score
VPULABUAM-13-NL	0.36	*0.66*	0.46
IACAS-51-NL	0.3	0.46	0.36
SS-NL (ours)	*0.5*	0.53	*0.52*
HMS-NL (ours)	*0.5*	0.63	*0.56*
VPULABUAM-13-10%	0.04	0.08	0.05
IACAS-51-NL-10%	0.03	0.04	0.03
SS-10% (ours)	*0.5*	0.53	0.52
HMS-10% (ours)	*0.5*	*0.63*	*0.56*

tains multiple actions then system should return the labels of all the actions present correctly. Even though our action detection framework is not specifically designed for this task, we still outperform the competition, as shown in Table 8.1.

8.4.2 Detection and Localisation Performance

This evaluation strategy takes localisation (space and time) information into account [27]. We use a 10% threshold quality level for the four thresholds (Sect. 4.2.5 of [19]), which is the same as that used in the LIRIS-HARL competition. In Table 8.1, we denote these results as "method-name-NL" (NL for no localisation) and "method-name-10%". In both cases (without localisation and with 10% overlap), our method outperforms existing approaches, achieving an improvement from 46% [22] to 56%, in terms of F1 score without localisation measures, and a improvement from 5% [22] to 56% (11.2 times better) gain in the F1-score when 10% localisation information *is* taken into account. In Table 8.2 we list the results we obtained using the overall integrated performance scores (Sect. 4.2.5 of [19])—our method yields significantly better quantitative and qualitative results with an improvement from 3% [22] to 43% (14.3% times better) in terms of F1 score, a relative gain across the spectrum of measures. Samples of qualitative instance segmentation results are shown in Fig. 8.7.

Table 8.2 Qualitative thresholds and integrated score on LIRIS HARL dataset

Method	I_{sr}	I_{sp}	I_{tr}	I_{tp}	IQ
VPULABUAM-13-IQ	0.02	0.03	0.03	0.03	0.03
IACAS-51-IQ	0.01	0.01	0.03	00.0	0.02
SS-IQ (ours)	0.52	0.22	0.41	0.39	0.38
HMS-IQ (ours)	0.49	0.35	0.46	0.43	*0.44*

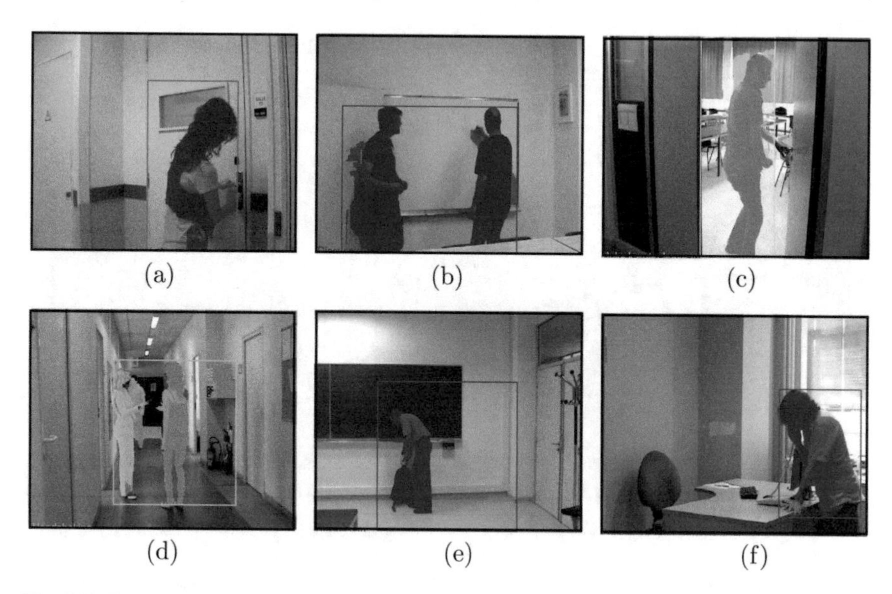

Fig. 8.7 Correct (**a–c**) and incorrect (**d–f**) instance segmentation results on the LIRIS-HARL dataset [26], the correct category is shown in brackets. **a** 'Try enter room unsuccessfully'. **b** 'Discussion'. **c** 'Unlock enter/leave room'. **d** 'Handshaking' (Give take object from person). **e** 'Discussion' (Leave bag unattended). **f** 'Put take object into/from desk' (Telephone conversation)

The pure classification accuracy of the HMS- and SS-based approaches are reflected in the Confusion Matrices shown in Fig. 8.9. Confusion matrices show the the complexity of the dataset. Some of the actions are wrongly classified, e.g., "*telephone-conversation*" is classified as "*put/take object to/from box/desk*", same can be observed for action "*unlock enter/leave room*" in SS approach.

8.4.3 Performance Versus Detection Quality Curves

The plots in Fig. 8.8 attest the robustness of our method, as they depict the curves corresponding to precision, recall and F1-score over varying quality thresholds.

When the threshold t_{tr} for temporal recall is considered (see Fig. 8.8 plot a) we achieved a highest recall of 50% for both HMS- and SS-based approaches and a highest precision of 65% for HMS-based approach at threshold value of $t_{tr} = 0$. As the threshold increases towards $t_{tr} = 1$, SS-based method shows a robust performance, with highest recall = 50% and precision = 52%, HMS-based method shows promising results with an acceptable drop in precision and recall. Note that when $t_{tr} = 1$, we assume that all frames of an activity instance need to be detected in order for the instance itself to be considered as detected.

As for the competing methods, IACAS-51 [11] yields the next competing recall of 2.4% and a precision of 3.7% with a threshold value of $t_{tr} = 1$.

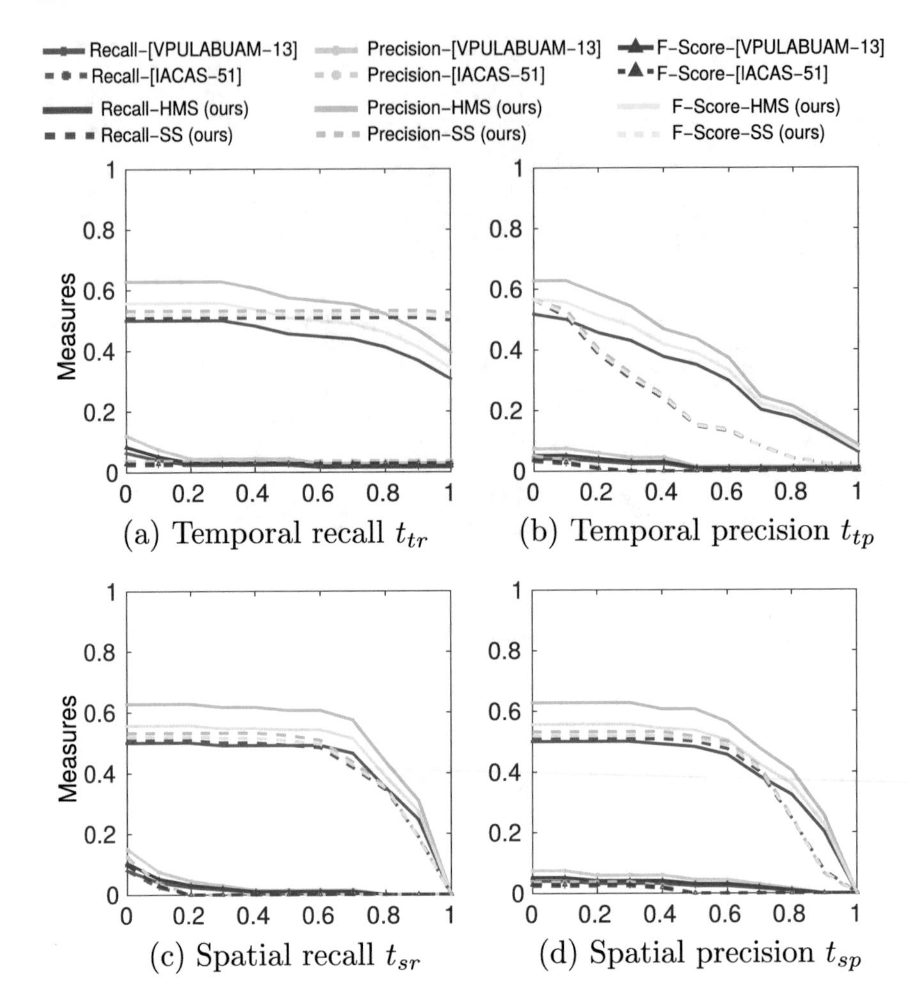

Fig. 8.8 Performance versus detection quality curves

When acting on the value of the temporal frame-wise precision threshold t_{tp} (see Fig. 8.8 plot b) we can observe that at $t_{tp} = 1$, when we assume that not a single spurious frame outside the ground truth temporal window is allowed, our HMS-based region proposal approach gives highest recall of 8% and precision 10.7%, where, as SS-based approach has significantly lower recall = 2% and precision = 2.4%, which is still significantly higher than the performance of the existing methods. Indeed, at $t_{tp} = 1$, VPULABUAM-13 has recall = 0.8% and precision = 1% where IACAS-51 yields both zero precision and zero recall. This results tell us that HMS-based approach performs superior in detecting temporal extent of an action and thus is suitable for action localisation in temporally untrimmed videos. The remaining two plots c, d of Fig. 8.8 illustrate the overall performance when spatial overlap is taken into

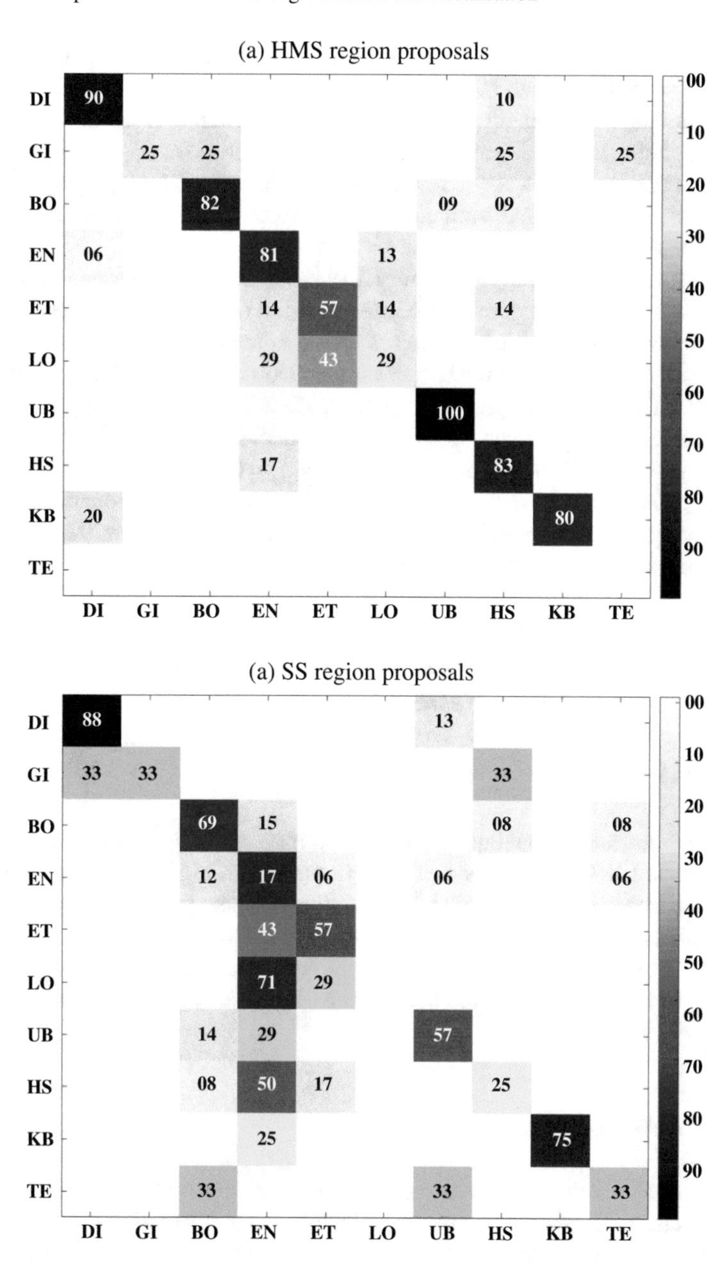

Fig. 8.9 Confusion matrix obtained by human motion segmentation (HMS) and selective search (SS) region proposal approach. They show the classification accuracy of HMS- and SS-based methods on LIRIS HARL human activity dataset. HMS region proposal based method provides better classification accuracy on the the complex LIRIS dataset [26]

account. Both plots show metrics approaching zero when the corresponding spatial thresholds (pixel-wise recall t_{sr} and pixel-wise precision t_{sp}) approach 1. Note that it is highly unlikely for a ground truth activity to be consistently (spatially) included in the corresponding detected activity over all the consecutive frames (spatial recall), as indicated in the plot c. It is also rare for a detected activity to be (spatially) included in the corresponding ground truth activity over all the frames (spatial precision) as indicated in plot d.

For the pixel-wise recall (plot c), our HMS based method shows consistent recall between 45 and 50% and precision between 59 and 65.5% up to a threshold value of $t_{sr} = 0.7$, where as, SS-based region proposal approach gives comparable recall between 48.3 and 50.8%, but relative lower precision between 43.5 and 53.2% up to $t_{sr} = 0.7$. For the pixel-wise precision (plot d), HMS and SS-based approaches give similar recall between 39 and 50%, where as HMS-method again outperforms in precision with 48–63% up to a threshold value of $t_{sp} = 0.7$, where as SS has precision 41–53% up to a threshold value $t_{sp} = 0.7$. Finally, we draw conclusion that our HMS-based region proposal approach shows superior qualitative and quantitative detection performance on the challenging LIRIS HARL dataset.

8.4.4 *Qualitative Action Instance Segmentation and Localisation Results*

8.4.4.1 LIRIS HARL Dataset

Figure 8.10 shows additional qualitative action instance segmentation and localisation results on LIRIS HARL dataset [26]. In particular, Fig. 8.10a, d show that the proposed approach can successfully detect action instances belonging to a same class or different classes at finer pixel-level. In (a), two action instances of a single action class (i.e. "*typing on keyboard*") are present, whereas in (d) two action instances belonging to two different action classes (i.e. "*handshaking*" and "*leave baggage unattended*") are present.

8.4.4.2 UCF-101-24 Dataset

To demonstrate that the proposed instance segmentation method generalises well on other datasets, we present here some sample instance segmentation results on UCF-101-24. We compute the binary segmentation masks for some selected UCF-101-24 test video clips, and apply the bounding-boxes predicted by our proposed action detection model [21] on the top of the binary masks to generate the final instance segmentation results which are shown in Figs. 8.11 and 8.12. Note that, the proposed approach can successfully localise multiple instances of the "*biking*" (Fig. 8.11b), "*fencing*" (Fig. 8.12a), and "*ice dancing*" (Fig. 8.12c) actions at finer pixel level in space and time.

Fig. 8.10 Qualitative action instance segmentation and localisation results on LIRIS HARL dataset. Ground-truth action labels: **TK**—typing on keyboard, **HS**—handshaking, **DC**—discussion, **LBU**—leave baggage unattended, **GOP**—give object to person, **POD**—put object into desk, **TERU**—try enter room unsuccessfully, **UER**—unlock enter room, **TC**—telephone conversation. Correct results: **a**, **b**, **c**, **d**, **e**, **f**, **g**, **h**, **j**; incorrect results: **h**, **i**, **k**, **l**. In **h**, out of two instances of **TK** action class, only one instance has been successfully detected. In **i**, the ground truth action class **GOP** has been misclassified as **HS** class. In **k**, the ground truth action classes **TK** and **HS** have been misclassified as **DC** class. In **l**, the ground truth action class **TC** has been misclassified as **POD** class

Fig. 8.11 Qualitative action instance segmentation and localisation results on UCF-101-24 test videos. The green boxes represent ground truth annotations, whereas the blue boxes denote the frame-level detections. Each row represents an UCF-101-24 test video clip where the 1st and 2nd rows in each set (i.e. set **a**–**c**) are the input video frames and their corresponding outputs respectively. From each clip 4 selected frames are shown. Predicted action labels: **a** "basketball"; **b** "biking"; **c** "cliffdiving"

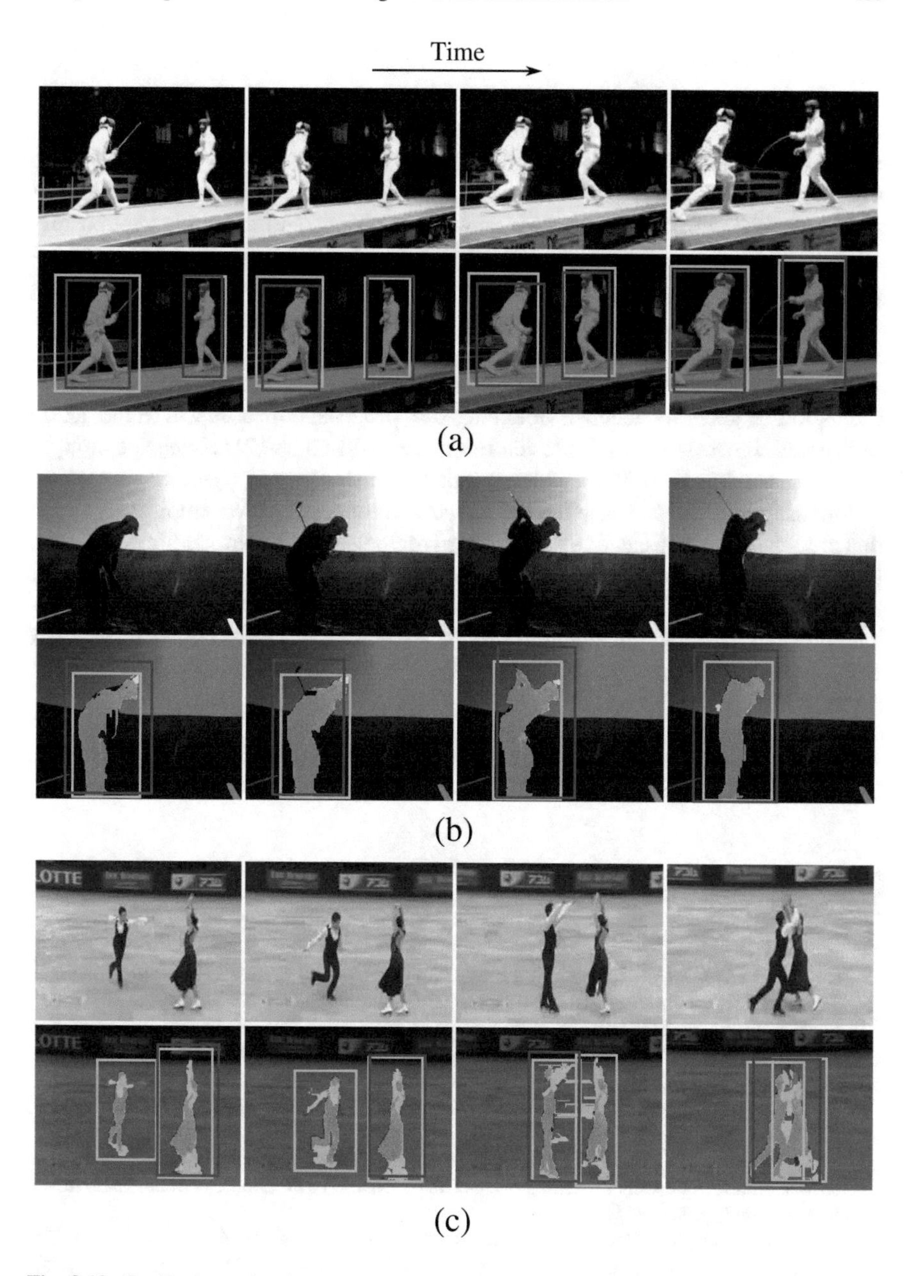

Fig. 8.12 Qualitative action instance segmentation and localisation results on UCF-101-24 test videos. The green boxes represent ground truth annotations, whereas the blue boxes denote the frame-level detections. Each row represents an UCF-101-24 test video clip where the 1st and 2nd rows in each set (i.e. set **a**–**c**) are the input video frames and their corresponding outputs respectively. From each clip 4 selected frames are shown. Predicted action labels: **a** "fencing"; **b** "golfswing"; **c** "icedancing"

8.5 Discussion

Unlike state-of-the-art supervised instance segmentation approaches (for objects) [9, 10] which require expensive ground-truth segmentation (i.e. per pixel class- and instance-aware labelling) to train their networks, the proposed framework does not require such expensive ground-truth annotations. Thanks to the human action segmentation [17] algorithm which computes human action binary masks using unsupervised learning, thus, does not require expensive ground-truth labels. However, the major drawback of [17] is that it is computationally expensive. For example, it takes several days to compute the binary masks for all frames in LIRIS HARL dataset. Another limitation is that the HMS (human motion segmentation) based region proposals fail to generate accurate bounding box proposals in cases where the action segmentations of two or multiple actors get merged into one 2D connected component, e.g., see Fig. 8.10 **(8)** in which out of two instances of "*typing on keyboard*" action class, only one instance has been successfully detected. We empirically found that in such instances Selective Search based region proposals work more effectively. Lastly, as there are no ground truth instance segmentation annotations available for LIRIS HARL and UCF-101-24 datasets, we could not perform an quantitative evaluation of the instance segmentation results. Also note, the J-HMDB-21 dataset has a single action instance per video, and thus, not suitable for evaluating instance segmentation methods.

Acknowledgements This work was partly supported by ERC grant ERC-2012-AdG 321162-HELIOS, EPSRC grant Seebibyte EP/M013774/1 and EPSRC/MURIgrant EP/N019474/1.

References

1. Boykov, Y., Veksler, O., & Zabih, R. (2001). Fast approximate energy minimization via graph cuts. *IEEE Transactions on Pattern Analysis and Machine Intelligence*, *23*(11), 1222–1239.
2. Brox, T., Bruhn, A., Papenberg, N., & Weickert, J. (2004). High accuracy optical flow estimation based on a theory for warping. *Computer Vision-ECCV, 2004*, 25–36.
3. Brox, T., & Malik, J. (2011). Large displacement optical flow: Descriptor matching in variational motion estimation. *IEEE Transactions on Pattern Analysis and Machine Intelligence*, *33*(3), 500–513.
4. Chen, W., Xiong, C., Xu, R., & Corso, J. J. (2014). Actionness ranking with lattice conditional ordinal random fields. In *Proceedings of the IEEE Conference on Computer Vision and Pattern Recognition* (pp. 748–755).
5. Everingham, M., Van Gool, L., Williams, C. K. I., Winn, J., & Zisserman, A. (2007). The PASCAL visual object classes challenge (VOC2007) results. Available at: http://www.pascal-network.org/challenges/VOC/voc2007/workshop/index.html.
6. Felzenszwalb, P. F., Girshick, R. B., McAllester, D., & Ramanan, D. (2010). Object detection with discriminatively trained part-based models. *IEEE Transactions on Pattern Analysis and Machine Intelligence*, *32*(9), 1627–1645.
7. Girshick, R., Donahue, J., Darrel, T., & Malik, J. (2014). Rich feature hierarchies for accurate object detection and semantic segmentation. In *IEEE International Conference on Computer Vision and Pattern Recognition*.

8. Gkioxari, G., & Malik, J. (2015). Finding action tubes. In *IEEE International Conference on Computer Vision and Pattern Recognition.*
9. Hariharan, B., Arbeláez, P., Girshick, R., & Malik, J. (2014). Simultaneous detection and segmentation. In *European Conference on Computer Vision* (pp. 297–312). Springer.
10. He, K., Gkioxari, G., Dollar, P., & Girshick, R. (2017). Mask r-cnn. In *The IEEE International Conference on Computer Vision (ICCV).*
11. He, Y., Liu, H., Sui, W., Xiang, S., & Pan, C. (2012). Liris harl competition participant. Institute of Automation, Chinese Academy of Sciences, Beijing. http://liris.cnrs.fr/harl2012/results.html.
12. Jain, S. D., & Grauman, K. (2014). Supervoxel-consistent foreground propagation in video. In *European Conference on Computer Vision* (pp. 656–671). Springer.
13. Jhuang, H., Gall, J., Zuffi, S., Schmid, C., & Black, M. J. (2013). Towards understanding action recognition. In *Proceedings of the IEEE International Conference on Computer Vision (ICCV)* (pp. 3192–3199).
14. Jia, Y., Shelhamer, E., Donahue, J., Karayev, S., Long, J., Girshick, R. B., et al. (2014). Caffe: Convolutional architecture for fast feature embedding. http://arxiv.org/abs/1408.5093.
15. Kohli, P., Torr, P. H., et al. (2009). Robust higher order potentials for enforcing label consistency. *International Journal of Computer Vision*, *82*(3), 302–324.
16. Krizhevsky, A., Sutskever, I., & Hinton, G. E. (2012). Imagenet classification with deep convolutional neural networks. *Advances in Neural Information Processing Systems*, 1097–1105.
17. Lu, J., Xu, R., & Corso, J. J. (2015). Human action segmentation with hierarchical supervoxel consistency. In *IEEE International Conference on Computer Vision and Pattern Recognition.*
18. Peng, X., & Schmid, C. (2016). Multi-region two-stream r-cnn for action detection. In *European Conference on Computer Vision* (pp. 744–759). Springer.
19. Saha, S. Spatio-temporal human action detection and instance segmentation in videos. Ph.D. thesis. Available at: https://tinyurl.com/y4py79cn.
20. Saha, S., Singh, G., Sapienza, M., Torr, P. H., & Cuzzolin, F. (2017). Spatio-temporal human action localisation and instance segmentation in temporally untrimmed videos. arXiv:1707.07213.
21. Saha, S., Singh, G., Sapienza, M., Torr, P. H. S., & Cuzzolin, F. (2016). Deep learning for detecting multiple space-time action tubes in videos. In *British Machine Vision Conference.*
22. SanMiguel, J. C., & Suja, S. (2012). Liris harl competition participant. Video Processing and Understanding Lab, Universidad Autonoma of Madrid, Spain, http://liris.cnrs.fr/harl2012/results.html.
23. Soomro, K., Zamir, A. R., & Shah, M. (2012). UCF101: A dataset of 101 human action classes from videos in the wild. Technical Report, CRCV-TR-12-01.
24. Uijlings, J. R., Van De Sande, K. E., Gevers, T., & Smeulders, A. W. (2013). Selective search for object recognition. *International Journal of Computer Vision*, *104*(2), 154–171.
25. Weinzaepfel, P., Harchaoui, Z., & Schmid, C. (2015). Learning to track for spatio-temporal action localization. *IEEE International Conference on Computer Vision and Pattern Recognition.*
26. Wolf, C., Mille, J., Lombardi, E., Celiktutan, O., Jiu, M., Baccouche, M., et al. The LIRIS Human activities dataset and the ICPR 2012 human activities recognition and localization competition. Technical Report, LIRIS UMR 5205 CNRS/INSA de Lyon/Université Claude Bernard Lyon 1/Université Lumière Lyon 2/École Centrale de Lyon (2012). http://liris.cnrs.fr/publis/?id=5498.
27. Wolf, C., Mille, J., Lombardi, E., Celiktutan, O., Jiu, M., Dogan, E., et al. (2014). Evaluation of video activity localizations integrating quality and quantity measurements. *Computer Vision and Image Understanding*, *127*, 14–30.
28. Xu, C., Xiong, C., & Corso, J. J. (2012). Streaming hierarchical video segmentation. In *European Conference on Computer Vision* (pp. 626–639). Springer.
29. Zeiler, M. D., & Fergus, R. (2013). Visualizing and understanding convolutional networks.

Chapter 9
Vision During Action: Extracting Contact and Motion from Manipulation Videos—Toward Parsing Human Activity

Konstantinos Zampogiannis, Kanishka Ganguly, Cornelia Fermüller, and Yiannis Aloimonos

Abstract When we physically interact with our environment using our hands, we touch objects and force them to move: contact and motion are defining properties of manipulation. In this paper, we present an active, bottom-up method for the detection of actor–object contacts and the extraction of moved objects and their motions in RGBD videos of manipulation actions. At the core of our approach lies non-rigid registration: we continuously warp a point cloud model of the observed scene to the current video frame, generating a set of dense 3D point trajectories. Under loose assumptions, we employ simple point cloud segmentation techniques to extract the actor and subsequently detect actor–environment contacts based on the estimated trajectories. For each such interaction, using the detected contact as an attention mechanism, we obtain an initial motion segment for the manipulated object by clustering trajectories in the contact area vicinity and then we jointly refine the object segment and estimate its 6DOF pose in all observed frames. Because of its generality and the fundamental, yet highly informative, nature of its outputs, our approach is applicable to a wide range of perception and planning tasks. We qualitatively evaluate our method on a number of input sequences and present a comprehensive robot imitation learning example, in which we demonstrate the crucial role of our outputs in developing action representations/plans from observation.

9.1 Introduction

A manipulation action, by its very definition, involves the handling of objects by an intelligent agent. Every such interaction requires physical contact between the actor and some object, followed by the exertion of forces on the manipulated object, which typically induce motion. When we open a door, pick up a coffee mug, or pull a chair, we invariably touch an object and cause it (or parts of it) to move. This obvious observation demonstrates that *contact* and *motion* are two fundamental aspects of manipulation.

K. Zampogiannis · K. Ganguly · C. Fermüller · Y. Aloimonos (✉)
University of Maryland, College Park, MD, USA
e-mail: yiannis@cs.umd.edu

N. Noceti et al. (eds.), *Modelling Human Motion*,
https://doi.org/10.1007/978-3-030-46732-6_9

Contact and motion information alone are often sufficient to describe manipulations in a wide range of applications, as they naturally encode crucial information regarding the performed action. Contact encodes *where* the affected object was touched/grasped, as well as *when* and for how long the interaction took place. Motion conveys *what* part of the environment (i.e., which object or object part) was manipulated and *how* it moved.

The ability to automatically extract contact and object motion information from video either directly solves or can significantly facilitate a number of common perception tasks. For example, in the context of manipulation actions, knowledge of the spatiotemporal extent of an actor–object contact automatically provides action *detection/segmentation* in the time domain, as well as *localization* of the detected action in the observed space [1, 2]. At the same time, motion information bridges the gap between the observation of an action and its semantic grounding. Knowing what part of the environment was moved effectively acts as an attention mechanism for the manipulated *object recognition* [3, 4], while the extracted motion profile provides invaluable cues for *action recognition*, in both "traditional" [1, 2, 5] and deep learning [6] frameworks.

Robot imitation learning is rapidly gaining attention. The use of robots in less controlled workspaces and even domestic environments necessitates the development of easily applicable methods for robot "programming": autonomous robots for manipulation tasks must efficiently *learn* how to manipulate. Exploiting contact and motion information can largely automate robot replication of a wide class of actions. As we will discuss later, the detected contact area can effectively bootstrap the grasping stage by guiding primitive fitting and grasp planning, while the extracted object and its motion capture the trajectory to be replicated as well as any applicable kinematic/collision constraints. Thus, the components introduced in this work are essential for building complex, hierarchical models of action (e.g., behavior trees, activity graphs) as they appear in the recent literature [7–13].

In this paper, we present an unsupervised, bottom-up method for estimating from RGBD video the contacts and object motions in manipulation tasks. Our approach is fully 3D and relies on dense motion estimation: we start by capturing a point cloud model of the observed scene and continuously warp/update it throughout the duration of the video. Building upon our estimated dense 3D point trajectories, we use simple concepts and common sense rules to segment the actor and detect actor–environment contact locations and time intervals. Subsequently, we exploit the detected contact to guide the motion segmentation of the manipulated object and, finally, estimate its 6DOF pose in all observed video frames. Our intermediate and final results are summarized in Table 9.1.

It is worth noting that we do not treat contact detection and object motion segmentation/estimation independently: we use the detected contact as an *attention mechanism* to guide the extraction of the manipulated object and its motion. This *active* approach provides an elegant and effective solution to our motion segmentation task. A passive approach to our problem would typically segment the whole observed scene into an *unknown* (i.e., to be estimated) number of motion clusters. By exploiting contact, we avoid having to solve a much larger and less constrained

Table 9.1 List of the inputs, intermediate results, and final outputs of our proposed system

Input	Intermediate results	Final outputs
RGBD video of manipulation	• **Dense 3D point trajectories** for the whole sequence duration • **Actor/background labels** for all model points at all times	• **3D trajectories** of detected actor–environment **contact points** • Manipulated **object segments** and their **6DOF poses** for every time point

problem, while gaining significant improvements in terms of both computational efficiency and segmentation/estimation accuracy.

The generality of our framework, combined with the highly informative nature of our outputs, renders our approach applicable to a wide spectrum of perception and planning tasks. In Sect. 9.3, we provide a detailed technical description of our method, while in Sect. 9.4, we demonstrate our intermediate results and final outputs for a number of input sequences. In Sect. 9.5, we present a comprehensive example of how our outputs were successfully used to facilitate a robot imitation learning task.

9.2 Related Work

We focus our literature review on recent works in four areas that are most relevant to our twofold problem, and the major processes/components upon which we build. We deliberately do not review works from the action recognition literature; while our approach may very appropriately become a component of a higher-level reasoning solution, the scope of this paper is the extraction of contacts, moving objects, and their motions.

Scene Flow Scene flow refers to the dense 3D motion field of an observed scene with respect to a camera; its 2D projection onto the image plane of the camera is the optical flow. Scene flow, analogously to optical flow, is typically computed from multiview frame pairs [14]. There have been a number of successful recent works on scene flow estimation from RGBD frame pairs, following both variational [15–19] and deep learning [20] frameworks. While being of great relevance in a number of motion reasoning tasks, plain scene flow cannot be directly integrated into our pipeline, which requires *model-to-frame* motion estimation: the scene flow motion field has a 2D support (i.e., the image plane), effectively warping the 2.5D geometry of an RGBD frame, while we need to appropriately warp a *full* 3D point cloud model.

Non-rigid Registration The non-rigid alignment of 3D point sets can be viewed as a generalization of scene flow, in the sense that the estimated motion field is supported by a 3D point cloud: the goal is to estimate point-wise transformations (usually rigid) that best align the point set to the target geometry under certain global prior constraints (e.g., "as-rigid-as-possible" [21]). The warp field estimation is performed either by iterating between correspondence estimation and motion optimization [22–25], or in a correspondence-free fashion, by aligning volumetric signed distance fields (SDFs) [26]. For this work, and due to lack of publicly available solutions, we have implemented a non-rigid registration algorithm similar to [24] and [25] (Sect. 9.3.2) and released it as part of our cilantro [27] library.

Contact Detection A CNN-based method for grasp recognition is introduced in [28]. A 2D approach for detecting "touch" interactions between a caregiver and an infant is presented in [29]. To the best of our knowledge, there is no prior work on explicitly determining the spatiotemporal extent of human–environment contact.

Motion Segmentation A very large volume of works on motion segmentation have casted the problem as subspace clustering of 2D point trajectories, assuming an affine camera model [30–35]. In [36], an active approach for the segmentation and kinematic modeling of articulated objects is proposed, which relies on the robot manipulation capabilities to induce object motion. In [37], object segmentation is performed from two RGBD frames, one before and one after the manipulation of the object, by rigidly aligning and "differencing" the two views and robustly estimating rigid motion between the "difference" regions. The same method is used in [15], where scene flow is used to obtain motion proposals, followed by an MRF inference step. In [38], joint tracking and reconstruction of multiple rigidly moving objects are achieved by combining two segmentation/grouping strategies with multiple surfel fusion [39] instances. A naive integration of a generic motion segmentation algorithm for the extraction of the manipulated object into our pipeline would be suboptimal in multiple ways. For instance, given the fact that there may exist an unknown number of other object motions that are irrelevant to the manipulation, we would be solving an unnecessarily hard problem. For the same reason, we would have little control over the segmentation granularity, which could cause the manipulated object to be over-/under-segmented. Instead, we leverage the detected contact and bootstrap our segmentation by an informed trajectory clustering approach that is similar to [40].

9.3 Our Approach

9.3.1 Overview

We present an automated system that, given a video of a human performing a manipulation task as input, *detects* and *tracks* the parts of the environment that participate in the manipulation. More specifically, our system is able to visually detect physical contact between the actor and their environment, and, using contact as an attention mechanism, eventually segment the manipulated object and estimate its 6DOF pose in every observed video frame. Our pipeline, as well as the interactions of the involved

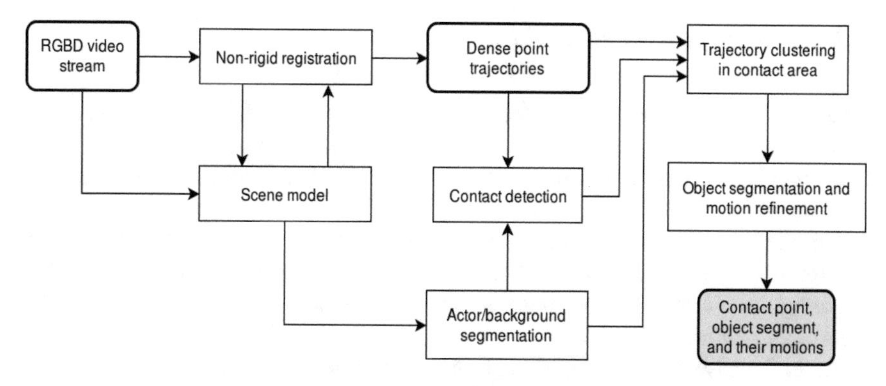

Fig. 9.1 A high-level overview of our modules and their connections in the proposed pipeline

processes, is sketched in Fig. 9.1 and followed by a more detailed description. An in-depth discussion of our core modules is provided in the following subsections.

The input to our system is an RGBD frame sequence, captured by a commodity depth sensor, of a human actor performing a task that involves the manipulation of objects in their environment. We assume that the input depth images are registered to and in sync with their color counterparts. Using estimates of the color camera intrinsics (e.g., from the manufacturer provided specifications), all input RGBD frames are back-projected to 3D point clouds (colored, with estimated surface normals), on which all subsequent processing is performed.

At the core of our method lies non-rigid point cloud registration, described in detail in Sect. 9.3.2. An initial point cloud model of the observed scene is built from the first observed frame and is then consecutively transformed to the current observation based on the estimated *model-to-frame* warp field at every time instance. This process generates a dense set of point trajectories, each associated with a point in the initial model. In order to keep the presentation clean, we opted to obtain the scene model from the first frame and keep it fixed in terms of its point set. Non-rigid reconstruction techniques for updating the model over time [24, 25] can be easily integrated into our pipeline if required.

To perform actor/background segmentation, we follow the semiautomatic approach described in Sect. 9.3.3. The obtained binary labeling is propagated to the whole temporal extent of the observed action via our estimated dense point trajectories and enables us to easily detect human–environment *contacts* as described in Sect. 9.3.4.

Given the dense scene point trajectories, the actor/background labels, and the (hand) contact interaction locations and time intervals, our final goal is, for each detected interaction, to *segment* the manipulated object and re-estimate its *motion* for every time instance, assuming it is rigid (i.e., fully defined by a 6DOF pose). Our contact-guided motion segmentation approach for this task is described in Sect. 9.3.5.

In Table 9.1, we summarize our proposed system's expected inputs, final outputs, and some useful generated intermediate results.

9.3.2 Non-rigid Registration

As described in the previous subsection, whenever a new RGBD frame (point cloud) becomes available, our scene model is non-rigidly warped from its previous state (that corresponds to the previous frame) to the new (current) observation. Since parts of the scene model may be invisible in the current state (e.g., because of self-occlusion), we cannot directly apply a traditional scene flow algorithm, as that would only provide us with motion estimates for (some of) the currently visible points. Instead, we adopt a more general approach, by implementing a non-rigid iterative closest point (ICP) algorithm, similar to [23–25].

As is the case with rigid ICP [41], our algorithm iterates between a correspondence search step and a warp field optimization step for the given correspondences. Our correspondence search typically amounts to finding the nearest neighbors of each point in the current frame to the model point cloud in its previous state. Correspondences that exhibit large point distance, normal angle, or color difference are discarded. Nearest neighbor searches are done efficiently by parallel kd-tree queries.

In the following, we will focus on the warp field optimization step of our scheme. It has been found that modeling the warp field using locally affine [23] or locally rigid [24] transformations provides better motion estimation results than adopting a simple translational local model, due to better regularization. In our implementation, for each point of the scene model in its previous state, we compute a full 6DOF rigid transformation that best aligns it to the current frame.

Let $X = \{x_i\}$ be the set of scene model points in the previous state that needs to be registered to the point set $Y = \{y_i\}$ of the current frame, whose surface normals we denote by $Y^n = \{n_i\}$. Let $S = \{s_i\} \subseteq \{1, \ldots, |X|\}$ and $D = \{d_i\} \subseteq \{1, \ldots, |Y|\}$ be the index sets of corresponding points in X and Y, respectively, such that $\left(x_{s_i}, y_{d_i}\right)$ is a pair of corresponding points. Let $T = \{T_i\}$ be the unknown warp field of rigid transformations, such that $T_i \in SE(3)$ and $|T| = |X|$, and $T_i(x_i)$ denotes the application of T_i to model point x_i. Local transformations are parameterized by 3 Euler angles (α, β, γ) for their rotational part and 3 offsets (t^x, t^y, t^z) for their translational part and are represented as 6D vectors $T_i = \left[\alpha_i\ \beta_i\ \gamma_i\ t_i^x\ t_i^y\ t_i^z\right]^{\mathrm{T}}$.

Our goal at this stage is to estimate a warp field T, of $6\,|X|$ unknown parameters, that maps model points in S as closely as possible to frame models in D. We formulate this property as the minimization of a weighted combination of sums of point-to-plane and point-to-point squared distances between corresponding pairs:

$$E_{\text{data}}(T) = \sum_{i=1}^{|S|} \left(n_{d_i}^T \left(T_{s_i}\left(x_{s_i}\right) - y_{d_i}\right)\right)^2 + w_{\text{point}} \sum_{i=1}^{|S|} \left\| T_{s_i}\left(x_{s_i}\right) - y_{d_i} \right\|^2. \tag{9.1}$$

Pure point-to-plane metric optimization generally converges faster and to better solutions than pure point-to-point [42] and is the standard trend in the state of the art for both rigid [39, 43] and non-rigid [24, 25] registrations. However, we have found that integrating a point-to-point term (second term in (9.1)) with a small weight (e.g., with $w_{\text{point}} \approx 0.1$) to the registration cost improves motion estimation on surfaces that lack geometric texture.

The set of estimated correspondences is only expected to cover a subset of X and Y, as not all model points are expected to be visible in the current frame, and the latter may suffer from missing data. Furthermore, even for model points with existing data terms (correspondences) in (9.1), analogously to the aperture problem in optical flow estimation, the estimation of point-wise transformation parameters locally is under-constrained. These reasons render the minimization of the cost function in (9.1) ill-posed. To overcome this, we introduce a “stiffness” regularization term that imposes an as-rigid-as-possible prior [21] by directly penalizing differences between transformation parameters of neighboring model points in a way similar to [23]. We fix a neighborhood graph on X, based on point locations, and use $N(i)$ to denote the indices of the neighbors of point x_i to formulate our stiffness prior term as:

$$E_{\text{stiff}}(T) = \sum_{i=1}^{|X|} \sum_{j \in \mathcal{N}(i)} w_{ij} \psi_\delta\big(T_i - T_j\big), \tag{9.2}$$

where $w_{ij} = \exp\Big(-\|x_i - x_j\| / \Big(2\sigma_{\text{reg}}^2\Big)\Big)$, σ_{reg} controls the radial extent of the regularization neighborhoods, “−” denotes regular matrix subtraction for the 6D vector representations of the local transformations, and ψ_δ denotes the sum of the Huber loss function values over the 6 residual components. Parameter δ controls the point at which the loss function behavior switches from quadratic (L^2-norm) to absolute linear (L^1-norm). Since L^1-norm regularization is known to better preserve solution discontinuities, we choose a small value of $\delta = 10^{-4}$.

Our complete registration cost function is a weighted combination of costs (9.1) and (9.2):

$$E(T) = E_{\text{data}}(T) + w_{\text{stiff}} E_{\text{stiff}}(T), \tag{9.3}$$

where w_{stiff} controls the overall regularization weight (set to $w_{\text{stiff}} = 200$ in our experiments). We minimize $E(T)$ in (9.3), which is nonlinear in the unknowns, by performing a small number of Gauss–Newton iterations. At every step, we linearize $E(T)$ around the current solution and obtain a solution increment $\hat{x}$ by solving the system of normal equations $J^T J \hat{x} = J^T r$, where J is the Jacobian matrix of the residual terms in E and r is the vector of residual values. We solve this sparse system iteratively, using the conjugate gradient algorithm with a diagonal preconditioner.

In Fig. 9.2, we show two sample outputs of our algorithm in an RGBD frame pair non-rigid alignment scenario. Our registration module accurately estimates deformations even for complex motions of significant magnitude.

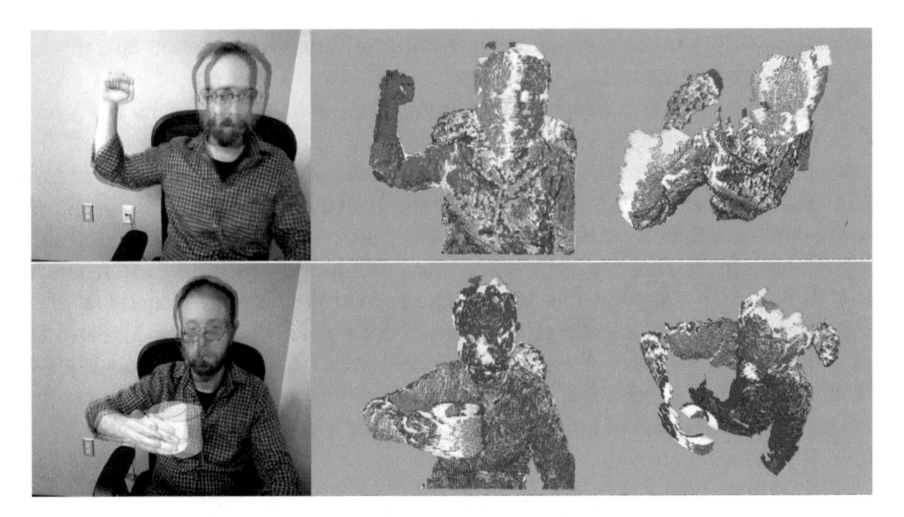

Fig. 9.2 Non-rigid registration: displacement vectors are depicted as white lines, aligning the source (red) to the target (blue) geometry

9.3.3 Human Actor Segmentation

We follow a semiautomatic approach to perform actor/background segmentation that relies on simple point cloud segmentation techniques.

We construct a proximity graph over the scene model points in the initial state, in which each node is a model point and two nodes are connected if and only if their Euclidean distance falls below a predefined threshold. Assuming that the actor is *initially* not in contact with any other part of the scene (i.e., the minimum distance of an actor point to a background point is at least our predefined distance threshold) and the observed actor points are not too severely disconnected in the initial state, the actor points will be exactly defined by one connected component of this proximity graph. The selection of the correct (actor) component can be automated by filtering all the extracted components based on context-specific criteria (e.g., rough size, shape, location, etc.) or by picking the component whose image projection exhibits maximum overlap with the output of a 2D human detector [44, 45]. Equivalently, we may begin by selecting a seed point known to belong to the actor and then perform region growing on the model point cloud until our distance threshold is no longer satisfied. Again, the selection of the seed point can be automated by resorting to standard 2D means (e.g., by picking the point with the strongest skin color response [46, 47] within a 2D human detector output [44, 45]).

We believe that the assumptions imposed by our Euclidean clustering-based approach for the actor segmentation task are not too restricting, as the main setting we focus on (representing human demonstrations for robot learning) is reasonably controlled in the first place.

We note that, since we opted to keep the scene model point set fixed and track it throughout the observed action, the obtained segmentation automatically becomes available at all time points.

9.3.4 Contact Detection

The outputs of the above two processes are a dense set of *point trajectories* and their respective actor/background *labels*. Given this information, it is straightforward to reason about *contact*, simply by examining whether the minimum distance between parts of the two clusters is small enough at any given time. In other words, we can easily infer both *when* the actor comes into/goes out of contact with part of the environment and *where* this interaction is taking place.

Some of the contact interactions detected using this criterion may, of course, be semantically irrelevant to the performed action. Since semantic reasoning is not part of our core framework, these cases have to be handled by a higher level module. However, under reasonably controlled scenarios, we argue that it is sufficient to simply assume that the detected contacts are established by the actor *hands*, with the goal of manipulating an *object* in their environment.

9.3.5 Manipulated Object Motion/Segmentation

Knowing the dense scene point trajectories, labeled as either actor or background, as well as the contact locations and intervals, our next goal is to infer what part of the environment is being manipulated, or, in other words, which object was moved. We assume that every contact interaction involves the movement of a *single* object, and that the latter undergoes *rigid* motion. In the following, we only focus on the *background* part of the scene around the contact point area, ignoring the human point trajectories. We propose the following two-step approach.

First, we bootstrap our segmentation task by finding a coarse/partial mask of the moving object, using standard unsupervised clustering techniques. Specifically, we cluster the point trajectories that are labeled as background and lie within a fixed radius of the detected contact point at the beginning of the interaction into two groups. We adopt a spectral clustering approach, using the "random walk" graph Laplacian [48] and a standard *k*-means last step. Our pairwise trajectory similarities are given by

$$s_{ij} = \exp\bigl(-(d_{\max} - d_{\min})^2/\bigl(2\sigma^2\bigr)\bigr),$$

where $d_{\min}$ and $d_{\max}$ are the minimum and maximum Euclidean point distance of trajectories i and j over the duration of the interaction, respectively. This similarity metric enforces similar trajectories to exhibit relatively constant point-wise distances;

i.e., it promotes clusters that undergo rigid motion. From the two output clusters, one is expected to cover (part of) the object being manipulated. Operating under the assumption that only interaction can cause motion in the scene, we pick the cluster that exhibits the largest average motion over the duration of contact as our object segment candidate.

In the above, we restricted our focus within a region of the contact point, in order to (1) avoid that our binary classification is influenced by other captured motions in the scene that are not related to the current interaction and (2) make the classification itself more computationally tractable. As long as these requirements are met, the choice of radius is not important.

Subsequently, we obtain a refined, more accurate segment of the moving object by requiring that the latter undergoes a rigid motion that is at every time point consistent with that of the previously found motion cluster. Let B^t denotes the background (nonactor) part the scene model point cloud at time t, for $t = 0, \ldots, T$, and $\widehat{M}^t \subseteq B^t$ be the initial motion cluster state at the same time instance. For all $t = 1, \ldots, T$, we robustly estimate the rigid motion between point sets $\widehat{M}^0$ and $\widehat{M}^t$ (i.e., relative to the first frame), using the closed-form solution of [49] under a RANSAC scheme and then find the set of points in *all* of B^t that are consistent with this motion model between B^0 and B^t. If we denote this set of motion inliers by I^t (which is a set of indices of points in B^t), we obtain our final object segment for this interaction as the intersection of inlier indices for all time instances $t = 1, \ldots, T$:

$$I \equiv \bigcap_{t=1}^{T} I^t \tag{9.4}$$

The subset of the background points indexed by I, as well as the per-frame RANSAC motion (pose) estimates of this last step, are the final outputs of our pipeline for the given interaction.

9.4 Experiments

9.4.1 Qualitative Evaluation

We provide a qualitative evaluation of our method for video inputs recorded in different settings, covering three different scenarios: (1) a tabletop object manipulation that involves flipping a pitcher, (2) opening a drawer, and (3) opening a room door. All videos were captured from a static viewpoint, using a standard RGBD sensor.

For each scenario, we depict (in Figs. 9.3, 9.4, and 9.5, respectively) the scene model point cloud state at three time snapshots: one right before, one during, and one right after the manipulation. For each time point, we show the corresponding color image and render the tracked point cloud from two viewpoints. The actor segment

Fig. 9.3 Flipping a pitcher: scene tracking, labeling, and contact detection

is colored green, the background is red, and the detected contact area is marked by blue. We also render the point-wise displacements induced by the estimated warp field (from the currently visible state to its next) as white lines (mostly visible in areas that exhibit large motion). The outputs displayed in these figures are in direct correspondence with the processes described in Sects. 9.3.2, 9.3.3, and 9.3.4.

Next, we demonstrate our attention-driven motion segmentation and 6DOF pose estimation of the manipulated object. In Fig. 9.6, we render the background part of the scene model in its initial state with the actor removed and show the two steps of our segmentation method described in Sect. 9.3.5. In the middle column, the blue segment corresponds to the initial motion segment, obtained by clustering trajectories in the vicinity of the contact point, which was propagated back to the initial model state and is highlighted in yellow. In the left column, we show the refined, final motion segment. We note that, because of our choice of the radius around the contact point in which we focus our attention in the first step, the initial segment in the first two cases is the same as the final one.

In Fig. 9.7, we show the estimated rigid motion (6DOF pose) of the segmented object. To more clearly visualize the evolution of object pose over time, we attach a local coordinate frame to the object, at the location of the contact point, whose axes were chosen as the principal components of the extracted object point cloud segment.

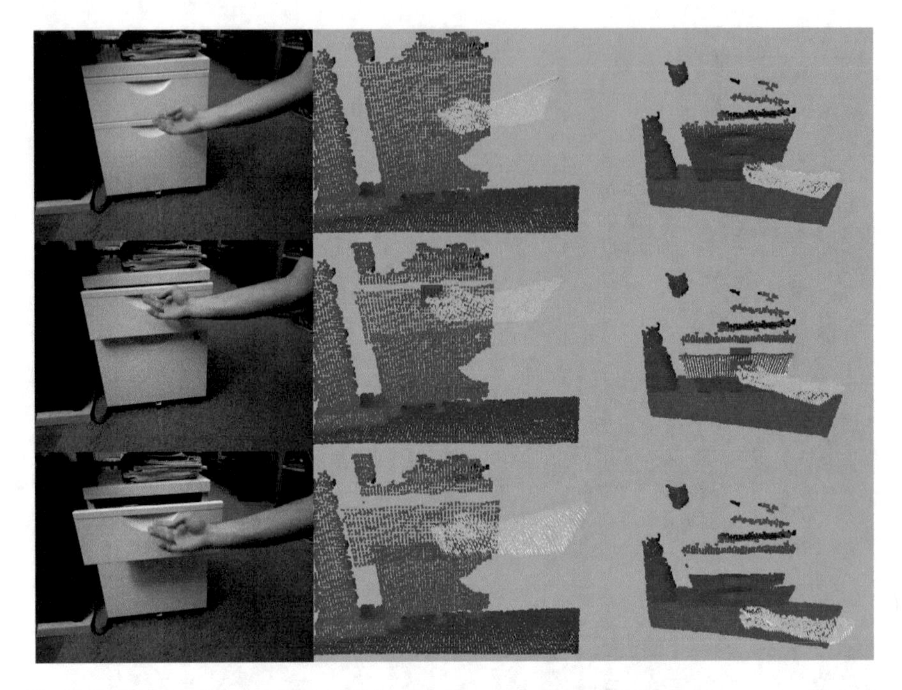

Fig. 9.4 Opening a drawer: scene tracking, labeling, and contact detection

The above illustrations provide a qualitative demonstration of the successful application of our proposed pipeline to three different manipulation videos. In all cases, contacts were detected correctly and the manipulated object was accurately segmented and tracked. A more thorough, quantitative evaluation of our contact and segmentation outputs on an extended set of videos is in our plans for the immediate future.

9.4.2 Implementation

Our pipeline is implemented using the cilantro [27] library, which provides a self-contained set of facilities for all of the computational steps involved.

9.5 Application: Replication from Observation by a Robot

For any human–environment task to be successful, there is a well-defined process involved, demarcated into phases depending on human–environment contact and consequent motion. This allows us to generate a graph representation for actions,

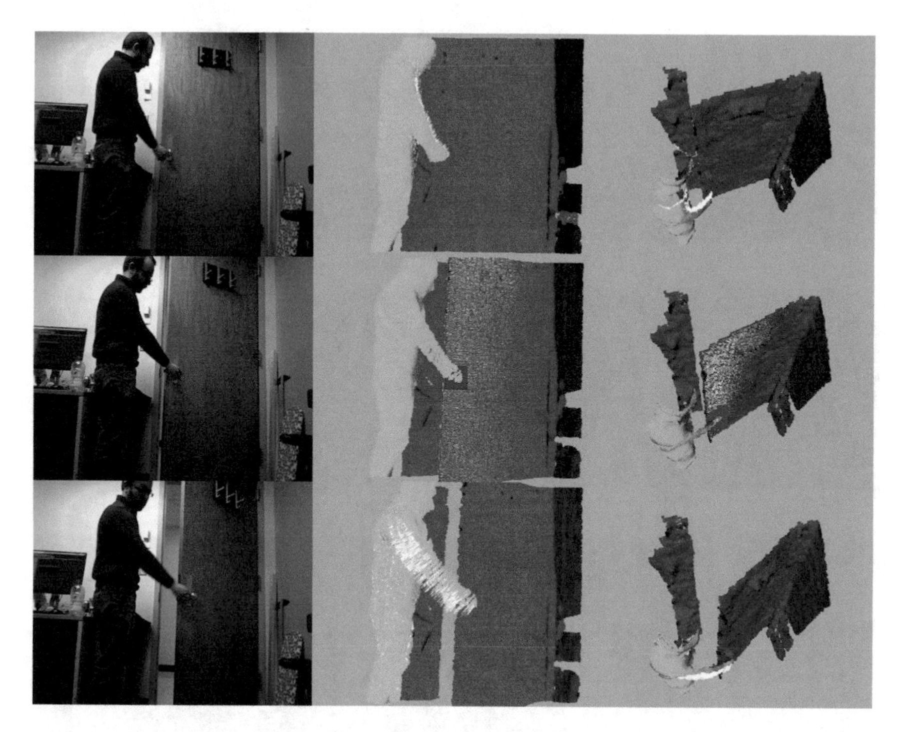

Fig. 9.5 Opening a door: scene tracking, labeling, and contact detection

such as that shown in Fig. 9.8, for the task of opening a refrigerator. Given this general representation of tasks, we demonstrate how our algorithm allows grounding of the *grasp* and *release* parts, based on contact detection, and also of the feedback loop for opening the door, based on motion analysis of segmented objects. Such a representation, featuring a tight coupling of planning and perception, is crucial for robots to observe and replicate human actions.

We now present a comprehensive application of our method to a real-world task, where a robot observes a human operator opening a refrigerator door and learns the process for replication. This can be seen in Fig. 9.9, where a RGBD sensor mounted to the robot's manipulator is used for observation. This process involves the segmentation of the human and the environment from the observed video input, analyzing the contact between the human agent and the environment (the refrigerator handle in this case), and finally performing 3D motion tracking and segmentation on the action of opening the door, using our methods elucidated in Sect. 9.3. These analyses, and the corresponding outputs, are then converted into an intermediate graph-like representation, which encodes both semantic labeling of regions of interest, such as doors and handles in our case, as well as motion trajectories computed from observing the human agent. The combination of these allows the robot to understand and generalize the action to be performed even in changing scenarios.

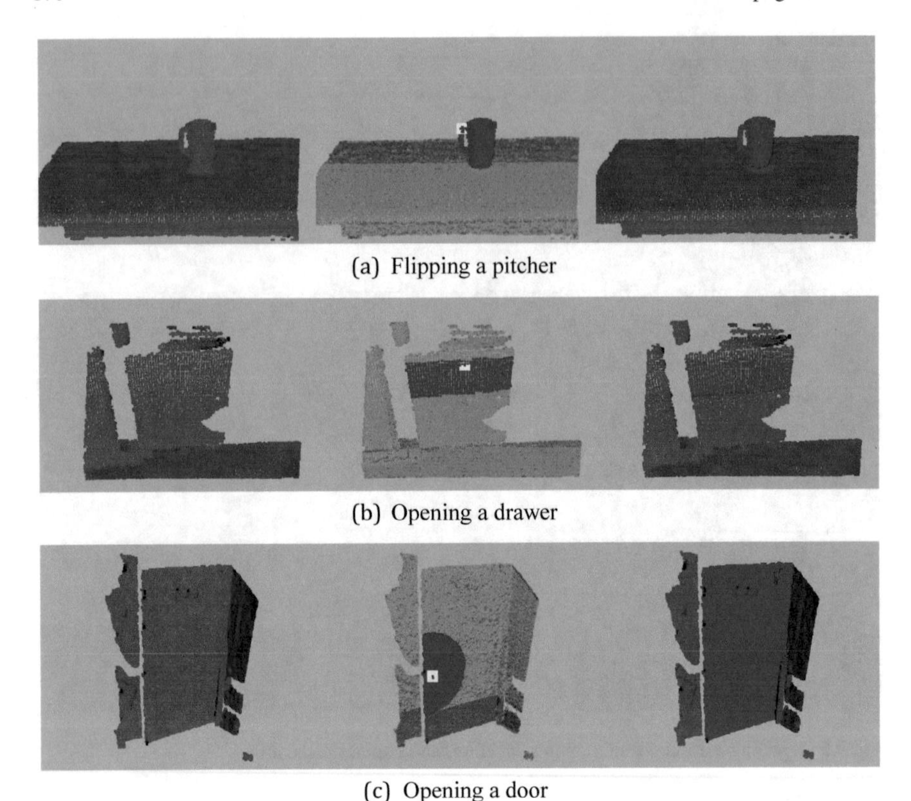

(a) Flipping a pitcher

(b) Opening a drawer

(c) Opening a door

Fig. 9.6 Motion segmentation of the manipulated object. First column: scene background points (the actor is removed). Second column: initial motion segment (blue) obtained by spectral clustering of point trajectories around contact area (yellow). Third column: final motion segment

We present a detailed explanation of each step involved in the process of a robot's replication of an action by observing a human. This entire process is visually described in Fig. 9.10, which separates our application into three phases, namely *preprocessing*, *planning*, and *execution*.

9.5.1 Preprocessing Stage

The preprocessing stage is responsible for taking the contact point, object segments, and their motion trajectories, as described in Fig. 9.1, and converting them into robot-specific trajectories for planning and execution. A visualization of this input can be seen in Fig. 9.11, where (a) depicts the RGB frame of the human performing the action. Subfigure (b) shows the contact point, highlighted in yellow, along with

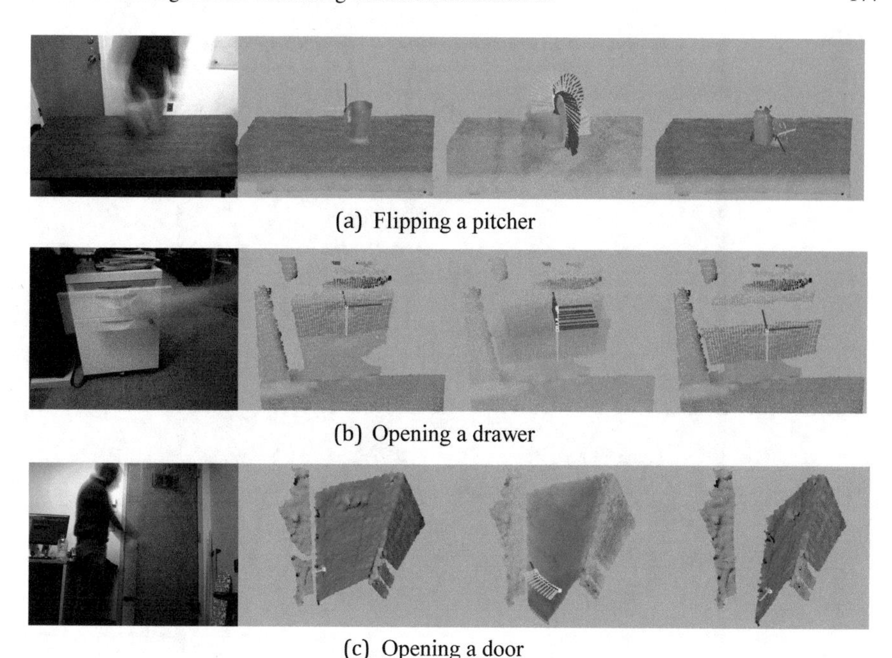

(a) Flipping a pitcher

(b) Opening a drawer

(c) Opening a door

Fig. 9.7 Estimated rigid motion of the manipulated object. A coordinate frame is attached to the object segment (blue) at the contact point location (yellow). First column: temporal accumulation of color frames for the whole action duration. Second column: object state before manipulation. Third column: object trajectory as a series of 6DOF poses. Fourth column: object state after manipulation

an initial object frame. Subfigure (c) demonstrates a dynamic view of the motion trajectory and segmentation of the door, along with the tracked contact point axes across time. Subfigure (d) shows the final pose of the door, after opening has finished.

In this stage, we exploit domain knowledge to semantically ground contact points and object segments, in order to assist affordance analysis and common sense reasoning for robot manipulation, since that provides us with task-dependent priors. For instance, since we know that our task involves opening a refrigerator door, we can make prior assumptions that the contact point between the human agent and the environment will happen at the handle and any consequent motion will be of the door and handle only.

9.5.2 Door Handle Detection

These priors allow us to robustly fit a plane to the points of the door (extracted object) using standard least squares fitting under RANSAC and obtain a set of points for

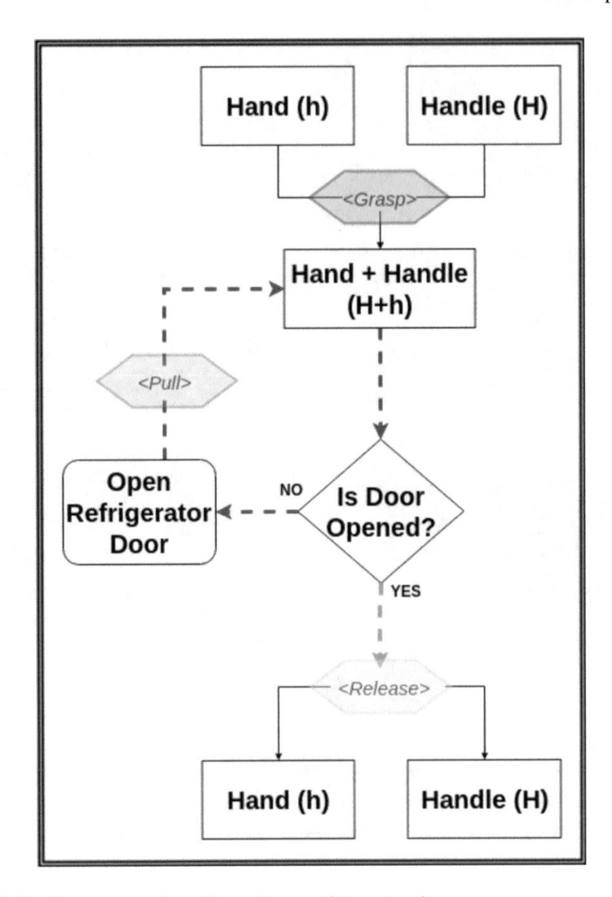

Fig. 9.8 High-level representation of opening a refrigerator door

the door handle (plane outliers). We then fit a cylinder to these points, in order to generate a grasp primitive with a 6DOF pose, for robot grasp planning. The estimated trajectories of the object segment, as mentioned in Table 9.1, are not directly utilized by the robot execution system, but must instead be converted to a robot-specific representation before replication can take place. Our algorithm outputs a series of 6DOF poses P_i for every time point $t_i \in T$. These are then converted to a series of robot-usable poses for the planning phase.

9.5.3 *Planning Stage*

The outputs from the preprocessing stage, namely the robot-specific 6DOF poses of the handle and the cylinder of specified radius and height depicting the handle are passed into the *planning* stage of our pipeline, for both grasp planning and trajectory

Fig. 9.9 Robot observing a human opening a door

planning. The robot visualizer (rviz) [50] package in ROS allows for simulation and visualization of the robot during planning and execution, via real-time feedback from the robot's state estimator. It also has point cloud visualization capabilities, which can be overlaid over primitive shapes. We use this tool for the planning stage, with the Baxter robot and our detected refrigerator (Fig. 9.12).

9.5.4 Grasp Planning

Given a primitive shape, such as a block or cylinder, we are able to use the MoveIt! Simple Grasps [51] package to generate grasp candidates for a parallel gripper (such as one mounted on the Baxter robot). The package integrates with the "MoveIt!" library's pick and place pipeline to simulate and generate multiple potential grasp candidates, i.e., approach poses (Fig. 9.13). There is also a grasp filtering stage, which uses task- and configuration-specific constraints to remove kinematically infeasible grasps, by performing feasibility tests via inverse kinematics solvers. At the end of the grasp planning pipeline, we have a set of candidate grasps, sorted by a grasp quality metric, of which one is chosen for execution in the next stage.

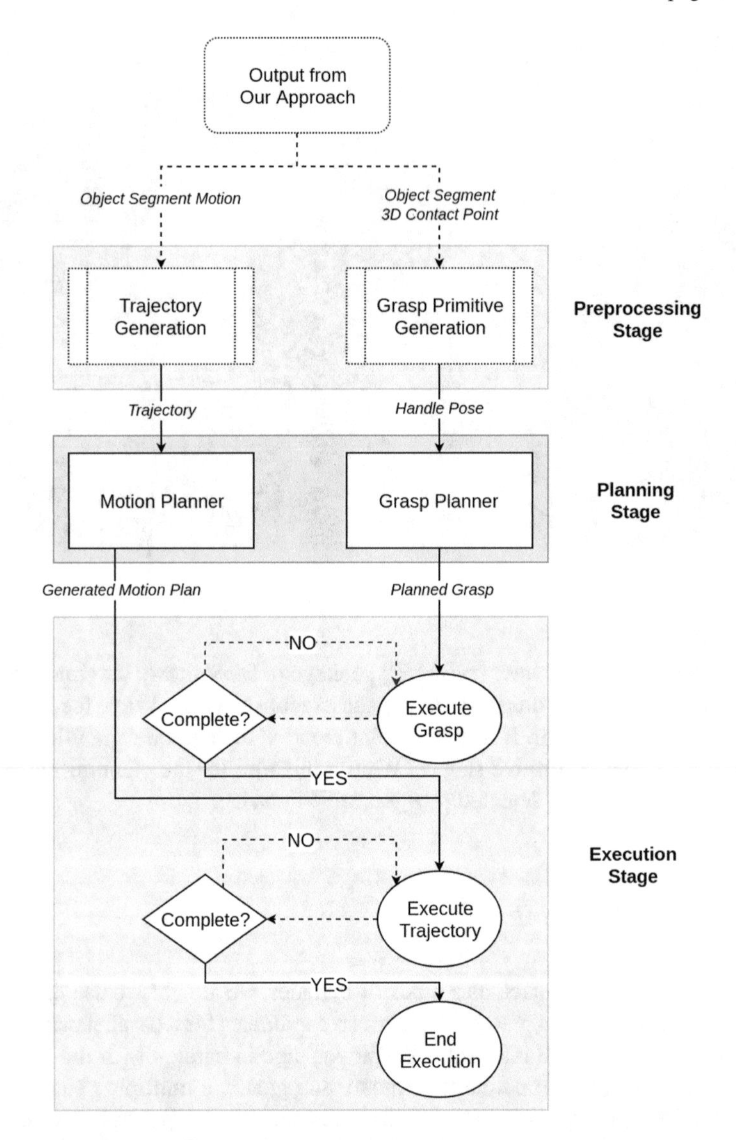

Fig. 9.10 State transition diagram of our process

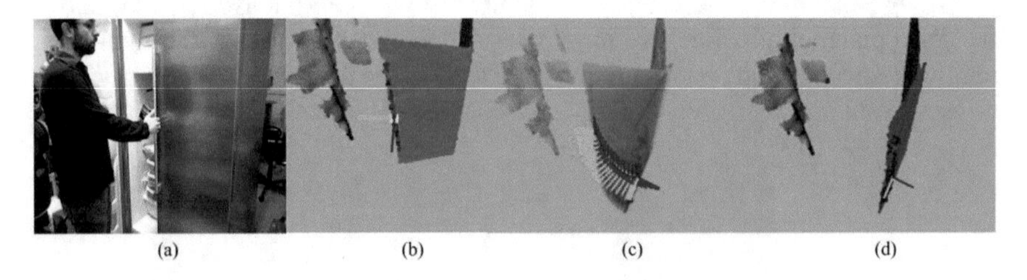

Fig. 9.11 Input to the *preprocessing* stage from our algorithm

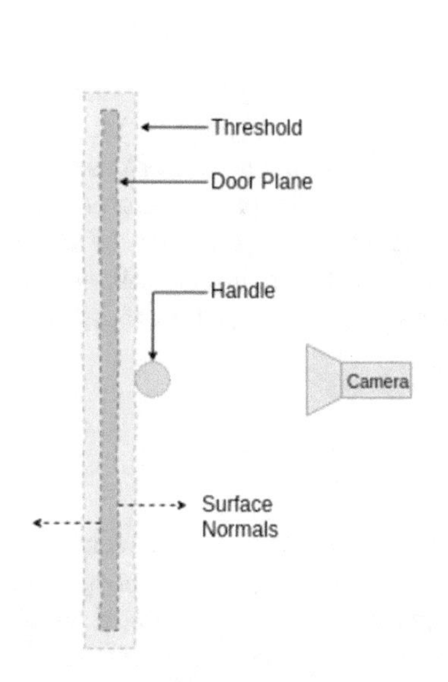

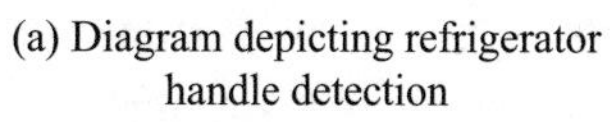

(a) Diagram depicting refrigerator handle detection

(b) Point cloud of refrigerator with detected handle and door

Fig. 9.12 Handle detection

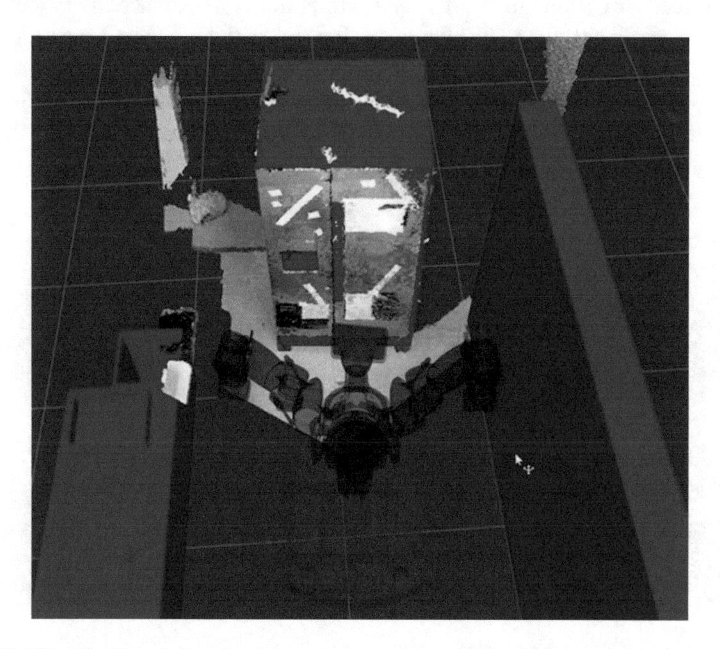

Fig. 9.13 Visualization of planning stage

9.5.5 Trajectory Planning

The ordered set of the poses over time obtained from the *preprocessing* stage is then used to generate a Cartesian path, using the robot operating system's "MoveIt!" [52] motion planning library. This abstraction allows us to input a set of poses through which the end-effector must pass, along with parameters for path validity and obstacle avoidance. "MoveIt!" then uses inverse kinematics solutions for the specified manipulator configuration combined with sampling-based planning algorithms, such as rapidly exploring random trees [53], to generate a trajectory for the robot to execute.

9.5.6 Execution Stage

The *execution* stage takes as input the grasp and trajectory plans generated in the *planning* stage and executes the plan on the robot. First, the generated grasp candidate is used to move the end-effector to a pre-grasp pose and the parallel gripper is aligned to the cylindrical shape of the handle. The grasp is executed based on a feedback control loop, with the termination condition decided by collision avoidance and force feedback. Upon successful grasp of the handle, our pipeline transitions into the trajectory execution stage, which attempts to follow the generated plan based on feedback from the robot's state estimation system. Once the trajectory has been successfully executed, the human motion replication pipeline is complete. This execution process is demonstrated by the robot in Fig. 9.14, beginning with the robot grasping the handle in the top-leftmost figure and ending with the robot releasing the handle in the bottom-leftmost figure, with intermediate frames showing the robot imitating the motion trajectory of the human.

Fig. 9.14 Robot replicating human by opening refrigerator

In future work, we plan to implement a dynamic motion primitives [54] (DMP)-based approach, which will allow more accurate and robust tracking of trajectories by the robot.

9.6 Conclusions

In this paper, we have introduced an active, bottom-up method for the extraction of two fundamental features of an observed manipulation, namely the contact points and motion trajectories of segmented objects. We have qualitatively demonstrated the success of our approach on a set of video inputs and described in detail its fundamental role in a robot imitation scenario. Owing to its general applicability and the manipulation defining nature of its output features, our method can effectively bridge the gap between observation and the development of action representations and plans.

There are many possible directions for future work. At a lower level, we plan to integrate *dynamic reconstruction* into our pipeline to obtain a more complete model for the manipulated object; at this moment, this can be achieved by introducing a step of static scene reconstruction before the manipulation happens, after which we run our algorithm. We also plan to extend our method so that it also can handle *articulated* manipulated objects, as well as objects that are *indirectly* manipulated (e.g., via the use of tools).

On the planning end, one of our future goals is to release a software component for the fully automated replication of *door-opening* tasks (Sect. 9.5), given only a single demonstration. This module will be hardware agnostic up until the final execution stage of the pipeline, such that the generated plan to be imitated can be handled by any robot agent, given the specific manipulator and end-effector configurations.

Acknowledgements The support of ONR under grant award N00014-17-1-2622 and the support of the National Science Foundation under grants SMA 1540916 and CNS 1544787 are greatly acknowledged.

References

1. Poppe, R. (2010). A survey on vision-based human action recognition. *Image and Vision Computing, 28*(6), 976–990.
2. Weinland, D., Ronfard, R., & Boyer, E. (2011). A survey of vision-based methods for action representation, segmentation and recognition. *Computer Vision and Image Understanding, 115*(2), 224–241.
3. Rutishauser, U., Walther, D., Koch, C., & Perona, P. (2004). Is bottom-up attention useful for object recognition? In *Proceedings of the 2004 IEEE Computer Society Conference on Computer Vision and Pattern Recognition, CVPR 2004* (Vol. 2, pp. II–II). IEEE.
4. Ba, J., Mnih, V., & Kavukcuoglu, K. Multiple object recognition with visual attention. arXiv preprint arXiv:1412.7755.

5. Wang, H., & Schmid, C. (2013). Action recognition with improved trajectories. In *2013 IEEE International Conference on Computer Vision (ICCV)* (pp. 3551–3558). IEEE.
6. Simonyan, K., & Zisserman, A. (2014). Two-stream convolutional networks for action recognition in videos. In *Advances in Neural Information Processing Systems* (pp. 568–576).
7. Krüger, N., Geib, C., Piater, J., Petrick, R., Steedman, M., Wörgötter, F., et al. (2011). Object–action complexes: Grounded abstractions of sensory–motor processes. *Robotics and Autonomous Systems, 59*(10):740–757.
8. Amaro, K. R., Beetz, M., & Cheng, G. (2014). Understanding human activities from observation via semantic reasoning for humanoid robots. In *IROS Workshop on AI and Robotics*.
9. Summers-Stay, D., Teo, C. L., Yang, Y., Fermüller, C., & Aloimonos, Y. (2012). Using a minimal action grammar for activity understanding in the real world. In *2012 IEEE/RSJ International Conference on Intelligent Robots and Systems (IROS)* (pp. 4104–4111). IEEE.
10. Yang, Y., Guha, A., Fermüller, C., & Aloimonos, Y. (2014). A cognitive system for understanding human manipulation actions. *Advances in Cognitive Systems, 3*, 67–86.
11. Yang, Y., Li, Y., Fermüller, C., & Aloimonos, Y. (2015). Robot learning manipulation action plans by "watching" unconstrained videos from the world wide web. In *AAAI* (pp. 3686–3693).
12. Aksoy, E. E., Abramov, A., Dörr, J., Ning, K., Dellen, B., & Wörgötter, F. (2011). Learning the semantics of object–action relations by observation. *The International Journal of Robotics Research, 30*(10), 1229–1249.
13. Zampogiannis, K., Yang, Y., Fermüller, C., & Aloimonos, Y. (2015). Learning the spatial semantics of manipulation actions through preposition grounding. In *2015 IEEE International Conference on Robotics and Automation (ICRA)* (pp. 1389–1396). IEEE.
14. Yan, Z., & Xiang, X. Scene flow estimation: A survey. arXiv preprint arXiv:1612.02590.
15. Herbst, E., Ren, X., & Fox, D. (2013). RGB-D flow: Dense 3-D motion estimation using color and depth. In *2013 IEEE International Conference on Robotics and Automation (ICRA)* (pp. 2276–2282). IEEE.
16. Quiroga, J., Brox, T., Devernay, F., & Crowley, J. (2014). Dense semi-rigid scene flow estimation from RGBD images. In *European Conference on Computer Vision* (pp. 567–582). Berlin: Springer.
17. Jaimez, M., Souiai, M., Stückler, J., Gonzalez-Jimenez, J., & Cremers, D. (2015). Motion cooperation: Smooth piece-wise rigid scene flow from RGB-D images. In *2015 International Conference on 3D Vision (3DV)* (pp. 64–72). IEEE.
18. Jaimez, M., Souiai, M., Gonzalez-Jimenez, J., & Cremers, D. (2015). A primal-dual framework for real-time dense RGB-D scene flow. In *2015 IEEE International Conference on Robotics and Automation (ICRA)* (pp. 98–104). IEEE.
19. Jaimez, M., Kerl, C., Gonzalez-Jimenez, J., & Cremers, D. (2017). Fast odometry and scene flow from RGB-D cameras based on geometric clustering. In *2017 IEEE International Conference on Robotics and Automation (ICRA)* (pp. 3992–3999). IEEE.
20. Mayer, N., Ilg, E., Hausser, P., Fischer, P., Cremers, D., Dosovitskiy, A., et al. (2016). A large dataset to train convolutional networks for disparity, optical flow, and scene flow estimation. In: *Proceedings of the IEEE Conference on Computer Vision and Pattern Recognition* (pp. 4040–4048).
21. Sorkine, O., & Alexa, M. (2007). As-rigid-as-possible surface modeling. In *Symposium on Geometry Processing* (Vol. 4, p. 30).
22. Tam, G. K., Cheng, Z.-Q., Lai, Y.-K., Langbein, F. C., Liu, Y., Marshall, D., et al. (2013). Registration of 3D point clouds and meshes: A survey from rigid to nonrigid. *IEEE Transactions on Visualization and Computer Graphics, 19*(7), 1199–1217.
23. Amberg, B., Romdhani, S., & Vetter, Y. (2007). Optimal step nonrigid ICP algorithms for surface registration. In *IEEE Conference on Computer Vision and Pattern Recognition, CVPR'07* (pp. 1–8). IEEE.
24. Newcombe, R. A., Fox, D., & Seitz, S. M. (2015). DynamicFusion: Reconstruction and tracking of non-rigid scenes in real-time. In *Proceedings of the IEEE Conference on Computer Vision and Pattern Recognition* (pp. 343–352).

25. Innmann, M., Zollhöfer, M., Nießner, M., Theobalt, C., & Stamminger, M. VolumeDeform: Real-time volumetric non-rigid reconstruction.
26. Slavcheva, M., Baust, M., Cremers, D., & Ilic, S. (2017). KillingFusion: Non-rigid 3D reconstruction without correspondences. In *IEEE Conference on Computer Vision and Pattern Recognition (CVPR)* (Vol. 3, p. 7).
27. Zampogiannis, K., Fermuller, C., & Aloimonos, Y. (2018). Cilantro: A lean, versatile, and efficient library for point cloud data processing. In *Proceedings of the 26th ACM International Conference on Multimedia, MM'18* (pp. 1364–1367). New York, NY, USA: ACM. https://doi.org/10.1145/3240508.3243655.
28. Yang, Y., Fermuller, C., Li, Y., & Aloimonos, Y. (2015). Grasp type revisited: A modern perspective on a classical feature for vision. In *Proceedings of the IEEE Conference on Computer Vision and Pattern Recognition* (pp. 400–408).
29. Chen, Q., Li, H., Abu-Zhaya, R., Seidl, A., Zhu, F., & Delp, E. J. (2016). Touch event recognition for human interaction. *Electronic Imaging, 2016*(11), 1–6.
30. Yan, J., & Pollefeys, M. (2006). A general framework for motion segmentation: Independent, articulated, rigid, non-rigid, degenerate and non-degenerate. In *European Conference on Computer Vision* (pp. 94–106). Berlin: Springer.
31. Tron, R., & Vidal, R. (2007). A benchmark for the comparison of 3-D motion segmentation algorithms. In *IEEE Conference on Computer Vision and Pattern Recognition, CVPR'07* (pp. 1–8). IEEE.
32. Costeira, J., & Kanade, T. (1995). A multi-body factorization method for motion analysis. In *Proceedings of the Fifth International Conference on Computer Vision* (pp. 1071–1076). IEEE.
33. Kanatani, K. (2001). Motion segmentation by subspace separation and model selection. In *Proceedings of the Eighth IEEE International Conference on Computer Vision, ICCV 2001* (Vol. 2, pp. 586–591). IEEE.
34. Rao, S., Tron, R., Vidal, R., & Ma, Y. (2010). Motion segmentation in the presence of outlying, incomplete, or corrupted trajectories. *IEEE Transactions on Pattern Analysis and Machine Intelligence, 32*(10), 1832–1845.
35. Vidal, R., & Hartley, R. (2004). Motion segmentation with missing data using power factorization and GPCA. In *Proceedings of the 2004 IEEE Computer Society Conference on Computer Vision and Pattern Recognition, CVPR 2004* (Vol. 2, pp. II–II). IEEE.
36. Katz, D., Kazemi, M., Bagnell, J. A., & Stentz, A. (2013). Interactive segmentation, tracking, and kinematic modeling of unknown 3D articulated objects. In *IEEE International Conference on Robotics and Automation (ICRA)* (pp. 5003–5010). IEEE.
37. Herbst, E., Ren, X., & Fox, D. (2012). Object segmentation from motion with dense feature matching. In *ICRA Workshop on Semantic Perception, Mapping and Exploration* (Vol. 2).
38. Rünz, M., & Agapito, L. (2017). Co-fusion: Real-time segmentation, tracking and fusion of multiple objects. In *2017 IEEE International Conference on Robotics and Automation (ICRA)* (pp. 4471–4478).
39. Whelan, T., Leutenegger, S., Salas-Moreno, R., Glocker, B., & Davison, A. (2015). ElasticFusion: Dense slam without a pose graph. In *Robotics: Science and Systems*.
40. Ochs, P., Malik, J., & Brox, T. (2014). Segmentation of moving objects by long term video analysis. *IEEE Transactions on Pattern Analysis and Machine Intelligence, 36*(6), 1187–1200.
41. Besl, P. J., & McKay, N. D. (1992). Method for registration of 3-D shapes. In *Sensor Fusion IV: Control Paradigms and Data Structures* (Vol. 1611, pp. 586–607). International Society for Optics and Photonics.
42. Rusinkiewicz, S., & Levoy, M. (2001). Efficient variants of the ICP algorithm. In *Proceedings of the Third International Conference on 3-D Digital Imaging and Modeling* (pp. 145–152). IEEE.
43. Newcombe, R. A., Izadi, S., Hilliges, O., Molyneaux, D., Kim, D., Davison, A. J., et al. (2011). KinectFusion: Real-time dense surface mapping and tracking. In *10th IEEE International Symposium on Mixed and Augmented Reality (ISMAR)* (pp. 127–136). IEEE.
44. Cao, Z., Simon, T., Wei, S.-E., & Sheikh, Y. (2017). Realtime multi-person 2D pose estimation using part affinity fields. In *CVPR*.

45. Dalal, N., & Triggs, B (2005). Histograms of oriented gradients for human detection. In *IEEE Computer Society Conference on Computer Vision and Pattern Recognition, CVPR 2005* (Vol. 1, pp. 886–893). IEEE.
46. Jones, M. J., & Rehg, J. M. (2002). Statistical color models with application to skin detection. *International Journal of Computer Vision, 46*(1), 81–96.
47. Vezhnevets, V., Sazonov, V., & Andreeva, A. (2003). A survey on pixel-based skin color detection techniques. In *Proceedings of Graphicon* (Vol. 3, pp. 85–92), Moscow, Russia.
48. Von Luxburg, U. (2007). A tutorial on spectral clustering. *Statistics and Computing, 17*(4), 395–416.
49. Umeyama, S. (1991). Least-squares estimation of transformation parameters between two point patterns. *IEEE Transactions on Pattern Analysis and Machine Intelligence, 13*(4), 376–380.
50. Hershberger, D., Gossow, D., & Faust, J. (2012). rviz, https://github.com/ros-visualization/rviz.
51. Coleman, D. T. (2016). "moveit!" simple grasps. https://github.com/davetcoleman/moveit_simple_grasps.
52. Chitta, S., Sucan, I., & Cousins, S. (2012). MoveIt! [ROS topics]. *IEEE Robotics Automation Magazine, 19*(1), 18–19. https://doi.org/10.1109/mra.2011.2181749.
53. Lavalle, S. M. (1998). *Rapidly-exploring random trees: A new tool for path planning*. Technical Report, Iowa State University.
54. Schaal, S. (2002). Dynamic movement primitives—A framework for motor control in humans and humanoid robotics.

Chapter 10
Human Action Recognition and Assessment Via Deep Neural Network Self-Organization

German I. Parisi

Abstract The robust recognition and assessment of human actions are crucial in human-robot interaction (HRI) domains. While state-of-the-art models of action perception show remarkable results in large-scale action datasets, they mostly lack the flexibility, robustness, and scalability needed to operate in natural HRI scenarios which require the continuous acquisition of sensory information as well as the classification or assessment of human body patterns in real time. In this chapter, I introduce a set of hierarchical models for the learning and recognition of actions from depth maps and RGB images through the use of neural network self-organization. A particularity of these models is the use of growing self-organizing networks that quickly adapt to non-stationary distributions and implement dedicated mechanisms for continual learning from temporally correlated input.

10.1 Introduction

Artificial systems for human action recognition from videos have been extensively studied in the literature, with a large variety of machine learning models and benchmark datasets [21, 66]. The robust learning and recognition of human actions are crucial in human-robot interaction (HRI) scenarios where, for instance, robots are required to efficiently process rich streams of visual input with the goal of undertaking assistive actions in a residential context (Fig. 10.1).

Deep learning architectures such as convolutional neural networks (CNNs) have been shown to recognize actions from videos with high accuracy through the use of hierarchies that functionally resemble the organization of earlier areas of the visual cortex (see [21] for a survey). However, the majority of these models are computationally expensive to train and lack the flexibility and robustness to operate in the above-described HRI scenarios. A popular stream of vision research has focused on the use of depth sensing devices such as the Microsoft Kinect and ASUS Xtion Live

G. I. Parisi (✉)
University of Hamburg, Hamburg, Germany
e-mail: parisi@informatik.uni-hamburg.de
URL: http://giparisi.github.io

N. Noceti et al. (eds.), *Modelling Human Motion*,
https://doi.org/10.1007/978-3-030-46732-6_10

Fig. 10.1 Person tracking and action recognition with a depth sensor on a humanoid robot in a domestic environment [61]

for human action recognition in HRI applications using depth information instead of, or in combination with, RGB images. Post-processed depth map sequences provide real-time estimations of 3D human motion in cluttered environments with increased robustness to varying illumination conditions and reducing the computational cost for motion segmentation and pose estimation (see [22] for a survey). However, learning models using low-dimensional 3D information (e.g. 3D skeleton joints) have often failed to show robust performance in real-world environments since this type of input can be particularly noisy and susceptible to self-occlusion.

In this chapter, I introduce a set of neural network models for the efficient learning and classification of human actions from depth information and RGB images. These models use different variants of growing self-organizing networks for the learning of action sequences and real-time inference. In Sect. 10.2, I summarize the fundamentals of neural network self-organization with focus on a particular type of growing network, the Grow When Required (GWR) model, that can grow and remove neurons in response to a time-varying input distribution, and the Gamma-GWR which extends the GWR with temporal context for the efficient learning of visual representations from temporally correlated input. Hierarchical arrangements of such networks, which I describe in Sect. 10.3, can be used for efficiently processing body pose and motion features and learning a set of training actions.

Understanding people's emotions plays a central role in human social interaction and behavior [64]. Perception systems making use of affective information can significantly improve the overall HRI experience, for instance, by triggering pro-active robot behavior as a response to the user's emotional state. An increasing corpus of research has been conducted in the recognition of affective states, e.g., through the processing of facial expressions [2], speech detection [46] and the combination of these multimodal cues [4]. While facial expressions can easily convey emotional states, it is often the case in HRI scenarios that a person is not facing the sensor or is standing far away from the camera, resulting in insufficient spatial resolution to extract facial features. The recognition of emotions from body motion, instead, has received less attention in the literature but has a great value in HRI domains. The main reason is that affective information is seen as harder to extrapolate from complex full-body expressions with respect to facial expressions and speech analysis. In Sect. 10.3.2, I introduce a self-organizing neural architecture for emotion recognition from 3D body motion patterns.

In addition to recognizing short-term behavior such as domestic daily actions and dynamic emotional states, it is of interest to learn the user's behavior over longer periods of time [85]. The collected data can be used to perform longer-term gait assessment as an important indicator for a variety of health problems, e.g., physical diseases and neurological disorders such as Parkinson's disease [1]. The analysis and assessment of body motion have recently attracted significant interest in the healthcare community with many application areas such as physical rehabilitation, diagnosis of pathologies, and assessment of sports performance. The correctness of postural transitions is fundamental during the execution of well-defined physical exercises since inaccurate movements may not only significantly reduce the overall efficiency of the movement and but also increase the risk of injury [29]. As an example, in the healthcare domain, the correct execution of physical rehabilitation routines is crucial for patients to improve their health condition [84]. Similarly, in weight-lifting training, correct postures improve the mechanical efficiency of the body and lead the athlete to achieve better results across training sessions. In Sect. 10.4, I introduce a self-organizing neural architecture for learning body motion sequences comprising weight-lifting exercise and assessing their correctness in real time.

State-of-the-art models of action recognition have mostly proposed the learning of a static batch of body patterns [21]. However, systems and robots operating in real-world settings are required to acquire and fine-tune internal representations and behavior in a continual learning fashion. Continual learning refers to the ability of a system to seamlessly learn from continuous streams of information while preventing *catastrophic forgetting*, i.e., a condition in which new incoming information strongly interferes with previously learned representations [38, 53]. Continual machine learning research has mainly focused on the recognition of static image patterns whereas the processing of complex stimuli such as dynamic body motion patterns has been overlooked. In particular, the majority of these models address supervised continual learning on static image datasets such as the MNIST [35] and the CIFAR-10 [34] and have not reported results on video sequences. In Sect. 10.5, I introduce the use of deep neural network self-organization for the continual learning of human actions from RGB video sequences. Reported results evidence that deep self-organization can mitigate catastrophic forgetting while showing competitive performance with state-of-the-art batch learning models.

Despite significant advances in artificial vision, learning models are still far from providing the flexibility, robustness, and scalability exhibited by biological systems. In particular, current models of action recognition are designed for and evaluated on highly controlled experimental conditions, whereas systems and robots in HRI scenarios are exposed to continuous streams of (often noisy) sensory information. In Sect. 10.6, I discuss a number of challenges and directions for future research.

10.2 Neural Network Self-Organization

10.2.1 Background

Input-driven self-organization is a crucial component of cortical processing which shapes topographic maps based on visual experience [45, 86]. Different artificial models of input-driven self-organization have been proposed to resemble the basic dynamics of Hebbian learning and structural plasticity [24], with neural map organization resulting from unsupervised statistical learning. The goal of the self-organizing learning is to cause different parts of a network to respond similarly to certain input samples starting from an initially unorganized state. Typically, during the training phase these networks build a map through a competitive process, also referred to as *vector quantization*, so that a set of neurons represent prototype vectors encoding a submanifold in the input space. Throughout this process, the network learns significant *topological relations* of the input without supervision.

A well-established model is the self-organizing map (SOM) [33] in which the number of prototype vectors (or neurons) that can be trained is pre-defined. However, empirically selecting a convenient number of neurons can be tedious, especially when dealing with non-stationary, temporally-correlated input distributions [78]. To alleviate this issue, a number of growing models have been proposed that dynamically allocate or remove neurons in response to sensory experience. An example is the Grow When Required (GWR) network [37] which grows or shrinks to better match the input distribution. The GWR has the ability to add new neurons whenever the current input is not sufficiently matched by the existing neurons (whereas other popular models, e.g. Growing Neural Gas (GNG) [16]), will add neurons only at fixed, pre-defined intervals). Because of their ability to allocate novel trainable resources, GWR-like models have the advantage of mitigating the disruptive interference of existing internal representations when learning from novel sensory observations.

10.2.2 Grow When Required (GWR) Networks

The GWR [37] is a growing self-organizing network that learns the prototype neural weights from a multi-dimensional input distribution. It consists of a set of neurons with their associated weight vectors and edges that create links between neurons. For a given input vector $\mathbf{x}(t) \in \mathbb{R}^n$, its best-matching neuron or unit (BMU) in the network, b, is computed as the index of the neural weight that minimizes the distance to the input:

$$b = \arg\min_{j \in A} \|\mathbf{x}(t) - \mathbf{w}_j\|, \tag{10.1}$$

where A is the set of neurons and $\| \cdot \|$ denotes the Euclidean distance.

The network starts with two randomly initialized neurons. Each neuron j is equipped with a habituation counter h that considers the number of times that the neuron has fired. Newly created neurons start with $h_j = 1$ and iteratively decreased towards 0 according to the habituation rule

$$\Delta h_i = \tau_i \cdot 1.05 \cdot (1 - h_i) - \tau_i, \tag{10.2}$$

where $i \in \{b, n\}$ and τ_i is a constant that controls the monotonically decreasing behavior. Typically, h_b is habituated faster than h_n by setting $\tau_b > \tau_n$.

A new neuron is added if the activity of the network computed as $a = \exp - \|\mathbf{x}(t) - \mathbf{w}_b\|$ is smaller than a given activation threshold a_T and if the habituation counter h_b is smaller than a given threshold h_T. The new neuron is created half-way between the BMU and the input. This mechanism leads to creating neurons only after the existing ones have been sufficiently trained.

At each iteration, the neural weights are updated according to:

$$\Delta \mathbf{w}_i = \epsilon_i \cdot h_i \cdot (\mathbf{x}(t) - \mathbf{w}_i), \tag{10.3}$$

where ϵ_i is a constant learning rate ($\epsilon_n < \epsilon_b$) and the index i indicates the BMU b and its topological neighbors. Connections between neurons are updated on the basis of neural co-activation, i.e. when two neurons fire together, a connection between them is created if it does not exist.

While the mechanisms for creating new neurons and connections in the GWR do not resemble biologically plausible mechanisms of neurogenesis (e.g., [11, 32, 43]), the GWR learning algorithm represents an efficient model that incrementally adapts to non-stationary input. A comparison between GNG and GWR learning in terms of the number of neurons, quantization error (average discrepancy between the input and its BMU), and parameters modulating network growth (average network activation and habituation rate) is shown in Fig. 10.2. This learning behavior is particularly convenient for incremental learning scenarios since neurons will be created to promptly distribute in the input space, thereby allowing a faster convergence through iterative fine-tuning of the topological map. The neural update rate decreases as the neurons become more habituated, which has the effect of preventing that noisy input interferes with consolidated neural representations.

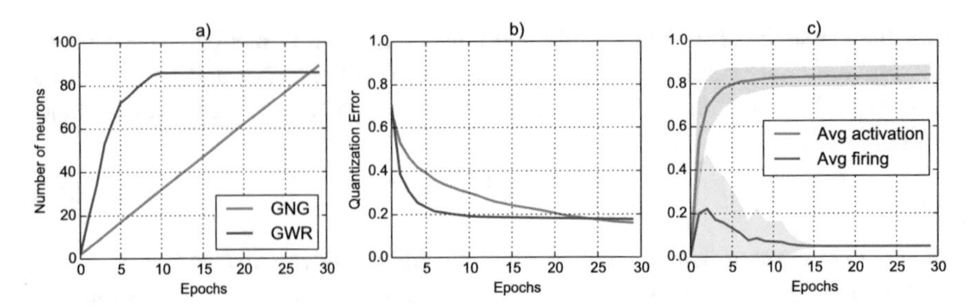

Fig. 10.2 Comparison of GNG and GWR training: **a** number of neurons, **b** quantization error, and **c** GWR average activation and habituation counter through 30 training epochs on the Iris dataset [56]

10.2.3 Gamma-GWR

The GWR model does not account for the learning of latent temporal structure. For this purpose, the Gamma-GWR [56] extends the GWR with temporal context. Each neuron consists of a weight vector $\mathbf{w}_j$ and a number K of context descriptors $\mathbf{c}_{j,k}$ ($\mathbf{w}_j, \mathbf{c}_{j,k} \in \mathbb{R}^n$).

Given the input $\mathbf{x}(t) \in \mathbb{R}^n$, the index of the BMU, b, is computed as:

$$b = \arg\min_{j \in A}(d_j), \tag{10.4}$$

$$d_j = \alpha_0 \|\mathbf{x}(t) - \mathbf{w}_j\| + \sum_{k=1}^{K} \alpha_k \|\mathbf{C}_k(t) - \mathbf{c}_{j,k}\|, \tag{10.5}$$

$$\mathbf{C}_k(t) = \beta \cdot \mathbf{w}_b^{t-1} + (1 - \beta) \cdot \mathbf{c}_{b,k-1}^{t-1}, \tag{10.6}$$

where $\| \cdot \|$ denotes the Euclidean distance, α_i and β are constant values that modulate the influence of the temporal context, $\mathbf{w}_b^{t-1}$ is the weight vector of the BMU at $t-1$, and $\mathbf{C}_k \in \mathbb{R}^n$ is the global context of the network with $\mathbf{C}_k(t_0) = 0$. If $K = 0$, then Eq. 10.5 resembles the learning dynamics of the standard GWR without temporal context. For a given input $\mathbf{x}(t)$, the activity of the network, $a(t)$, is defined in relation to the distance between the input and its BMU (Eq. 10.4) as follows:

$$a(t) = \exp(-d_b), \tag{10.7}$$

thus yielding the highest activation value of 1 when the network can perfectly match the input sequence ($d_b = 0$).

The training of the existing neurons is carried out by adapting the BMU b and its neighboring neurons n:

$$\Delta \mathbf{w}_i = \epsilon_i \cdot h_i \cdot (\mathbf{x}(t) - \mathbf{w}_i), \tag{10.8}$$

$$\Delta \mathbf{c}_{i,k} = \epsilon_i \cdot h_i \cdot (\mathbf{C}_k(t) - \mathbf{c}_{i,k}), \tag{10.9}$$

where $i \in \{b, n\}$ and ϵ_i is a constant learning rate ($\epsilon_n < \epsilon_b$). The habituation counters h_i are updated according to Eq. 10.2.

Empirical studies with large-scale datasets have shown that Gamma-GWR networks with additive neurogenesis show a better performance than a static network with the same number of neurons, thereby providing insights into the design of neural architectures in incremental learning scenarios when the total number of neurons is fixed [50].

10.3 Human Action Recognition

10.3.1 Self-Organizing Integration of Pose-Motion Cues

Human action perception in the brain is supported by a highly adaptive system with separate neural pathways for the distinct processing of body pose and motion features at multiple levels and their subsequent integration in higher areas [13, 83]. The ventral pathway recognizes sequences of body form snapshots, while the dorsal pathway recognizes optic-flow patterns. Both pathways comprise hierarchies that extrapolate visual features with increasing complexity of representation [23, 36, 81]. It has been shown that while early visual areas such as the primary visual cortex (V1) and the motion-sensitive area (MT+) yield higher responses to instantaneous sensory input, high-level areas such as the superior temporal sulcus (STS) are more affected by information accumulated over longer timescales [23]. Neurons in higher levels of the hierarchy are also characterized by gradual invariance to the position and the scale of the stimulus [47]. Hierarchical aggregation is a crucial organizational principle of cortical processing for dealing with perceptual and cognitive processes that unfold over time [14]. With the use of extended models of neural network self-organization, it is possible to obtain progressively generalized representations of sensory inputs and learn inherent spatiotemporal dependencies of input sequences.

In Parisi et al. [60], we proposed a learning architecture consisting of a two-stream hierarchy of GWR networks that processes extracted pose and motion features in parallel and subsequently integrates neuronal activation trajectories from both streams. This integration network functionally resembles the response of STS model neurons encoding sequence-selective prototypes of action segments in the joint pose-motion domain. An overall overview of the architecture is depicted in Fig. 10.3. The hierarchical arrangement of the networks yields progressively specialized neurons encoding latent spatiotemporal dynamics of the input. We process the visual input under the assumption that action recognition is selective for temporal order [18, 23]. Therefore, the recognition of an action occurs only when neural trajectories are activated in the correct temporal order with respect to the learned action template.

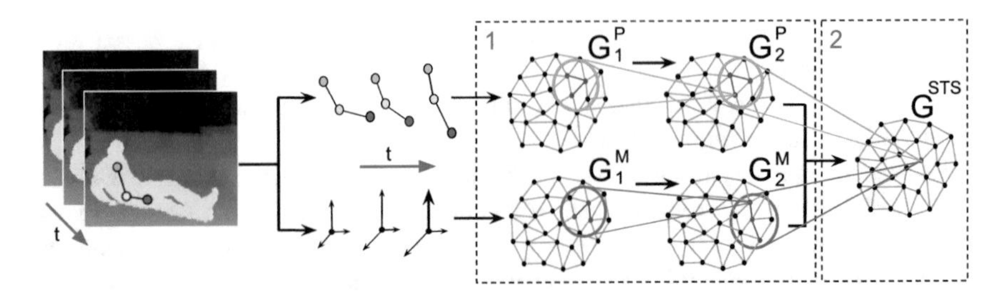

Fig. 10.3 GWR-based architecture for pose-motion integration and action classification: **a** hierarchical processing of pose-motion features in parallel; **b** integration of neuron trajectories in the joint pose-motion feature space [60]

Following the notation in Fig. 10.3, G_1^P and G_1^M are trained with pose and motion features respectively. After this step, we train G_2^P and G_2^M with concatenated trajectories of neural activations in the previous network layer. The STS stage integrates pose-motion features by training G^{STS} with the concatenation of vectors from G_2^P and G_2^M in the pose-motion feature space. After the training of G^{STS} is completed, each neuron will encode a sequence-selective prototype action segment, thereby integrating changes in the configuration of a person's body pose over time. For the classification of actions, we extended the standard implementation of the GWR in which an associative matrix stores the frequency-based distribution of sample labels, i.e. each neuron stores the number of times that a given sample label has been associated to its neural weight. This labeling strategy does not require a predefined number of action classes since the associative matrix can be dynamically expanded when a novel label class is encountered.

We evaluated our approach both on our Knowledge Technology (KT) full-body action dataset [59] and the public action benchmark CAD-60 [80]. The KT dataset is composed of 10 full-body actions performed by 13 subjects with a normal physical condition. The dataset contains the following actions: *standing, walking, jogging, picking up, sitting, jumping, falling down, lying down, crawling,* and *standing up*. Videos were captured in a home-like environment with a Kinect sensor installed 1, 30 m above the ground. Depth maps were sampled with a VGA resolution of 640×480 and an operation range from 0.8 to 3.5 m at 30 frames per second. From the raw depth map sequences, 3D body joints were estimated on the basis of the tracking skeleton model provided by OpenNI SDK. Snapshots of full-body actions are shown in Fig. 10.4 as raw depth images, segmented body silhouettes, skeletons, and body centroids. We proposed a simplified skeleton model consisting of three centroids and two body slopes. The centroids were estimated as the centers of mass that follow the distribution of the main body masses on each posture. As can be seen in Fig. 10.5, three centroids are sufficient to represent prominent posture characteristics while maintaining a low-dimensional feature space. Such low-dimensional representation increases tracking robustness for situations of partial occlusion with respect to a skeleton model comprising a larger number of body joints. Our experiments showed

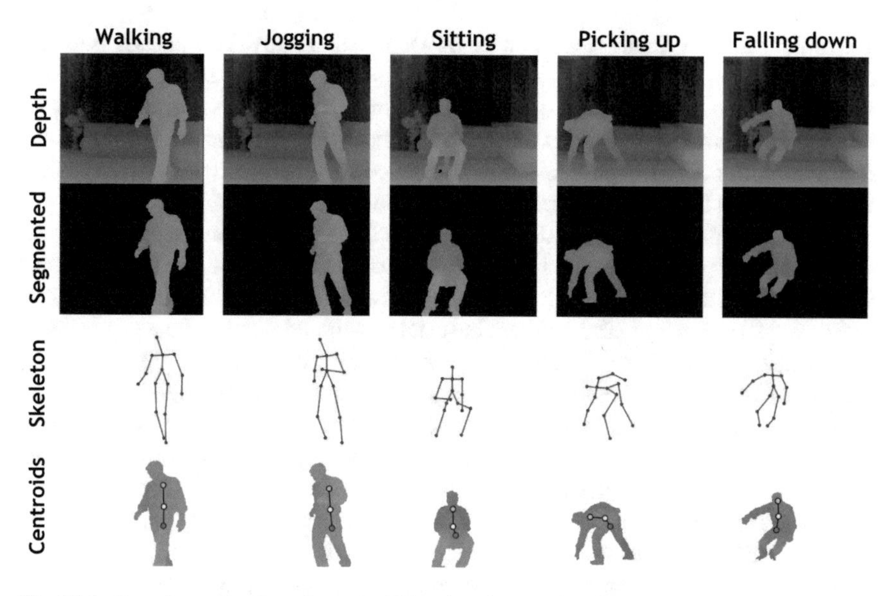

Fig. 10.4 Snapshots of actions from the KT action dataset visualized as raw depth images, segmented body, skeleton, and body centroids

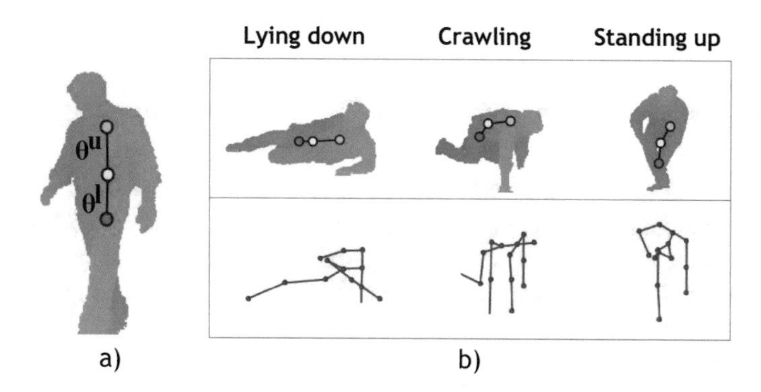

Fig. 10.5 Full-body action representations: **a** three centroids with body slopes θ^u and θ^l, and **b** comparison of body centroids (top) and noisy skeletons (bottom)

that a GWR-based approach outperforms the same type of architecture using GNG networks with an average accuracy rate of 94% (5% higher than GNG-based).

The Cornell activity dataset CAD-60 [80] is composed of 60 RGB-D videos of four subjects (two males, two females, one left-handed) performing 12 activities: *rinsing mouth, brushing teeth, wearing contact lens, talking on the phone, drinking water, opening pill container, cooking (chopping), cooking (stirring), talking on couch, relaxing on couch, writing on whiteboard, working on computer*. The activities were performed in 5 different environments: office, kitchen, bedroom, bathroom, and living room. The videos were collected with a Kinect sensor with distance ranges from

1.2 to 3.5 m and a depth resolution of 640×480 at 15 fps. The dataset provides raw depth maps, RGB images, and skeleton data. We used the set of 3D positions without the *feet*, leading to 13 joints (i.e., 39 input dimensions). Instead of using world coordinates, we encoded the joint positions using the center of the hips as the frame of reference to obtain translation invariance. We computed joint motion as the difference of two consecutive frames for each pose transition.

For our evaluation on the CAD-60, we adopted the same scheme as [80] using all the 12 activities plus a random action with a *new person* strategy, i.e. the first 3 subjects for training and the remaining for test purposes. We obtained 91.9% precision, 90.2% recall, and 91% F-score. The reported best state-of-the-art result is 93.8% precision, 94.5% recall, and 94.1% F-score [75], where they estimate, prior to learning, a number of key poses to compute spatiotemporal action templates. Here, each action must be segmented into atomic action templates composed of a set of n key poses, where n depends on the action's duration and complexity. Furthermore, experiments with real-time inference have not been reported. The second-best approach achieves 93.2% precision, 91.9% recall, and 91.5% F-score [12], in which they used a dynamic Bayesian Mixture Model to classify motion relations between body poses. However, the authors estimated their own skeleton model from raw depth images and did not use the one provided by the CAD-60 benchmark dataset. Therefore, differences in the tracked skeleton exist that hinder a direct quantitative comparison with our approach.

10.3.2 Emotion Recognition from Body Expressions

The recognition of emotions plays an important role in our daily life and is essential for social communication and it can be particularly useful in HRI scenarios. For instance, a socially-assistive robot may be able to strengthen its relationship with the user if it can understand whether that person is bored, angry, or upset. Body expressions convey an additional social cue to reinforce or complement facial expressions [65, 71]. Furthermore, this approach can complement the use of facial expressions when the user is not facing the sensor or is too distant from it for facial features to be computed. Despite its promising applications in HRI domains, emotion recognition from body motion patterns has received significantly less attention with respect to facial expressions and speech analysis.

Movement kinematics such velocity and acceleration represent significant features when it comes to recognizing emotions from body patterns [65, 71]. Similarly, using temporal features in terms of body motion resulted in higher recognition rates than pose features alone [62]. Schindler et al. [73] presented an image-based classification system for recognizing emotion from images of body postures. The overall recognition accuracy of his system resulted in 80% for six basic emotions. Although these systems show a high recognition rate, they are limited to postural emotions, which are not sufficient for a real-time interactive situation between humans and robots in a domestic environment. Piana et al. [63] proposed a real-time emotion recognition system using postural, kinematic, and geometrical features extracted

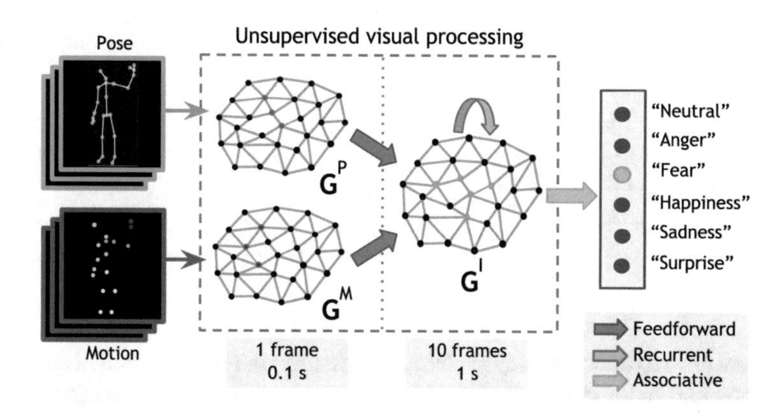

Fig. 10.6 Proposed learning architecture with a hierarchy of self-organizing networks. The first layer processes pose and motion features from individual frames, whereas in the second layer a Gamma-GWR network learns the spatiotemporal structure of the joint pose-motion representations [10]

from sequences of 3D skeletons videos. However, they only considered a reduced set of upper-body joints, i.e., head, shoulders, elbows, hands, and torso (Fig. 10.6).

In Elfaramawy et al. [10], we proposed a self-organizing architecture to recognize emotional states from body motion patterns. The focus of our study was to investigate whether full-body expressions from depth map videos convey adequate affective information for the task of emotion recognition. The overall architecture, shown in Fig. 10.4, consists of a hierarchy of self-organizing networks for learning sequences of 3D body joint features. In the first layer, two GWR networks [37], G^P and G^M, learn a dictionary of prototype samples of pose and motion features respectively. Motion features are obtained by computing the difference between two consecutive frames containing pose features. In the second layer, a Gamma-GWR [56], G^I, is used to learn prototype sequences and associate symbolic labels to unsupervised visual representations of emotions for the purpose of classification. While in the model presented in Sect. 10.3.1, networks were trained with concatenated trajectories of neural activations from a previous network layer, in this case we use the recurrent Gamma-GWR. This is because sequences of bodily expressions comprising emotions require a larger temporal window to be processed and, by explicitly concatenating neural activations from previous layers, the dimensionality of the input increases [60]. Here, instead, the temporal context of the Gamma-GWR is used to efficiently process larger temporal windows and reduce quantization error over time. During the inference phase, unlabeled novel samples are processed by the hierarchical architecture, yielding patterns of neural weight activations. One best-matching neuron in G^I will activate for every 10 processed input frames.

For the evaluation of our system, we collected a dataset named the Body Expressions of Emotion (BEE), with nineteen participants performing six different emotional states: *anger, fear, happiness, neutral, sadness*, and *surprise*. The dataset was acquired in an HRI scenario consisting of a humanoid robot Nao extended with a

Table 10.1 A comparison of overall recognition of emotions between our system and human performance

	System (%)	Human (%)
Accuracy	88.8	90.2
Precision	66.3	70.1
Recall	68	70.7
F-score	66.8	68.9

depth sensor to extract 3D body skeleton information in real time. Nineteen participants took part in the data recordings (fourteen male, five female, age ranging from 21 to 33). The participants were students at the University of Hamburg and they declared not to have suffered any physical injury resulting in motor impairments. To compare the performance of our system to human observers, we performed an additional study in which 15 raters that did not take part in the data collection phase had to label depth map sequences as one of the six possible emotions.

For our approach, we used the full 3D skeleton model except for the *feet*, leading to 13 joints (i.e., 39 input dimensions). To obtain translation invariance, we encoded the joint positions using the center of the hips as the frame of reference. We then computed joint motion as the difference of two consecutive frames for each pose transition. Experimental results showed that our system successfully learned to classify the set of six training emotions and that its performance was very competitive with respect to human observers (see Table 10.1). The overall accuracy of emotions recognized by human observers was 90.2%, whereas our system showed an overall accuracy of 88.8%.

As additional future work, we could investigate the development of a multimodal emotion recognition scenario, i.e., by taking into account auditory information that complements the use of visual cues [4]. The integration of audio-visual stimuli for emotion recognition has been shown to be very challenging but also strongly promising for a more natural HRI experience.

10.4 Body Motion Assessment

10.4.1 Background

The correct execution of well-defined movements plays a key role in physical rehabilitation and sports. While the goal of action recognition approaches is to categorize a set of distinct classes by extrapolating inter-class differences, action assessment requires instead a model to capture intra-class dissimilarities that allow expressing a measurement on how much an action follows its learned template. The quality of actions can be computed in terms of how much a performed movement matches the

Fig. 10.7 Isual feedback for correct squat sequence (top), and a sequence containing *knees in* mistake (bottom; joints and limbs in red) [58]

correct continuation of a learned motion sequence template. Visual representations can then provide useful qualitative feedback to assist the user in the correct performance of the routine and the correction of mistakes (Fig. 10.7). The task of assessing the quality of actions and providing feedback in real time for correcting inaccurate movements represents a challenging visual task.

Artificial systems for the visual assessment of body motion have been previously investigated for applications mainly focused on physical rehabilitation and sports training. For instance, Chan et al. [5] proposed a physical rehabilitation system using a Kinect sensor for young patients with motor disabilities. The idea was to assist the users while performing a set of simple movements necessary to improve their motor proficiency during the rehabilitation period. Although experimental results have shown improved motivation for users using visual hints, only movements involving the arms at constant speed were considered. Furthermore, the system does not provide real-time feedback to enable the user to timely spot and correct mistakes. Similarly, Su et al. [79] proposed the estimation of feedback for Kinect-based rehabilitation exercises by comparing tracked motion with a pre-recorded execution by the same person. The comparison was carried out on sequences using dynamic time warping and fuzzy logic with the Euclidean distance as a similarity measure. The evaluation of the exercises was based on the degree of similarity between the current sequence and a correct sequence. The system provided qualitative feedback on the similarity of body joints and execution speed, but it did not suggest the user how to correct the movement.

10.4.2 Motion Prediction and Correction

In Parisi et al. [54], we proposed a learning architecture that consists of two hierarchically arranged layers with self-organizing networks for human motion assessment in real time (Fig. 10.8). The first layer is composed of two GWR networks, G^P and G^M, that learn a dictionary of posture and motion feature vectors respectively. This hierarchical scheme has the advantage of using a fixed set of learned features to compose more complex patterns in the second layer, where the Gamma-GWR G^I with $K = 1$ is trained with sequences of posture-motion activation patterns from the first layer to learn the spatiotemporal structure of the input.

The underlying idea for assessing the quality of a sequence is to measure how much the current input sequence differs from a learned sequence template. Provided that a trained model G^I represents a training sequence with a satisfactory degree of accuracy, it is then possible to quantitatively compute how much a novel sequence differs from such expected pattern. We defined a function $\mathfrak{f}_\Omega$ that computes the difference of a current input sequence, Ω_t, from its expected input, i.e. the prediction of the next element of the sequence given Ω_{t-1}:

$$\mathfrak{f}_\Omega(t) = \|\Omega_t - \mathfrak{p}(\Omega_{t-1})\|, \tag{10.10}$$

$$\mathfrak{p}(\Omega_{t-1}) = \mathbf{w}_p \text{ with } p = \arg\min_{j \in A} \|\mathbf{c}_j - \Omega_{t-1}\|, \tag{10.11}$$

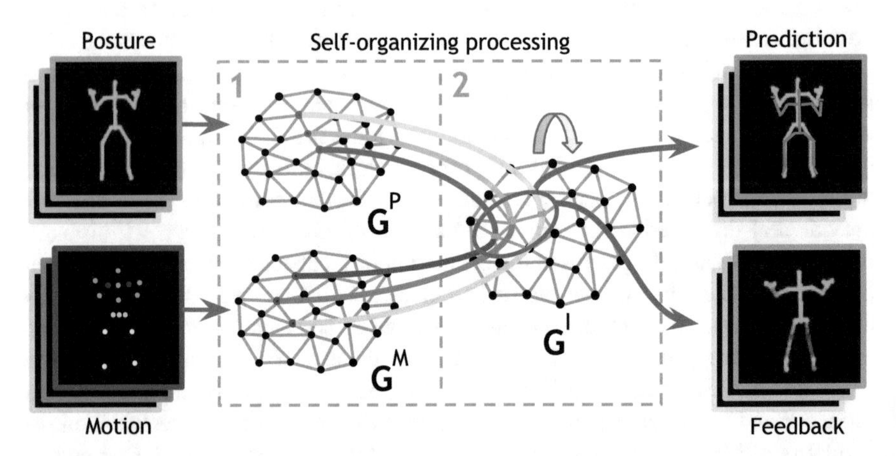

Fig. 10.8 Learning architecture with growing self-organizing networks. In layer 1, two GWR networks learn posture and motion features respectively. In layer 2, a Gamma-GWR learns spatiotemporal dynamics of body motion. This mechanism allows predicting the template continuation of a learned sequence and computing feedback as the difference between its current and its expected execution [54]

Fig. 10.9 Visual hints for the correct execution of the *finger to nose* routine. Progressively fading violet lines indicate the learned action template [58]

where A is the set of neurons and $\|\cdot\|$ denotes the Euclidean distance. Since the weight and context vectors of the prototype neurons lie in the same feature space as the input ($\mathbf{w}_i, \mathbf{c}_i \in \mathbb{R}^{|\Omega|}$), it is possible to provide joint-wise feedback computations. The recursive prediction function $\mathfrak{p}$ can be applied an arbitrary number of timesteps into the future. Therefore, after the training phase is completed, it is possible to compute $\mathfrak{f}_\Omega(t)$ in real time with linear computational complexity $\mathcal{O}(|A|)$.

The visual effect of this prediction mechanism is shown in Fig. 10.9. For this example, the architecture was trained with the *Finger to nose* routine which consists of keeping your arm bent at the elbow and then touching your nose with the tip of your finger. As soon the person starts performing the routine, we can see progressively fading violet lines representing the next 30 time steps which lead to visual assistance for successful execution. The value 30 was empirically determined to provide a substantial reference to future steps while limiting visual clutter. To compute visual feedback, we used the $\mathfrak{p}$ predictions as hints on how to perform a routine over 100 timesteps into the future, and then use $\mathfrak{f}_\Omega(t)$ to spot mistakes on novel sequences that do not follow the expected pattern for individual joint pairs. Execution mistakes are detected if $\mathfrak{f}_\Omega(t)$ exceeds a given threshold $\mathfrak{f}_T$ over i timesteps. Visual representations of these computations can then provide useful qualitative feedback to correct mistakes during the performance of the routine (Fig. 10.7). Our approach learns also motion intensity to better detect temporal discrepancies. Therefore, it is possible to provide accurate feedback on posture transitions and the correct execution of lockouts.

10.4.3 Dataset and Evaluation

We evaluate our approach with a data set containing 3 powerlifting exercises performed by 17 athletes: High bar back squat, Deadlift, and Dumbbell lateral raise. The data collection took place at the Kinesiology Institute of the University of Hamburg, Germany, where 17 volunteering participants (9 male, 8 female) performed 3 different powerlifting exercises. We captured body motion of correct and incorrect executions with a Kinect v2 sensor and estimated body joints using Kinect SDK 2.0 that provides a set of 25 joint coordinates at 30 frames per second. The participants executed the routines frontal to the sensor placed at 1 m from the ground. We extracted the 3D joints for *head, neck, wrists, elbows, shoulders, spine, hips, knees*, and *ankles*, for a total of 13 3D-joints (39 dimensions). We computed motion intensity from posture sequences as the difference between consecutive joint pairs. The Kinect's skeleton model (Fig. 10.7), although not faithful to human anatomy, provides reliable estimations of the joints' position over time when the user is facing the sensor. We manually segmented single repetitions for all exercises. In order to obtain translation invariance, we subtracted the *spine_base* joint (the center of the hips) from all the joints in absolute coordinates.

We evaluated our method for computing feedback with individual and multiple subjects. We divided the correct body motion data with threefold cross-validation into training and test sets and trained the models with data containing correct motion sequences only. For the inference phase, both the correct and incorrect movements were used with feedback threshold $\mathfrak{f}^T = 0.7$ over 100 frames. Our expectation was that the output of the feedback function would be higher for sequences containing mistakes. We observed true positives (TP), false negatives (FN), true negatives (TN), and false positives (FP) as well as the measures true positive rate (TPR or sensitivity), true negative rate (TPR or specificity), and positive predictive value (PPV or precision). Results for single- and multiple-subject data on E1, E2, and E3 routines are displayed in Tables 6.1 and 6.2 respectively, along with a comparison with the best-performing feedback function $\mathfrak{f}_b$ from [58] in which we used only pose frames without explicit motion information.

The evaluation on single subjects showed that the system successfully provides feedback on posture errors with high accuracy. GWR-like networks allow reducing the temporal quantization error over longer timesteps, so that more accurate feedback can be computed and thus reduce the number of false negatives and false positives. Furthermore, since the networks can create new neurons according to the distribution of the input, each network can learn a larger number of possible executions of the same routine, thus being more suitable for training sessions with multiple subjects. Tests with multiple-subject data showed significantly decreased performance, mostly due to a large number of false positives. This is not exactly a flaw due to the learning mechanism but rather a consequence people having different body configurations and, therefore, slightly different ways to perform the same routine. To attenuate this issue, we can set different values for the feedback threshold $\mathfrak{f}_T$. For larger values, the system would tolerate more variance in the performance. However, one must consider whether a higher degree of variance is not desirable in some application domains. For instance, rehabilitation routines may be tailored to a specific subject based on their specific body configuration and health condition (Tables 10.2 and 10.3).

Our results encourage further work in embedding this type of real-time system into an assistive robot that can interact with the user and motivate the correct performance of physical rehabilitation routines and sports training. The positive effects of having a

Table 10.2 Single-subject evaluation

		TP	FN	TN	FP	TPR	TNR	PPV
E1	$\mathfrak{f}_{\mathfrak{b}}$	35	10	33	0	0.77	1	1
	$\mathfrak{f}_{\Omega}$	35	2	41	0	0.97	1	1
E2	$\mathfrak{f}_{\mathfrak{b}}$	24	0	20	0	1	1	1
	$\mathfrak{f}_{\Omega}$	24	0	20	0	1	1	1
E3	$\mathfrak{f}_{\mathfrak{b}}$	63	0	26	0	1	1	1
	$\mathfrak{f}_{\Omega}$	63	0	26	0	1	1	1

Table 10.3 Multi-subject evaluation. Best results in italics

		TP	FN	TN	FP	TPR	TNR	PPV
E1	$\mathfrak{f}_{\mathfrak{b}}$	326	1	7	151	0.99	0.04	0.68
	$\mathfrak{f}_{\Omega}$	328	1	13	143	0.99	*0.08*	*0.70*
E2	$\mathfrak{f}_{\mathfrak{b}}$	127	2	0	121	0.98	0	0.51
	$\mathfrak{f}_{\Omega}$	139	0	0	111	*1*	0	*0.56*
E3	$\mathfrak{f}_{\mathfrak{b}}$	123	0	8	41	1	0.16	0.75
	$\mathfrak{f}_{\Omega}$	126	0	15	31	1	*0.33*	*0.80*

motivational robot for health-related tasks has been shown in a number of studies [9, 30, 44]. The assessment of body motion plays a role not only for the detection of mistakes on training sequences but also in the timely recognition of gait deterioration, e.g., linked to age-related cognitive declines. Growing learning architectures are particularly suitable for this task since they can adapt to the user through longer periods of time while still detecting significant changes in their motor skills.

10.5 Continual Learning of Human Actions

10.5.1 Background

Deep learning models for visual tasks typically comprise a set of convolution and pooling layers trained in a hierarchical fashion for yielding action feature representations with increasing degree of abstraction (see [21] for a recent survey). This processing scheme is in agreement with neurophysiological studies supporting the presence of functional hierarchies with increasingly large spatial and temporal receptive fields along cortical pathways [18, 23] However, the training of deep learning models for action sequences has been proven to be computationally expensive and requires an adequately large number of training samples for the successful learning of spatiotemporal filters. Consequently, the question arises whether traditional deep learning models for action recognition can account for real-world learning scenarios, in which the number of training samples may not be sufficiently high and system may be required to learn from novel input in a continual learning fashion.

Continual learning refers to the ability of a system to continually acquire and fine-tune knowledge and skills over time while preventing *catastrophic forgetting* (see [6, 53] for recent reviews). Empirical evidence shows that connectionists architectures are in general prone to catastrophic forgetting, i.e., when learning a new class or task, the overall performance on previously learned classes and tasks may abruptly decrease due to the novel input interfering with or completely overwriting existing representations [15, 39]. To alleviate catastrophic forgetting in neural networks, researchers have studied how to address the *plasticity-stability dilemma* [20], i.e. how which extent networks should adapt to novel knowledge without forgetting previously learned knowledge. Specifically for self-organizing networks such as the GWR, catastrophic forgetting is modulated by the conditions of map plasticity, the available resources to represent information, and the similarity between new and old knowledge [56, 69]. While the vast majority of the proposed continual learning models are designed for processing i.d.d. data from datasets of static images such as MNIST and CIFAR (e.g. [31, 68, 77, 87]), here I introduce deep self-organization for the continual learning of non-stationary, non-i.d.d. data from videos comprising human actions.

The approaches described in Sects. 10.3 and 10.4 rely on the extraction of a simplified 3D skeleton model from which low-dimensional pose and motion features can be computed to process actor-independent action dynamics. The use of such models is in line with biological evidence demonstrating that human observers are very proficient in learning and recognizing complex motion underlying a skeleton structure [25, 26]. These studies show that the presence of a holistic structure improves the learning speed and accuracy of action patterns, also for non-biologically relevant motion such as artificial complex motion patterns. However, skeleton models are susceptible to sensor noise and situations of partial occlusion and self-occlusion (e.g. caused by body rotation). In this section, I describe how self-organizing architectures can be extended to learning and recognize actions in a continual learning fashion from raw RGB image questions.

10.5.2 Deep Neural Network Self-Organization

In Parisi et al. [56], we proposed a self-organizing architecture consisting of a series of hierarchically arranged growing networks for the continual learning of actions from high-dimensional input streams (Fig. 10.10). Each layer in the hierarchy comprises a Gamma-GWR and a pooling mechanism for learning action features with increasingly large spatiotemporal receptive fields. In the last layer, neural activation patterns from distinct pathways are integrated. The proposed deep architecture is composed of two distinct processing streams for pose and motion features, and their subsequent integration in the STS layer. Neurons in the G^{STS} network are activated by the latest $K + 1$ input samples, i.e. from time t to $t - K$.

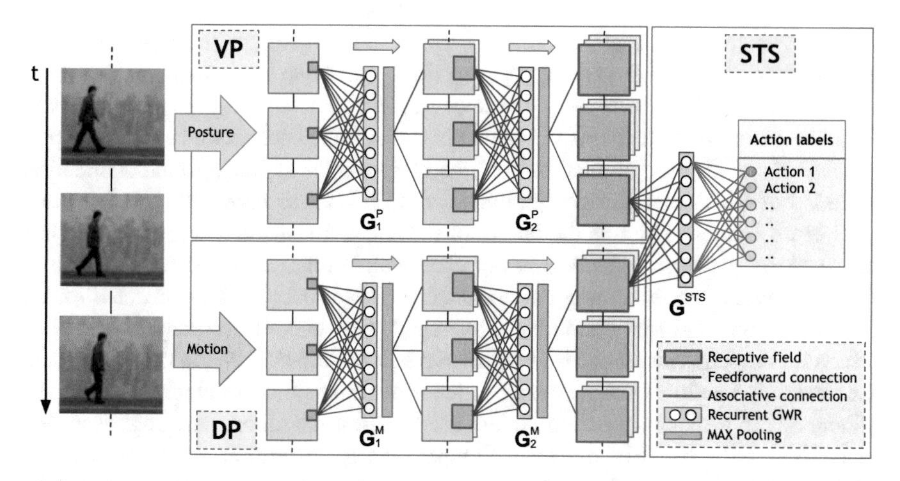

Fig. 10.10 Diagram of our deep neural architecture with Gamma-GWR networks for continual action recognition. Posture and motion action cues are processed separately in the ventral (VP) and the dorsal pathway (DP) respectively. At the STS stage, the recurrent GWR network learns associative connections between prototype action representations and symbolic labels [56]

Deep architectures obtain invariant responses by alternating layers of feature detectors and nonlinear pooling neurons using, e.g., the maximum (MAX) operation, which has been shown to achieve higher feature specificity and more robust invariance with respect to linear summation [21]. Robust invariance to translation has been obtained via MAX and average pooling, with the MAX operator showing faster convergence and improved generalization [72]. In our architecture, we implemented MAX-pooling layers after each Gamma-GWR network (see Fig. 10.10). For each input image patch, a best-matching neuron $\mathbf{w}_b^{(n-1)} \in \mathbb{R}^m$ is be computed in layer $n-1$ and only its maximum weight value $\widetilde{\mathbf{w}}^{(n)} \in \mathbb{R}$ will be forwarded to the next layer n:

$$\widetilde{\mathbf{w}}^{(n)} = \max_{0 \le i \le m} \mathbf{w}_{b,i}^{(n-1)}, \tag{10.12}$$

where b is computed according to Eq. 10.4 and the superscript on $\widetilde{\mathbf{w}}^{(n)}$ indicates that this value is not an actual neural weight of layer n, but rather a pooled activation value from layer $n-1$ that will be used as input in layer n. Since the spatial receptive field of neurons increases along the hierarchy, this pooling process will yield scale and position invariance.

10.5.3 Datasets and Evaluation

We conducted experimental results with two action benchmarks: the Weizmann [19] and the KTH [74] datasets.

The Weizmann dataset contains 90 low-resolution image sequences with 10 actions performed by 9 subjects. The actions are *walk, run, jump, gallop sideways, bend, one-hand wave, two-hands wave, jump in place, jumping jack*, and *skip*. Sequences are sampled at 180 × 144 pixels with a static background and are about 3 seconds long. We used aligned foreground body shapes by background subtraction included in the dataset. For compatibility with [73], we trimmed all sequences to a total of 28 frames, which is the length of the shortest sequence, and evaluated our approach by performing *leave-one-out* cross-validation, i.e., 8 subjects were used for training and the remaining one for testing. This procedure was repeated for all 9 permutations and the results were averaged. Our overall accuracy was 98.7%, which is competitive with the best reported result of 99.64% [19]. In their approach, they extracted action features over a number of frames by concatenating 2D body silhouettes in a space-time volume and used nearest neighbors and Euclidean distance to classify. Notably, our results outperform the overall accuracy reported by [28] with three different deep learning models: convolutional neural network (CNN, 92.9%), multiple spatiotemporal scales neural network (MSTNN, 95.3%), and 3D CNN (96.2%). However, a direct comparison of the above-described methods with ours is hindered by the fact that they differ in the type of input and number of frames per sequence used during the training and the test phase.

The KTH action dataset contains 25 subjects performing 6 different actions: *walking, jogging, running, boxing, hand-waving* and *hand-clapping*, for a total of 2391 sequences. Action sequences were performed in 4 different scenarios: indoor, outdoor, variations in scale, and changes in clothing. Videos were collected with a spatial resolution of 160 × 120 pixels taken over homogeneous backgrounds and sampled at 25 frames per second. Following the evaluation schemes from the literature, we trained our model with 16 randomly selected subjects and used the other 9 subjects for testing. The overall classification accuracy averaged across 5 trials achieved by our model was 98.7%, which is competitive with the two best reported results: 95.6% [67] and 95.04% [17]. In the former approach, they used a hierarchical CNN model to capture sub-actions from complex ones. Key frames were extracted using binary coding of each frame in a video which helps to improve the performance of the hierarchical model (from 94.1 to 95.6%). In the latter approach, they computed handcrafted interest points with substantial motion, which requires high computational requirements for the estimation of ad-hoc interest points. Our model outperforms other hierarchical models that do not rely on handcrafted features, such as 3D CNN (90.2%, [27]) and 3D CNN in combination with long short-term memory (94.39%, [3]).

10.6 Conclusions and Open Challenges

The underlying neural mechanisms for action perception have been extensively studied, comprising cortical hierarchies for processing body motion cues with increasing complexity of representation [23, 36, 81], i.e. higher-level areas process information accumulated over larger temporal windows with increasing invariance to the position and the scale of stimuli. Consequently, the study of the biological mechanisms for action perception is fundamental for the development of artificial systems aimed to address the robust recognition of actions and learn in a continual fashion in HRI scenarios [52].

Motivated by the process of input-driven self-organization exhibited by topographic maps in the cortex [42, 45, 86], I introduced learning architectures hierarchically arranged growing networks that integrate body posture and motion features for action recognition and assessment. The proposed architectures can be considered a further step towards more flexible neural network models for learning robust visual representations on the basis of visual experience. Successful applications of deep neural network self-organization include human action recognition [10, 59, 60], gesture recognition [49, 51], body motion assessment [54, 58], human-object interaction [40, 41], continual learning [56, 57], and audio-visual integration [55].

Models of hierarchical action learning are typically feedforward. However, neurophysiological studies have shown that the visual cortex exhibits significant feedback connectivity between different cortical areas [13, 70]. In particular, action perception demonstrates strong top-down modulatory influences from attentional mechanisms [82] and higher-level cognitive representations such as biomechanically plausible motion [76]. Spatial attention allows animals and humans to process relevant environmental stimuli while suppressing irrelevant information. Therefore, attention as a modulator in action perception is also desirable from a computational perspective, thereby allowing the suppression of uninteresting parts of the visual scene and thus simplifying the detection and segmentation of human motion in cluttered environments.

The integration of multiple sensory modalities such as vision and audio is crucial for enhancing the perception of actions, especially in situations of uncertainty, with the aim to reliably operate in highly dynamic environments [48]. Experiments in HRI scenarios have shown that the integration of audio-visual cues significantly improves performance with respect to unimodal approaches for sensory-driven robot behavior [7, 8, 61]. The investigation of biological mechanisms of multimodal action perception is an important research direction for the development of learning systems exposed to rich streams of information in real-world scenarios.

Acknowledgements The author would like to thank Pablo Barros, Doreen Jirak, Jun Tani, and Stefan Wermter for great discussions and feedback.

References

1. Aerts, M., Esselink, R., Post, B., van de Warrenburg, B., & Bloem, B. (2012). Improving the diagnostic accuracy in parkinsonism: A three-pronged approach. *Practical Neurology, 12*(1), 77–87.
2. Alonso-Martin, F., Malfaz, M., Sequeira, J., Gorostiza, J. F., & Salichs, M. A. (2013). A multimodal emotion detection system during human-robot interaction. *Sensors, 13*(11), 15549–15581.

3. Baccouche, M., Mamalet, F., Wolf, C., Garcia, C., & Baskurt, A. (2011). Sequential deep learning for human action recognition. In *Human Behavior Understanding (HBU): Second International Workshop* (pp. 29–39). Springer Berlin Heidelberg.
4. Barros, P., & Wermter, S. (2016). Developing crossmodal expression recognition based on a deep neural model. *Adaptive Behavior*, *24*(5), 373–396.
5. Chang, Y.-J., Chen, S.-F., & Huang, J.-D. (2011). A Kinect-based system for physical rehabilitation: A pilot study for young adults with motor disabilities. *Research in Developmental Disabilities*, *32*(6), 2566–2570. ISSN 08914222. https://doi.org/10.1016/j.ridd.2011.07.002.
6. Chen, Z., & Liu, B. (2018). Lifelong machine learning. *Synthesis Lectures on Artificial Intelligence and Machine Learning*, *12*(3), 1–207.
7. Cruz, F., Parisi, G., Twiefel, J., & Wermter, S. (2016). Multi-modal integration of dynamic audiovisual patterns for an interactive reinforcement learning scenario. In *Proceedings of the IEEE/RSJ International Conference on Intelligent Robots and Systems (IROS)* (pp. 759–766).
8. Cruz, F., Parisi, G., Twiefel, J., & Wermter, S. (2018). Multi-modal integration of dynamic audiovisual patterns for an interactive reinforcement learning scenario. In *Proceedings of the IEEE/RSJ International Conference on Intelligent Robots and Systems (IROS)* (pp. 759–766).
9. Dautenhahn, K. (1999). Robots as social actors: Aurora and the case of autism. In *Third Cognitive Technology Conference*.
10. Elfaramawy, N., Barros, P., Parisi, G. I., & Wermter, S. Emotion recognition from body expressions with a neural network architecture. In *Proceedings of the International Conference on Human Agent Interaction (HAI'17)* (pp. 143–149). Bielefeld, Germany.
11. Eriksson, P. S., Perfilieva, E., Bjork-Eriksson, T., Alborn, A.-M., Nordborg, C., Peterson, D. A., & Gage, F. H. (1998). Neurogenesis in the adult human hippocampus. *Nature Medicine*, *4*(11), 1313–1317. ISSN 1078-8956. https://doi.org/10.1038/3305.
12. Faria, D. R., Premebida, C., & Nunes, U. (2014). A probabilistic approach for human everyday activities recognition using body motion from RGB-D images. In *Proceedings of the IEEE International Symposium on Robot and Human Interactive Communication (RO-MAN)* (pp. 842–849).
13. Felleman, D., & Van Essen, D. (1991). Distributed hierarchical processing in the primate cerebral cortex. *Cerebral Cortex*, *1*(1), 1–47.
14. Fonlupt, P. (2003). Perception and judgement of physical causality involve different brain structures. *Cognitive Brain Research*, *17*(2), 248–254. ISSN 0926-6410. https://doi.org/10.1016/S0926-6410(03)00112-5.
15. French, R. M. (1999). Catastrophic forgetting in connectionist networks. *Trends in Cognitive Sciences*, *3*(4), 128–135.
16. Fritzke, B. (1995). A growing neural gas network learns topologies. In *Advances in neural information processing systems* (Vol. 7, pp. 625–632). MIT Press.
17. Gao, Z., Chen, M.-Y., Hauptmann, A. G., & Cai, A. (2010). *Comparing Evaluation Protocols on the KTH Dataset* (pp. 88–100). Springer, Berlin, Heidelberg.
18. Giese, M. A., & Poggio, T. (2003 March). Neural mechanisms for the recognition of biological movements. *Nature Reviews Neuroscience*, *4*(3), 179–192. https://doi.org/10.1038/nrn1057.
19. Gorelick, L., Blank, M., Shechtman, E., Irani, M., & Basri, R. (2005). Actions as space-time shapes. In *Proceedings of the International Conference on Computer Vision (ICCV)* (pp. 1395–1402).
20. Grossberg, S. (1980). How does a brain build a cognitive code? *Psychological Review*, *87*, 1–51.
21. Guo, Y., Liu, Y., Oerlemans, A., Lao, S., Wu, S., & Lew, M. S. (2016). Deep learning for visual understanding: A review. *Neurocomputing*, *187*, 27–48.
22. Han, J., Shao, L., Xu, D., & Shotton, J. (2013). Enhanced computer vision with Microsoft Kinect sensor. *IEEE Transactions on Cybernetics*, *43*(5), 1318–1334.
23. Hasson, U., Yang, E., Vallines, I., Heeger, D. J., & Rubin, N. (2008). A hierarchy of temporal receptive windows in human cortex. *The Journal of Neuroscience*, *28*(10), 2539–2550. ISSN 1529-2401.

24. Hebb, D. O. (1949). *The organization of behavior: A neuropsychological theory*. New York: Wiley.
25. Hiris, E. (2007). Detection of biological and nonbiological motion. *Journal of Vision*, *7*(12), 1–16.
26. Jastorff, J., Kourtzi, Z., & Giese, M. A. (2006). Learning to discriminate complex movements: Biological versus artificial trajectories. *Journal of Vision*, *6*(8), 791–804.
27. Ji, S., Xu, W., Yang, M., & Yu, K. (2013). 3d convolutional neural networks for human action recognition. *IEEE Transactions on Pattern Analysis and Machine Intelligence*, *35*(1), 221–231.
28. Jung, M., Hwang, J., & Tani, J. (2015). Self-organization of spatio-temporal hierarchy via learning of dynamic visual image patterns on action sequences. *PLoS ONE*, *10*(7), e0131214, 07.
29. Kachouie, R., Sedighadeli, S., Khosla, R., & Chu, M. (2014). Socially assistive robots in elderly care: A mixed-method systematic literature review. *The International Journal of Human-Computer Interaction*, *30*(5), 369–393. https://doi.org/10.1080/10447318.2013.873278.
30. Kidd, C. D., & Breazeal, C. (2007). A robotic weight loss coach. In *Proceedings of the AAAI Conference on Artificial Intelligence* (pp. 1985–1986).
31. Kirkpatrick, J., Pascanu, R., Rabinowitz, N., Veness, J., Desjardins, G., Rusu, A. A., et al. (2017). Overcoming catastrophic forgetting in neural networks. In *Proceedings of the National Academy of Sciences*.
32. Knoblauch, A. (2017). Impact of structural plasticity on memory formation and decline. In A. van Ooyen & M. Butz (Eds.), *Rewiring the Brain: A Computational Approach to Structural Plasticity in the Adult Brain*. Elsevier, Academic Press.
33. Kohonen, T. (1991). Self-organizing maps: Optimization approaches. *Artificial Neural Networks*, *II*, 981–990.
34. Krizhevsky, A. (2009). *Learning multiple layers of features from tiny images*. Master's thesis, University of Toronto.
35. LeCun, Y., Bottou, L., Bengio, Y., & Haffner, P. (1998). *Gradient-based learning applied to document recognition*. In *Proceedings of the IEEE*.
36. Lerner, Y., Honey, C. J., Silbert, L. J., & Hasson, U. (2011). Topographic mapping of a hierarchy of temporal receptive windows using a narrated story. *The Journal of Neuroscience*, *31*(8), 2906–2915. https://doi.org/10.1523/jneurosci.3684-10.2011.
37. Marsland, S., Shapiro, J., & Nehmzow, U. (2002). A self-organising network that grows when required. *Neural Networks*, *15*(8–9), 1041–1058.
38. Mermillod, M., Bugaiska, A., & Bonin, P. (2013a). The stability-plasticity dilemma: Investigating the continuum from catastrophic forgetting to age-limited learning effects. *Frontiers in Psychology*, *4*(504).
39. Mermillod, M., Bugaiska, A., & Bonin, P. (2013). The stability-plasticity dilemma: Investigating the continuum from catastrophic forgetting to age-limited learning effects. *Frontiers in Psychology*, *4*, 504. ISSN 1664-1078.
40. Mici, L., Parisi, G. I., & Wermter, S. (2017). An incremental self-organizing architecture for sensorimotor learning and prediction. arXiv:1712.08521.
41. Mici, L., Parisi, G. I., & Wermter, S. (2018). A self-organizing neural network architecture for learning human-object interactions. *Neurocomputing*, *307*, 14–24.
42. Miikkulainen, R., Bednar, J. A., Choe, Y., & Sirosh, J. (2005). *Computational maps in the visual cortex*. Springer. ISBN 978-0-387-22024-6. https://doi.org/10.1007/0-387-28806-6.
43. Ming, G.-L., & Song, H. (2011). Adult neurogenesis in the mammalian brain: Significant answers and significant questions. *Neuron*, *70*(4), 687–702. https://doi.org/10.1016/j.neuron.2011.05.001. http://dx.doi.org/10.1038/nrn2147.
44. Nalin, M., Baroni, I., Sanna, A., & Pozzi, C. (2012). Robotic companion for diabetic children: Emotional and educational support to diabetic children, through an interactive robot. In *ACM SIGCHI* (pp. 260–263).
45. Nelson, C. A. (2000). Neural plasticity and human development: The role of early experience in sculpting memory systems. *Developmental Science*, *3*(2), 115–136.

46. Nwe, T. L., Foo, S. W., & Silva, L. C. D. (2003). Speech emotion recognition using hidden Markov models. *Speech Communication, 41*(4), 603–623.
47. Orban, G., Lagae, L., Verri, A., Raiguel, S., Xiao D., Maes, H., & Torre, V. (1982). First-order analysis of optical flow in monkey brain. *Proceedings of the National Academy of Sciences, 89*(7), 2595–2599.
48. Parisi, G. I., Barros, P., Fu, D., Magg, S., Wu, H., Liu, X., & Wermter, S. (2018). A neurorobotic experiment for crossmodal conflict resolution in complex environments. arXiv:1802.10408.
49. Parisi, G. I., Barros, P., & Wermter, S. (2014). FINGeR: Framework for interactive neural-based gesture recognition. In *Computational Intelligence and Machine Learning (ESANN), Bruges, Belgium: Proceedings of the European Symposium on Artificial Neural Networks* (pp. 443–447).
50. Parisi, G. I., Ji, X., & Wermter, S. (2018). On the role of neurogenesis in overcoming catastrophic forgetting. In *NIPS'18, Workshop on Continual Learning*, Montreal, Canada.
51. Parisi, G. I., Jirak, D., & Wermter, S. (2014). HandSOM—Neural clustering of hand motion for gesture recognition in real time. In *Proceedings of the IEEE International Symposium on Robot and Human Interactive Communication (RO-MAN)* (pp. 981–986). Edinburgh, Scotland, UK.
52. Parisi, G. I., & Kanan, C. (2019). Rethinking continual learning for autonomous agents and robots. arXiv:1907.01929.
53. Parisi, G. I., Kemker, R., Part, J. L., Kanan, C., & Wermter, S. (2019). Continual lifelong learning with neural networks: A review. *Neural Networks, 113*, 54–71.
54. Parisi, G. I., Magg, S., & Wermter, S. (2016a). Human motion assessment in real time using recurrent self-organization. In *Proceedings of the IEEE International Symposium on Robot and Human Interactive Communication (RO-MAN)* (pp. 71–76).
55. Parisi, G. I., Tani, J., Weber, C., & Wermter, S. (2016). Emergence of multimodal action representations from neural network self-organization. *Cognitive Systems Research*.
56. Parisi, G. I., Tani, J., Weber, C., & Wermter, S. (2017). Lifelong learning of humans actions with deep neural network self-organization. *Neural Networks, 96*, 137–149.
57. Parisi, G. I., Tani, J., Weber, C., & Wermter, S. (2018). Lifelong learning of spatiotemporal representations with dual-memory recurrent self-organization. arXiv:1805.10966.
58. Parisi, G. I., von Stosch, F., Magg, S., & Wermter, S. (2015). Learning human motion feedback with neural self-organization. In *Proceedings of International Joint Conference on Neural Networks (IJCNN)* (pp. 2973–2978).
59. Parisi, G. I., Weber, C., & Wermter, S. (2014). Human action recognition with hierarchical growing neural gas learning. In *Proceedings of the International Conference on Artificial Neural Networks (ICANN)* (pp. 89–96).
60. Parisi, G. I., Weber, C., & Wermter, S. (2015b). Self-organizing neural integration of pose-motion features for human action recognition. *Frontiers in Neurorobotics, 9*(3).
61. Parisi, G. I., Weber, C., & Wermter, S. (2016). A neurocognitive robot assistant for robust event detection. *Trends in ambient intelligent systems: Role of computational intelligence. Studies in computational intelligence* (pp. 1–28). Springer.
62. Patwardhan, A., & Knapp, G. (2016). Multimodal affect recognition using kinect. arXiv:1607.02652.
63. Piana, S., Stagliano, A., Odone, F., Verri, A., & Camurri, A. (2014). Real-time automatic emotion recognition from body gestures. arXiv:1402.5047.
64. Picard, R. W. (1997). *Affective computing*. Cambridge, MA, USA: MIT Press.
65. Pollick, F. E., Paterson, H. M., Bruderlin, A., & Sanford, A. J. (2001). Perceiving affect from arm movement. *Cognition, 82*(2), B51–B61.
66. Poppe, R. (2010). A survey on vision-based human action recognition. *Image and Vision Computing, 28*, 976–990.
67. Ravanbakhsh, M., Mousavi, H., Rastegari, M., Murino, V., & Davis, L. S. (2015). Action recognition with image based cnn features. arXiv:1512.03980.
68. Rebuffi, S., Kolesnikov, A., Sperl, G., & Lampert, C. H. (2017 July). Icarl: Incremental classifier and representation learning. In *2017 IEEE Conference on Computer Vision and Pattern Recognition (CVPR)* (pp. 5533–5542).

69. Richardson, F. M., & Thomas, M. S. (2008). Critical periods and catastrophic interference effects in the development of self-organizing feature maps. *Developmental Science*, *11*(3), 371–389.
70. Salin, P., & Bullier, J. (1995). Corticocortical connections in the visual system: Structure and function. *Physiological Reviews*, *75*(1), 107–154.
71. Sawada, M., Suda, K., & Ishii, M. (2003). Expression of emotions in dance: Relation between arm movement characteristics and emotion. *Perceptual and Motor Skills*, *97*(3), 697–708.
72. Scherer, D., Müller, A., & Behnke, S. (2010). Evaluation of pooling operations in convolutional architectures for object recognition. In *Proceedings of the International Conference on Artificial Neural Networks (ICANN)* (pp. 92–101). Berlin, Heidelberg: Springer. ISBN 3-642-15824-2, 978-3-642-15824-7.
73. Schindler, K., & Van Gool, L. J. (2008). Action snippets: How many frames does human action recognition require? In *Proceedings of the Conference on Computer Vision and Pattern Recognition (CVPR)*. IEEE Computer Society.
74. Schuldt, C., Laptev, I., & Caputo, B. (2004). Recognizing human actions: A local SVM approach. In *Proceedings of the International Conference on the Pattern Recognition (ICPR)* (pp. 2–36). Washington, DC, USA: IEEE Computer Society.
75. Shan, J., & Akella, S. (2014). 3D human action segmentation and recognition using pose kinetic energy. In *Workshop on advanced robotics and its social impacts (IEEE)*, pp. 69–75.
76. Shiffrar, M., & Freyd, J. J. (1990). Apparent motion of the human body. *Psychological Science*, *1*, 257–264.
77. Shin, H., Lee, J. K., Kim, J., & Kim, J. (2017). Continual learning with deep generative replay. In *Advances in neural information processing systems* (pp. 2990–2999).
78. Strickert, M., & Hammer, B. (2005). Merge SOM for temporal data. *Neurocomputing*, *64*, https://doi.org/10.1016/j.neucom.2004.11.014.
79. Su, C.-J. (2013). Personal rehabilitation exercise assistant with Kinect and dynamic time warping. *International Journal of Information and Education Technology*, *3*(4), 448–454. https://doi.org/10.7763/IJIET.2013.V3.316.
80. Sung, J., Ponce, C., Selman, B., & Saxena, A. (2012). Unstructured human activity detection from RGBD images. In *Proceedings of the International Conference on Robotics and Automation (ICRA)* (pp. 842–849).
81. Taylor, P., Hobbs, J. N., Burroni, J., & Siegelmann, H. T. (2015). The global landscape of cognition: Hierarchical aggregation as an organizational principle of human cortical networks and functions. *Scientific Reports*, *5*(18112).
82. Thornton, I. M., Rensink, R. A., & Shiffrar, M. (2002). Active versus passive processing of biological motion. *Perception*, *31*, 837–853.
83. Ungerleider, L., & Mishkin, M. (1982). Two cortical visual systems. *Analysis of visual behavior* (pp. 549–586). Cambridge: MIT press.
84. Velloso, E., Bulling, A., Gellersen, G., Ugulino, W., & Fuks, G. (2013). Qualitative activity recognition of weight lifting exercises. In *Augmented Human International Conference (ACM)* (pp. 116–123).
85. Vettier, B., & Garbay, C. (2014). Abductive agents for human activity monitoring. *International Journal on Artificial Intelligence Tools*, *23*.
86. Willshaw, D. J., & von der Malsburg, C. (1976). How patterned neural connections can be set up by self-organization. *Proceedings of the Royal Society of London B: Biological Sciences*, *194*(1117), 431–445.
87. Zenke, F., Poole, B., & Ganguli, S. (2017 Aug 06–11). Continual learning through synaptic intelligence. In *Proceedings of the 34th International Conference on Machine Learning*, volume 70 of *Proceedings of Machine Learning Research (PMLR)* (pp. 3987–3995). International Convention Centre, Sydney, Australia.

Chapter 11
Movement Expressivity Analysis: From Theory to Computation

Giovanna Varni and Maurizio Mancini

Abstract Movement expressivity, on which we focus in this chapter, has been widely studied and described by psychologists, sociologists, and neuroscientists. More recently, movement expressivity is receiving increasing attention also by computer scientists with the aim to develop machines with social and emotional intelligence. This chapter, after providing a definition of expressive movement, describes qualitative and quantitative methods, frameworks and algorithms for movement expressivity analysis.

11.1 Introduction

Movement Expressivity is the whole-body motor component of emotional episodes [36]. It is sometimes described as the *unintentional* action component of emotion expression, as argued by the leading models defining emotion intentional and unintentional action components (e.g., [21, 34, 35, 85, 86]). Movement expressivity can also be defined as the dynamic *movement* component in affect perception, in contrast to the static *form* component [49]. Pioneering studies on movement expressivity focused on identifying body movement patterns and postures associated to emotions (e.g., [24, 98]). More recently, movement expressivity was investigated in non-emblematic movements, that is, daily actions, such as walking or knocking at a door, performed with different emotions [43, 77]. To demonstrate that movement expressivity plays a central role in emotion communication, De Gelder [23] describes several intuitive examples: "an angry face is more menacing when accompanied by a fist, and a fearful face more worrisome when the person is in flight (that is, running away)"; "when a frightening event occurs, there might not be time to look for the

G. Varni (✉)
LTCI, Télécom Paris, Institut polytechnique de Paris, Paris, France
e-mail: giovanna.varni@telecom-paris.fr

M. Mancini
School of Computer Science and Information Technology,
University College Cork, Cork, Ireland
e-mail: m.mancini@cs.ucc.ie

N. Noceti et al. (eds.), *Modelling Human Motion*,
https://doi.org/10.1007/978-3-030-46732-6_11

fearful contortions in an individual's face, but a quick glance at the body may tell us all we need to know".

In the last two decades, due to the decreasing cost and increasing reliability of sensing technology, movement expressivity has started to be investigated also by computer scientists. More specifically, their research aimed to develop machines able to understand and express a certain degree of Social and Emotional Intelligence [7]. In humans, it corresponds to the capacity of being aware during interaction of one's own and others' feelings and integrate that information to communicate and interact in an individualised way with the others. This has resulted, for example, in the development of systems for automatically recognising bodily expression of emotions (e.g., [18, 76, 83]), computational models for the design of affective artificial agents, such embodied conversational agents and robots (e.g., [17, 37, 50, 51, 55, 60]), analytic techniques for measuring the interpersonal emotion dynamics expressed through the body during interactions (e.g., [94, 101]).

The goal of this chapter is to provide an overview of the literature on movement expressivity analysis. The chapter begins by defining movement expressivity and movement expressivty analysis from a psychological and sociological point of view (Sects. 11.2 and 11.3). Section 11.4 moves to a survey to computational approaches of movement expressivity: first, a review of devices allowing machines to "sense" human movement is presented in Sect. 11.4.1; then, data-sets of human expressive movements collected through these devices follow in Sect. 11.4.2; finally, computational frameworks adopted to address movement expressivity analysis are presented in Sect. 11.4.3. Algorithms for the extraction of expressive movement features are described in Sect. 11.5. We conclude the paper by briefly discussing how such models could be integrated into robots in Sect. 11.6.

11.2 From Gesture to Expressive Movement

Gesture constitutes a relevant source of information in human-human communication. Indeed, it encodes not only a denotative meaning (that is, "what" is communicated), but also an expressive information concerning the manner (that is, "how" meaning is communicated) through the gesture's execution. Several complementary definitions of gesture were proposed by Psychology in the past. Although Kendon [48] first claimed that "for an action to be treated as a gesture it must have features which make it stand out as such", the most traditional definitions of gesture stem from the studies performed on the alignment between speech and gesture during interactions [61]. Here, gesture is exclusively approached as a support to speech and intended only for communicating denotative meanings. McNeill [61] mainly focused on arms and hand gestures, distinguishes four classes: *iconic*, *metaphoric*, *deictic*, and *beats*.

Argyle [3] and Ekman and Friesen [31] stressed the fact that gesture plays a more relevant role in communication than supporting speech only: humans, indeed, continuously communicate meanings, feeling and emotions through their body movements, even when words are not used. Therefore, a broader definition of gesture taking into account all the facets of nonverbal communication was needed. At the present, the most adopted definition of gesture is the one provided by Kurtenbach [53], stating that a gesture is "a movement of the body that contains information". Focusing on "how" information is conveyed trough gesture, Allport and Vernon [1] first defined expressive movement as "individual differences in the manner of performing adaptive acts, considered as dependent less upon external and temporary conditions than upon enduring qualities of personality". Although this definition dates back to the 1930s, a new classification of gesture taking into account expressive information was provided only many years later by Buck [8]. He focused on bodily emotion communication, arguing the existence of two components: the *propositional* one and the *non-propositional* one. Propositional movements have a denotative meaning and they include also specific movements corresponding to emotion stereotypes (e.g., a clenched fist to show anger); non-propositional ones do not have denotative meaning and refer more to the quality of movement, like, for example, lightness or heaviness [11].

11.3 Qualitative Analysis of Movement Expressivity

The first studies on movement expressivity rooted in Psyhchology and Sociology. Movement expressivity was studied by detailing the movement features associated with emotion expression or by focusing on the specific variations that make any movement produced with the emotion recognizable.

Wallbott [98] directly illustrated the concept of expressive movement linking movement features such as speed, amplitude and fluidity with hot anger. He measured displacement of hand in psychiatric patients behavior and found four main movement characteristics: space, which describes the extension of movement; hastiness, which is related to speed and acceleration; intensity, which describes the energy of a movement; fluency-course, which is related to the flow between the end of a movement and the beginning of the following one. Similarly, Boone and Cunningham [6], taking inspiration by the research of De Meijer [25], found out that children are able to recognize anger, fear, grief and happiness looking at the following six expressive movement features only: the "frequency of upward arm movement, the duration of time arms were kept close to the body, the amount of muscle tension, the duration of time an individual leaned forward, the number of directional changes in face and torso, and the number of tempo changes an individual made in a given action sequence". Gross et al. [43], Montepare et al. [62], and Pollick et al. [77] conducted studies focusing on single non-emblematic movements such as knocking, drinking and walking. Neff and colleagues, with the aim to endow animated characters with expressive nonverbal capabilities and by reviewing arts and literature, such as theater and dance, found that body and movement characteristics such as balance, body silhouette (contour of the body), position of torso and shoulder influence the way in which people perceive others [63–65]. Other studies prove that persons using open-handed movement are perceived positively [74], and that the pace of nodding is a cue of (presence or lack of) patience. As reported by Noroozi et al. [70], it is important to note that, generally, to correctly interpret body expression of emotion, several parts of body must be taken into account at the same time.

The most ancient methods to address movement expressivity rely on the definition of *coding schemes* (e.g., [20, 24, 98]). They were the first enabling a discrimination among emotions through movement features. However, the existence of many different coding schemes made hard to build a unique comprehensive systematic description of movement expressivity. Another large class of methods is the *subtractive* one. Here, information is progressively subtracted from a stimulus in order to minimize or totally remove bias due to gesture shape and to identify features that are mainly involved in expressivity communication. The most famous approach grounds on the *Point Light Display* stimulus technique. Such technique was originally conceived by Johansson [46] to study the human sensitivity to biological movements (i.e., movements made by a biological organism). Through these movements, humans can identify and understand actions related to empathy and other's intentions. In the tradition of the work by Dittrich et al. [27] and Walk and Homan [96] found out that humans can perceive emotions in dance performances from point light displays movement only. Pollick et al. [77], grounding on the same methodology, studied the expression of emotion in everyday actions such as knocking at the door and drinking. He combined this approach with another more quantitative one based on correlation analysis. He found out significant correlation between movement features such as speed and the arousal dimension of emotion (i.e., activation), according to the circumplex model of affect [82].

Other qualitative methods took inspiration from performative arts, such as dance. As noticed by Chi et al. [19], these approaches rely on providing movement features related to the gesture shape and execution, more suitable to capture movement naturalness than those ones detected by adopting the psychological notion of gesture only. The most adopted movement system to formalize these movement features, and therefore expressivity, is the Laban Movement Analysis (LMA), developed by the choreographer Laban [54]. Through this analysis it is possible describing every type of movement performed in a variety of tasks. LMA provides models for interpreting movement, its functions and its expressions through the following 4 components: *Body* (what part of body is moved), *Effort* (how this part is moved), *Space* (where the movement is directed) and *Shape* (illustrating the relation of the movement with the surrounding environment). Following research focused on the Effort and Shape components (e.g., [24, 43, 56]). Dell et al. [26] proposes the Effort-Shape Analysis, assuming that conscious/unconscious personal inner attitudes towards efforts are observable during movement. Effort has 4 factors thought as a continuum with two opposite ends, and they enable a description of how exertion occurs in movement. The factors are:

- *space*: it refers to the direction of movement; it can be direct, with an identifiable target or indirect (flexible), without a clear target;

- *weight*: it describes how much strength is exerted by a movement; it can be light or heavy, with low or high impact;
- *time*: it is related to the "urgency" and the impulsiveness of the movement; it can be sudden and unpredictable suddenly or sustained and prepared;
- *flow*: it expresses how much energy is used to control (bound) the movement or is expressed through movement (free).

By combining the Laban's Effort factors Space, Time and Weight, it is possible to describe a large variety of movements and activities called *Basic Effort Actions*: for example, *punching* is a direct, strong and sudden action, while *floating* is an indirect, light and sustained one. Smooth movements, as reported in [66], are direct, light, sustained and bound actions. Shape describes how movement varies its shape according to (i) the distance from the body center, (ii) its path, (iii) the relation of the body with the surrounding environment.

11.4 Quantitative Movement Expressivity

11.4.1 How Computers Can "Sense" Human Movement

Human movement can be "sensed" by machines to extract features related to expressivity and use them to infer emotions by exploiting a variety of devices: cameras, range imaging devices, motion capture systems, and inertial sensors. A machine can be endowed with a single type of device (e.g., cameras) or with several ones (e.g., cameras and inertial sensors). In the latter case, a "multi-modal" movement analysis is possible through data fusion. Usually, this process guarantees to achieve a more reliable analysis.

Cameras are probably the first devices that have been exploited in human movement expressivity analysis. Computer vision is the area of Computer Science dealing with how machines can have a high-level comprehension of images and video data. A classical computer vision algorithm is *Optical Flow*, that computes the amount and direction of movement in a sequence of video frames. Traditional computer vision algorithms were generally used to extract full-body features, such as movement kinetic energy or the trajectory of the center of mass of a blob (Binary Large Object). The extraction of finer-grained features was made difficult due to the small size of the video frames and the low frame rate of the cameras (e.g., 25 fps). Moreover, the extraction of features from multiple people was particularly challenging due to the risk of occlusions.

In the last few years, approaches to video-based movement analysis grounded on deep learning methods, such as *Convolutional Neural Networks*, were developed. For example, [13, 14] designed an approach to efficiently detect the body configuration of multiple people in images and videos. For each person the algorithm jointly detects body, hands, face, and feet.

Motion capture systems exploit different technologies for "measuring an object's position and orientation in a physical space, then recording that information in a computer-usable form. Objects of interest include human and non-human bodies, facial expressions, camera and light positions, and other element in a scene" [30]. Motion capture idea is based on the evidence that the human visual system is able to perceive biological movement focusing only on a limited numbers of moving points (see the work of Johansson [46] described in Sect. 11.3). Motion capture systems can be classified in two classes: optical systems and non-optical systems. The former ones use data captured from cameras calibrated to provide overlapping fields of view and triangulate the 3D position of bodies on which special markers are attached. If the markers are LED emitting their own light, the system is called *active*, otherwise if the markers just reflect the light emitted by cameras, the system is called *passive*. In both the cases, the capture area can be increased by adding more cameras. At the present, passive systems are the most used ones. The latter class of motion capture systems includes inertial-magnetic systems, magnetic systems, and mechanical systems. In this last years, inertial-magnetic systems are increasingly used, as they are highly portability and do not require a time-consuming calibration process. However, they can be used only to detect and track human bodies. They are based on a set of small and light inertial sensors and magnetometers that are attached on the person's body and adopt biomechanical models and sensors fusion algorithms to provide full-body reconstruction.

A large variety of devices, exploting the same principles of inertial-magnetic sensors, can "sense" movement-related features. For example, inertial sensors can detect human body joints' velocity. While this information alone (i.e., not used in conjunction with magnetometers) can not be exploited to extract full-body posture, it can provide enough data to reliably detect movement features, such as kinetic energy.

Recently, a new generation of devices, called range imaging devices, has been also developed. Range imaging is the name for a collection of methods to produce a 2D image also embedding the distance between the points and the camera. The most known range imaging devices use the Structured Light and the Time-of-Flight methods, sometimes in combination with machine learning techniques to detect and track body joints (e.g., [90]). These devices can be considered as a trade-off between the ease of use, affordability and poor precision of computer vision approaches and the higher precision, but also higher cost and difficulty of use, of motion capture systems.

11.4.2 Expressive Movement Data-Sets

The increasing interest on movement expressivity, resulted in the collection of a plethora of data-sets. This is due to the fact that a single data-set could not be expected to address all the open questions in this research field. Following the path of Douglas-Cowie et al. [28], in this section we detail some data-sets focusing on: (i) the type and the number of participants involved (actors or not), (ii) the type of recorded data (audiovisual, motion capture and so on), (iii) the number of emotions they focused on, (iv) the way bodily expressive responses emerges (spontaneously, acted, or as a result of an induction or conditioning method). For the sake of clarity, we present the corpora according to the number of involved participants: only one or more than one.

11.4.2.1 Single Person Data-Sets

The pioneering studies on movement expressivity relied on bodily emotional expressions portrayed by professional actors or dancers acting emotions alone, e.g., [11, 47, 98]. The FP6-IST Network of Excellence HUMAINE delivered some interesting data-sets, as, for example, the one described in [18], involving 10 persons acting 8 emotions equally distributed in the valence-arousal space. The data-set consists of 240 video excerpts. Other data-sets addressed movement expressivity focusing on non-emblematic actions, such as daily actions. HUMAINE also supported the creation of the GEMEP (GEneva Multimodal Emotion Portrayals) data-set. It consists of 145 audio-video recordings of 10 professional actors portraying 15 affective states under the direction of a professional stage director [4].

Ma et al. [58] collected via motion capture 4080 movements from 30 non-professional actors performing actions like walking, knocking, lifting, throwing, and their combinations with different emotional intents (angry, happy, neutral, and sad), elicited through a scenario-based induction approach. Similarly, Emilya [33] is a synchronized multimodal data-set (audio, video and motion capture data) containing data of 11 non professional actors guided by a professional stage director to express 8 emotions during 7 movement tasks. The emotions, elicited through a scenario-based approach, were selected to cover the valence-arousal dimensions. A broader range of daily actions than [58] was used: walking, sitting down, knocking at the door, lifting and throwing objects with one hand, and moving objects on a table with two hands.

The EU-Emotion Stimulus Set is a collection of 418 dynamic multimodal (facial expressions, vocal expressions, body gestures) emotion and mental state expressions [71]. Twenty emotions and mental states plus a neutral state were portrayed by child and adult actors. The data-set contains 82 body gesture scenes acted by 8 actors. Some of these scenes concern emotions/mental states depending on a social interaction for their expression, for this reason they were acted with the involvement of a second actor.

11.4.2.2 Multi-party Data-Sets

Some more recent data-sets were collected to address movement expressivity in social contexts. However, their number is still quite small, due to the technological challenges that have to be faced to collect them (e.g., different persons data streams synchronization, occlusions management, and so on). The IEMOCAP (Interactive

Emotional dyadic MOtion CAPture) data-set [9] was recorded from 10 actors in 12 hours of dyadic scripted and spontaneous spoken communication sessions, designed to elicit specific types of emotions (happiness, anger, sadness, frustration and neutral). It contains audiovisual data, motion capture data (hands, head and face), and text transcriptions. The MMLI (Multimodal and Multiperson corpus of Laughter in Interaction) data-set is a multimodal data-set focusing on full body movements and different laughter types [67]. It contains 500 episodes of both induced and interactive laughs from human triads playing social games in a well-established scenario to elicit rich and natural non-verbal expressive behavior. The data consists of motion capture, facial tracking, multiple audio and video channels as well as physiological data. During the EU-FP7-ICT FET SIEMPRE Project,[1] focusing on the analysis of creative bodily expressive communication within groups of people, several multimodal data-sets of audiovisual and motion capture data were recorded. Data were collected in three musical scenarios: string quartet, orchestra, and audience.

11.4.3 Movement Expressivity Computational Frameworks

Although researchers in Computer Science and Robotics have an increasing interest on movement expressivity for developing movement-based interactive systems and applications, there is a scarcity of computational frameworks to address it. Many studies have been aimed at exploring how to use the LMA [54] observational system to define a computational framework e.g., [32, 91]. However, most of these works mainly result in systems' prototypes and sets of guidelines to open research perspectives where LMA can be a basic brick in conceiving movement-based interactive systems. An effort towards a clear formalization of LMA in terms of a general computational framework is still needed.

Some researchers addressed movement expressivity through classes of models that can be written as dynamical, discrete-time, state-space systems. For example, Caramiaux [15] in his work devoted to show how variations in movement is a way to understand expressivity, and to find computational solutions to capture and use such variations in interactive systems, proposed two models and their possible application in interactive scenarios. The first model takes into account temporal (slow-fast) and geometrical (small-big, tilt) movement variations; the second one focuses on the dynamical variations of the movement. However, these approaches are more focused on modeling a movement and its expressive content than to provide a general analytical framework to address it.

To our knowledge, the most adopted framework for addressing expressive movement analysis is the one proposed by Camurri et al. [12]. The authors grounds their four-layer framework on some of the theoretical models previously described in this chapter, e.g., [54, 99]. The main motivation of their work is to provide researchers with an analytic approach that can work independently of the considered modality

[1]http://www.infomus.org/siempre.

conveying expressivity (movement extracted from a large palette of sensors, audio and so on). The framework counts four layers. The first one is the *Physical signals* layer that pre-processes the raw data captured by sensors and send data to the *Low-level features* layer. This one computes low-level features, that is quantitative representations of the descriptors that psychologists, musicologists, researchers on music perception, researchers on human movement, and artists deem important for conveying expressivity. The third layer is the *Mid-level features and maps* layer devoted to compute features describing the qualities of a gesture. Generally, here, the first step is the movement unitizing to identify gestures on which to extract these features. When a clear unitizing cannot be found, this step consists of segmenting data on fixed-length windows. Examples of mid-level features are the values of the 4 Effort's Factors. Finally, such features are mapped onto conceptual structures by the *High-level features* layer. Concepts can include basic emotions or their dimensions, such as the well-known valence-arousal space. Other possible outputs include the 8 Basic Effort Actions of Laban (pressing, flicking, punching, floating, wringing, dabbing, slashing, gliding).

Despite the framework was originally conceived for the analysis, authors envisage its use also for synthesizing expressive behavior. The framework has been implemented in the Gesture Processing Library of the EyesWeb XMI research platform [10].

Recently, according to the intuition of Camurri et al. [12] that movement conveys expressivity more through its spatial and temporal features than through its syntactic meaning, Jessop [45] proposed a new four-layer framework specifically devoted to analyze and recognize expressive movement performances. The *Input data* layer corresponds to the *Physical signals* layer of the Camurri et al. [12] framework. The second layer, the *Expressive Features* layer, is devoted to extract temporal features conveying expressive content. Such features are then mapped onto high-level parametric spaces of expressivity. This association is done in the *High-level Parametric Spaces* layer. Finally, the *Output Control Parameters* layer enable a further mapping of these spaces onto parameters controlling the output media used in the performance. According to the author, the novelty of this framework is twofold: it enables to represent movement expressivity via trajectories in continuous expressive spaces, and it allows researchers to work at high levels of abstraction. However, this framework lacks in generalization with respect to the one of Camurri et al. [12], because it is specifically designed for movement analysis and does not enable a multimodal approach to expressivity.

There exist other examples of computational frameworks that can be used to perform movement expressivity analysis, but they are not specifically conceived to address this issue e.g., [2, 88, 95]. A notable example is the Social Signal Interpretation (SSI) framework aimed at recognizing social signals in real-time and is based on the concept of pipeline, see [95]. However, it is more technology-oriented than the other frameworks: its motivation is to provide researchers in Computer Science and Robotics with a *unique* tool able to overcome the most common technological bottlenecks that can occur from data capturing to classification (e.g., streams synchronization, feature extraction and fusion of data from different modalities).

11.5 Multi-scale Movement Expressivity Analysis Algorithms

We now provide a detailed description of a set of algorithms for movement expressivity analysis that appear in several publications co-authored by the authors of this chapter. The algorithms partially implement the movement expressivity framework of Camurri et al. [12], and have been successfully applied in a number of research works and EU ICT projects: [38–40, 59, 69, 75]. Their implementation has been re-written and adapted in this chapter by adopting a unified notation and a consistent approach to their description.

Algorithms are organized by time scale, that is, the amount of time needed for them to generate an output. Time scales range from *fine-grain*, in which algorithms generate output almost instantaneously, to *medium-grain*, in which algorithm's output is generated after a few seconds, to *coarse-grain* algorithms, that need more time (many seconds to minutes or hours) to provide an output.

11.5.1 Fine-Grain Time Scale

At the fine-grain time scale we find instantaneous features that can be directly computed from low-level sensors data. Sometimes, they can be computed at the hardware level, see for example the output of devices like the Inertial Movement Units (IMUs).

Psychologists, like De Meijer [24] and Wallbott [98], investigated the low-level expressive movement features that characterize emotion communication in humans. For example, movement *energy* and *expansiveness* are significantly different when expressing emotional states with opposite degrees of physical activation: e.g., hot anger versus sadness, joy versus boredom.

11.5.1.1 Kinetic Energy

Kinetic Energy (KE) can be computed from motion capture and inertial data, as demonstrated in previous works, such as [59, 68, 75]. Full-body KE is computed as follows:

$$KE = \frac{1}{2}\sum_{i=0}^{n} m_i v_i^2 \qquad (11.1)$$

where m_i is the mass of the ith user's body joint (e.g., head, right/left shoulder, right/left elbow and so on) and v_i is the velocity of that joint. The mass values can be obtained from anthropometric studies (e.g. [100]). If joints are tracked by a motion capture device, then v_i is the result of differencing the joint position, that is, by subtracting the position of the joint at the current data frame from its position at the previous frame.

If inertial sensors like IMU are used, then v_i can be obtained by integrating the acceleration data, that is, by summing the acceleration of the joint at the current data frame to its acceleration at the previous frame.

11.5.1.2 Bounding Volume

Bounding Volume (BV) indicates the body level of expansiveness/contraction [75]. It corresponds to the volume of the smallest cuboid enclosing the user's body. The BV can also be seen as a measure of body "openness": when a person stretches their arms and legs outward, then BV increases. It is computed by extracting the highest and lowest values of the 3D body joints coordinates. The cuboid corresponding to those coordinates is then generated:

$$BV = (\max_{\forall i \in N} x_i - \min_{\forall i \in N} x_i) * (\max_{\forall i \in N} y_i - \min_{\forall i \in N} y_i) * (\max_{\forall i \in N} z_i - \min_{\forall i \in N} z_i) \tag{11.2}$$

where $i \in N$ is the i-th body joint J_i expressed as the 3D position (x_i, y_i, z_i).

11.5.2 Medium-Grain Time Scale

Medium-grain expressive features describe qualities of gestures. Therefore such features are neither as "simple" to be computed as the fine-grain ones, nor "complex" and abstract as the coarse-grain ones. The expressive movement analysis framework defined by Piana et al. [75], for example, provides an algorithm for measuring impulsivity of movement (i.e., sudden movements, executed without pre-planning) as by looking at the variation of kinetic energy and bounding volume. Similarly, dynamic symmetry of movement is measured as the similarity of body joints' positions and accelerations.

11.5.2.1 Smoothness

As we reported in Sect. 11.3, Wallbott [97] states that smoothness is a possible cue of the fluency-course characteristics of psychiatric patients' movements. Todorov and Jordan [92] demonstrated a correspondence between (i) smooth trajectories performed by human arms, (ii) minimization of the third-order derivative (i.e., jerk) of the hand position and (iii) correlation between hand trajectory curvature and velocity. Glowinski and Mancini [39] defined an algorithm for hand smoothness extraction based on (iii). The hand position (P_x, P_y) is stored in a 1 second long timeseries. Curvature k and velocity v are computed as:

$$k = \left| \frac{P'_x P''_y - P'_y P''_x}{(P'^2_x + P'^2_y)^{\frac{3}{2}}} \right| \qquad v(P_x, P_y) = \sqrt{P'^2_x + P'^2_y} \tag{11.3}$$

where P'_x, P'_y, P''_x and P''_y are the first and second order derivatives of P_x and P_y. As demonstrated by Glowinski and Mancini [39], derivatives can be efficiently computed by applying a Savitzky-Golay filter [84] providing as output both the filtered signal and an approximation of the nth order smoothed derivatives.

Then, the Pearson correlation between $\log(k)$ and $\log(v)$ is calculated:

$$\rho_h(k, v) = \frac{\sigma_{\log(k),\log(v)}}{\sigma_{\log(k)}\sigma_{\log(v)}} \tag{11.4}$$

However, k and v are computed over a relatively "short" time window, so the covariance $\sigma_{\log(k),\log(v)}$ can be approximated by 1, as the k and v variate (or not) approximately at the same time:

$$\rho'_h(k, v) = \frac{1}{\sigma_{\log(k)}\sigma_{\log(v)}} \tag{11.5}$$

Finally, the Smoothness Index SmI is equal to $\rho'(k, v)$.

11.5.2.2 Suddenness

Suddenness can be a cue of various psychological disorders: drugs use, bipolarity, anti-social personality, and so on. For example, Heiser et al. [44] made objective measurements of impulsivity in children with hyperkinetic disorders through an infrared motion analysis system combined with a continuous performance test. Barratt [5] defined the Barratt Impulsiveness Scale (BIS), one of the most widely used measures of personality traits.

Niewiadomski et al. [69] exploited the characteristics of alpha-stable distributions to detect sudden movements. Alpha-stable distributions [57] can be modeled by Probability Density Functions (PDFs). These are characterised by four parameters $(\alpha, \beta, \gamma, \delta)$:

- $\alpha \in (0, 2]$ is the characteristic exponent that defines whether the distribution includes impulses;
- $\beta \in [-1, 1]$ determines the skewness of the pdf;
- $\gamma > 0$ corresponds to variance in Gaussian distributions;
- $\delta \in (-\infty, \infty)$ corresponds to the mean value in Gaussian distributions.

Starting from motion captured hand position, they compute the absolute velocity by differencing it. Then, they apply the $stblfit$ function, a C++ implementation of the stable fit Matlab algorithm.[2]

Suddenness S is equal to the resulting α parameter, varying in $(0, 2]$, scaled and multiplied by γ. This process implies 2 consequences: (i) when α tends to zero, the scaled value of α tends to one and vice-versa; (ii) movements exhibiting *low* (resp., *high*) velocity will correspond to *low* (resp., *high*) values of γ. That is, S will be high for sudden movements (α low) with large velocity variability (γ high). Also, the sign of β is tested: sudden movement exhibiting a fast deceleration of the hand will generate negative values of β; in this case the value of S is set to zero.

11.5.2.3 Entropy

Sample Entropy is a non-linear entropy extraction technique that was developed to quantify behavior regularity by Richman and Moorman [79] and improved by Govindan et al. [41]. It has been applied to a variety of physiological data (heart rate, EMG, see [87]) and was successfully exploited in the high-level analysis of movement expressivity by Glowinski et al. [38, 40]. As with the usual entropy measures, higher values of Sample Entropy are associated to high disorder, while smaller values indicate regularity.

Given a standardized one-dimensional discrete time series of length N, $X = \{x_1, \ldots, x_i, \ldots, x_N\}$, the Sample Entropy algorithm works as follows (the algorithm is taken from the paper [40]):

1. construct vectors of length m:

$$u_i(m) = \{x_i, \ldots, x_{i+m-1}\},\ 1 \leq i \leq N - m \tag{11.6}$$

2. compute the correlation sum $U_i^m(r)$ to estimate similar subsequences (or *template vectors*) of length m within the time series:

$$U_i^m(r) = \frac{1}{(N - m - 1)} \sum_{i=1, i \neq j}^{N-m} \Theta(r - \parallel u_i(m) - u_j(m) \parallel_\infty) \tag{11.7}$$

   where $u_i(m)$ and $u_j(m)$ are the template vectors of length m built from the standardized time series, at time i and j respectively, N is the number of samples in the time series, r is the tolerance (or *radius*), Θ is the Heaviside function, and $\parallel \parallel_\infty$ is the maximum norm defined by $\parallel u_i(m) - u_j(m) \parallel_\infty = \max_{0 \leq k \leq m-1} \mid x_{j+k} - x_{i+k} \mid$
3. calculate the average of U_i^m, i.e., the probability that two vectors will match in the m-dimensional reconstructed state space

[2] http://www.mathworks.com/matlabcentral/fileexchange/37514-stbl--alpha-stable-distributions-for-matlab.

$$U^m(r) = \frac{1}{(N-m)} \sum_{i=1}^{N-m} U_i^m(r) \tag{11.8}$$

4. set $m = m + 1$ and repeat steps 1–4
5. calculate the sample entropy of X:

$$SampEn(X, m, r) = -ln \frac{U^{m+1}(r)}{U^m(r)} \tag{11.9}$$

To sum up, SampEn is the negative natural logarithm of the conditional probability that sub-sequences of m points in a time series remain similar (as defined in Eq. 11.8) when extra point $(m + 1)$ is added to the sub-sequences. So, small values of SampEn indicate regularity. Ramdani et al. [78] suggest to set $m = 3$ and r (tolerance) $= 0.20$ when analysing human expressive movement.

11.5.3 Coarse-Grain Time Scale

The last level of abstraction of movement expressivity analysis is related to high-level concepts and messages, such as emotions. For example, Piana et al. [76], using high-level features, infer emotional states to help children with Autism Spectrum Conditions to learn how to express and understand emotional states. Another example of how computers can deal with emotion expression is the EU-ICT FET Project ILHAIRE,[3] that aimed to show how machines can effectively encode and decode human laughter [93]. In the following section we present one of the computational models of laughter detection developed during the Project.

11.5.3.1 Laughter Detection

Laughter can not only be the visible expression of some emotional states like joy and happiness, but it can also be a strong trigger for social interaction. For example, [42, 72] shown that it could communicate interest and reduce the sense of threat in a group. Morphology of laughter has been studied since [22], as well as its function in human interaction (e.g., laughing during a conversation) and its occurrence along with emotions [80].

In this section we focus on automated laughter decoding, and, in particular, on the computation of a high-level feature called Body Laughter Index (BLI, [59]). The feature is computed on the position and movement of a person's shoulders, that can be detected, for example, through motion capture, videocameras, and range imaging devices. As described above, high-level features are based on low and mid-level features. In particular, BLI is computed from the following ones:

---

[3] http://www.ilhaire.eu.

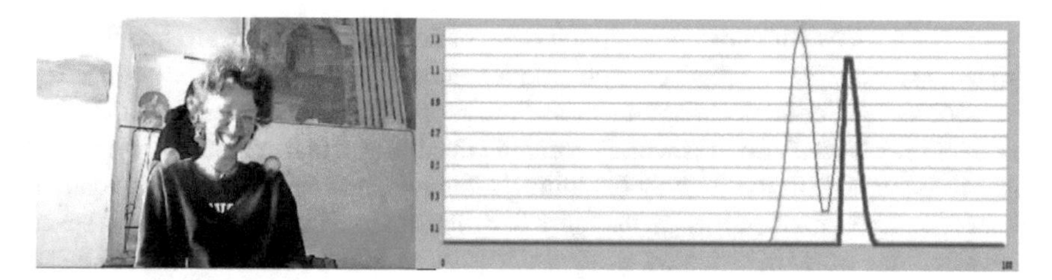

Fig. 11.1 An example of body laughter index (BLI) computation. On the left a person with green markers on her shoulders to track trajectories; on the right the BLI

- *Kinetic Energy (KE)* (low-level): see the description in Sect. 11.5.1.1.
- *Shoulders Correlation (SC)* (mid-level): It is the Pearson correlation ρ between the person's shoulders vertical positions.
- *Periodicity Index (PI)* (mid-level): It is the Periodicity Transform [89] of KE. The description of PT is out of the scope of this chapter. The output of PT is a description of the input signal in terms of periodic components. For example, if the person's trunk or limbs are oscillating during laughter, we expect a greater contribution of a periodic component around 4–5 Hz, as reported by Ruch and Ekman [81].

BLI is the result of the following equation:

$$BLI = \alpha\bar{\rho} + \beta\bar{KE} \tag{11.10}$$

In order to take into account the rhythmicity of movement extracted by PI, the computed BLI value is acknowledged only if the mean Periodicity Index belongs to the arbitrary range 2.5–8 Hz.

Figure 11.1 shows an example of analysis of a laughing person, taken from Mancini et al. [59]. BLI is plotted in red when it is acknowledged, in blue otherwise. Mancini et al. [59] presented a preliminary study in which BLI was validated on a laughter video corpus.

11.6 Conclusion and Implications on Robotics

This chapter focused on movement expressivity analysis, providing a survey of qualitative and quantitative approaches studied by researchers in Psychology, Sociology and Computer Science. In particular, we provided a survey of computational approaches to movement expressivity and a set of algorithms to extract it at different time scales.

It is well established that theories and computational models we previously illustrated in the chapter are embodiment-independent, that is, they can be applied to computer devices with any embodiment type, including the robotic ones, as demonstrated, for example, by the works on emotion and robots of [16, 73].

Dourish [29] defines Embodiment as "the property of being manifest in and of the everyday world". So, it is widely accepted that both virtual characters and robots able to interact with the user by exhibiting affective and social intelligence can be considered embodied entities, in the sense that they "unfold in real time and real space as a part of the world in which we are situated" [29]. Moreover, several authors, like Le et al. [55], demonstrated that it is possible to extend virtual agent frameworks to adapt them to work with robots, while Kriegel et al. [52] defined a software architecture to migrate artificial characters from virtual to physical bodies and vice-versa.

As reported by "A Roadmap for US Robotics: From Internet to Robotics[4]", that presents the long-term research roadmap on HRI, robots will have to "perceive, model and adapt to complex user behaviors, actions, and intent in semistructured tasks and environments, and transfer learned models across domains and environments" and "perceive, model and adapt to complex user behaviors, actions, and intent in semistructured tasks and environments, and transfer learned models across domains and environments". In Sects. 11.4 and 11.5, we describe quantitative approaches to movement expressivity analysis, that align with the need for addressing the issues of "perception" and "modeling" user behavior in human-robot interaction, as outlined in the above roadmap.

In the near future, research in this area will have to explore sensing and perception techniques for human social and emotional non-verbal behavior, in single as well as in multi-party interaction, without neglecting human variability across individuals, culture and time.

Acknowledgements Authors would like to acknowledge their colleagues who participated in the definition and implementation of the algorithms presented in Sect. 11.5: Antonio Camurri, Donald Glowinski, Radoslaw Niewiadomski, Stefano Piana, Gualtiero Volpe.

References

1. Allport, G., & Vernon, P. (1933). *Studies in expressive movement*. New York: Macmillan.
2. Anzalone, S. M., Avril, M., Salam, H., & Chetouani, M. (2014). Imi2s: A lightweight framework for distributed computing. In *International Conference on Simulation, Modeling, and Programming for Autonomous Robots* (pp. 267–278). Springer.
3. Argyle, M. (1988). *Bodily communication* (2nd ed.). London: Methuen & Co.
4. Bänziger, T., Pirker, H., & Scherer, K. (2006). Gemep-geneva multimodal emotion portrayals: A corpus for the study of multimodal emotional expressions. *Proceedings of LREC*, *6*, 15–019.
5. Barratt, E. S. (1994). Impulsiveness and aggression. *Violence and Mental Disorder: Developments in Risk Assessment*, *10*, 61–79.
6. Boone, R. T., & Cunningham, J. G. (1998). Children's decoding of emotion in expressive body movement: The development of cue attunement. *Developmental Psychology*, *34*(5), 1007.
7. Breazeal, C. (2003). Toward sociable robots. *Robotics and Autonomous Systems*, *42*(3–4), 167–175.

[4]https://cra.org/ccc/wp-content/uploads/sites/2/2016/11/roadmap3-final-rs-1.pdf.

8. Buck, R. (1984). The communication of emotion. Guilford press.
9. Busso, C., Bulut, M., Lee, C. C., Kazemzadeh, A., Mower, E., Kim, S., et al. (2008). Lemocap: Interactive emotional dyadic motion capture database. *Language Resources and Evaluation*, *42*(4), 335.
10. Camurri, A., Coletta, P., Varni, G., & Ghisio, S. (2007). Developing multimodal interactive systems with eyesweb xmi. In *Proceedings of the 7th International Conference on New Interfaces for Musical Expression* (pp. 305–308). ACM.
11. Camurri, A., Lagerlöf, I., & Volpe, G. (2003). Recognizing emotion from dance movement: Comparison of spectator recognition and automated techniques. *International Journal of Human-Computer Studies*, *59*(1–2), 213–225.
12. Camurri, A., Mazzarino, B., Ricchetti, M., Timmers, R., & Volpe, G. (2004). Multimodal analysis of expressive gesture in music and dance performances. In *Gesture-Based Communication in Human-computer Interaction* (pp. 20–39). Springer.
13. Cao, Z., Hidalgo, G., Simon, T., Wei, S. E., & Sheikh, Y. (2018). Openpose: Realtime multi-person 2d pose estimation using part affinity fields. arXiv:1812.08008.
14. Cao, Z., Simon, T., Wei, S. E., & Sheikh, Y. (2017). Realtime multi-person 2d pose estimation using part affinity fields. In *Proceedings of the IEEE Conference on Computer Vision and Pattern Recognition* (pp. 7291–7299).
15. Caramiaux, B. (2014). Motion modeling for expressive interaction: A design proposal using Bayesian adaptive systems. In *Proceedings of the 2014 International Workshop on Movement and Computing* (pp. 76–81).
16. Castellano, G., Leite, I., Pereira, A., Martinho, C., Paiva, A., & Mcowan, P. W. (2014). Context-sensitive affect recognition for a robotic game companion. *ACM Transactions on Interactive Intelligent Systems (TiiS)*, *4*(2), 10.
17. Castellano, G., Mancini, M., Peters, C., & McOwan, P. W. (2012). Expressive copying behavior for social agents: A perceptual analysis. *IEEE Transactions on Systems, Man, and Cybernetics-Part A: Systems and Humans*, *42*(3), 776–783.
18. Castellano, G., Villalba, S. D., & Camurri, A. (2007). Recognising human emotions from body movement and gesture dynamics. In *International Conference on Affective Computing and Intelligent Interaction* (pp. 71–82). Springer.
19. Chi, D., Costa, M., Zhao, L., & Badler, N. (2000). The emote model for effort and shape. In *Proceedings of the 27th Annual Conference on Computer Graphics and Interactive Techniques* (pp. 173–182). ACM Press/Addison-Wesley Publishing Co.
20. Dahl, S., & Friberg, A. (2007). Visual perception of expressiveness in musicians' body movements. *Music Perception: An Interdisciplinary Journal*, *24*(5), 433–454.
21. Damasio, A. R. (1999). The feeling of what happens: Body and emotion in the making of consciousness. Houghton Mifflin Harcourt.
22. Darwin, C. (1872). *The expression of the emotions in man and animals*. London: John Murray.
23. De Gelder, B. (2006). Towards the neurobiology of emotional body language. *Nature Reviews Neuroscience*, *7*(3), 242.
24. De Meijer, M. (1989). The contribution of general features of body movement to the attribution of emotions. *Journal of Nonverbal behavior*, *13*(4), 247–268.
25. De Meijer, M. (1991). The attribution of aggression and grief to body movements: The effect of sex-stereotypes. *European Journal of Social Psychology*, *21*(3), 249–259.
26. Dell, C. (1977). A primer for movement description using effort-shape and supplementary concepts. Princeton Book Company Pub.
27. Dittrich, W. H., Troscianko, T., Lea, S. E., & Morgan, D. (1996). Perception of emotion from dynamic point-light displays represented in dance. *Perception*, *25*(6), 727–738.
28. Douglas-Cowie, E., Campbell, N., Cowie, R., & Roach, P. (2003). Emotional speech: Towards a new generation of databases. *Speech Communication*, *40*(1–2), 33–60.
29. Dourish, P. (1999). Embodied interaction: Exploring the foundations of a new approach to hci. Unpublished paper, http://www.ics.uci.edu/jpd/publications/misc/embodied.pdf.
30. Dyer, S., Martin, J., & Zulauf, J. (1995). Motion capture white paper.

31. Ekman, P., & Friesen, W. (1974). Detecting deception from the body or face. *Journal of Personality and Social Psychology*, *29*, 288–298.
32. Fdili Alaoui, S., Carlson, K., Cuykendall, S., Bradley, K., Studd, K., & Schiphorst, T. (2015). How do experts observe movement? In *Proceedings of the 2nd International Workshop on Movement and Computing* (pp. 84–91).
33. Fourati, N., & Pelachaud, C. (2014). Emilya: Emotional body expression in daily actions database. In *LREC*(pp. 3486–3493).
34. Frijda, N. H. (2010). Impulsive action and motivation. *Biological Psychology*, *84*(3), 570–579.
35. Frijda, N. H. (2010). Not passion's slave. *Emotion Review*, *2*(1), 68–75.
36. Giraud, T., Focone, F., Isableu, B., Martin, J. C., & Demulier, V. (2016). Impact of elicited mood on movement expressivity during a fitness task. *Human Movement Science*, *49*, 9–26.
37. Glas, N., & Pelachaud, C. (2018). Topic management for an engaging conversational agent. *International Journal of Human-Computer Studies*, *120*, 107–124.
38. Glowinski, D., Coletta, P., Volpe, G., Camurri, A., Chiorri, C., & Schenone, A. (2010). Multi-scale entropy analysis of dominance in social creative activities. In *Proceedings of the 18th ACM International Conference on Multimedia* (pp. 1035–1038). ACM.
39. Glowinski, D., & Mancini, M. (2011). Towards real-time affect detection based on sample entropy analysis of expressive gesture. In *International Conference on Affective Computing and Intelligent Interaction* (pp. 527–537). Springer.
40. Glowinski, D., Mancini, M., Cowie, R., Camurri, A., Chiorri, C., & Doherty, C. (2013). The movements made by performers in a skilled quartet: A distinctive pattern, and the function that it serves. *Frontiers in Psychology*, *4*, 841.
41. Govindan, R., Wilson, J., Eswaran, H., Lowery, C., & Preißl, H. (2007). Revisiting sample entropy analysis. *Physica A: Statistical Mechanics and Its Applications*, *376*, 158–164.
42. Grammer, K. (1990). Strangers meet: Laughter and nonverbal signs of interest in opposite-sex encounters. *Journal of Nonverbal Behavior*, *14*(4), 209–236.
43. Gross, M. M., Crane, E. A., & Fredrickson, B. L. (2010). Methodology for assessing bodily expression of emotion. *Journal of Nonverbal Behavior*, *34*(4), 223–248.
44. Heiser, P., Frey, J., Smidt, J., Sommerlad, C., Wehmeier, P., Hebebrand, J., et al. (2004). Objective measurement of hyperactivity, impulsivity, and inattention in children with hyperkinetic disorders before and after treatment with methylphenidate. *European Child & Adolescent Psychiatry*, *13*(2), 100–104.
45. Jessop, E. (2015). Capturing the body live: A framework for technological recognition and extension of physical expression in performance. *Leonardo*, *48*(1), 32–38.
46. Johansson, G. (1973). Visual perception of biological motion and a model for its analysis. *Perception & Psychophysics*, *14*(2), 201–211.
47. Kapur, A., Kapur, A., Virji-Babul, N., Tzanetakis, G., & Driessen, P. F. (2005). Gesture-based affective computing on motion capture data. In *International Conference on Affective Computing and Intelligent Interaction* (pp. 1–7). Springer.
48. Kendon, A. (1980). Gesticulation and speech: Two aspects of the. The relationship of verbal and nonverbal communication (25), 207.
49. Kleinsmith, A., & Bianchi-Berthouze, N. (2013). Affective body expression perception and recognition: A survey. *IEEE Transactions on Affective Computing*, *4*(1), 15–33.
50. Knight, H., & Simmons, R. (2016). Laban head-motions convey robot state: A call for robot body language. In *2016 IEEE International Conference on Robotics and Automation (ICRA)* (pp. 2881–2888). IEEE.
51. Kolkmeier, J., Lee, M., & Heylen, D. (2017). Moral conflicts in VR: Addressing grade disputes with a virtual trainer. In *International Conference on Intelligent Virtual Agents* (pp. 231–234). Springer.
52. Kriegel, M., Aylett, R., Cuba, P., Vala, M., & Paiva, A. (2011). Robots meet Ivas: A mind-body interface for migrating artificial intelligent agents. In *International Workshop on Intelligent Virtual Agents* (pp. 282–295). Springer.
53. Kurtenbach, G. (1990). Gestures in human-computer communication. *The Art of Human Computer Interface Design*, *309*.

54. Laban, R., & Lawrence, F. C. (1947). *Effort*. USA: Macdonald & Evans.
55. Le, Q. A., Hanoune, S., & Pelachaud, C. (2011). Design and implementation of an expressive gesture model for a humanoid robot. In *2011 11th IEEE-RAS International Conference on Humanoid Robots* (pp. 134–140). IEEE.
56. Levy, J. A., & Duke, M. P. (2003). The use of laban movement analysis in the study of personality, emotional state and movement style: An exploratory investigation of the veridicality of "body language". *Individual Differences Research, 1*(1).
57. Lévy, P. (1925). Calcul des probabilités (Vol. 9). Gauthier-Villars Paris.
58. Ma, Y., Paterson, H. M., & Pollick, F. E. (2006). A motion capture library for the study of identity, gender, and emotion perception from biological motion. *Behavior Research Methods, 38*(1), 134–141.
59. Mancini, M., Varni, G., Glowinski, D., & Volpe, G. (2012). Computing and evaluating the body laughter index. In *International Workshop on Human Behavior Understanding* (pp. 90–98). Springer.
60. Masuda, M., Kato, S., & Itoh, H. (2010). Laban-based motion rendering for emotional expression of human form robots. In *Pacific Rim Knowledge Acquisition Workshop* (pp. 49–60). Springer.
61. McNeill, D. (1992). Hand and mind: What gestures reveal about thought. University of Chicago press.
62. Montepare, J. M., Goldstein, S. B., & Clausen, A. (1987). The identification of emotions from gait information. *Journal of Nonverbal Behavior, 11*(1), 33–42.
63. Neff, M., & Fiume, E. (2004). Artistically based computer generation of expressive motion. In *Proceedings of the AISB Symposium on Language, Speech and Gesture for Expressive Characters* (pp. 29–39).
64. Neff, M., & Fiume, E. (2005). AER: Aesthetic exploration and refinement for expressive character animation. In *Proceedings of the 2005 ACM SIGGRAPH/Eurographics Symposium on Computer Animation* (pp. 161–170). ACM Press, New York, NY, USA.
65. Neff, M., & Kim, Y. (2009). Interactive editing of motion style using drives and correlations. In *SCA '09: Proceedings of the 2009 ACM SIGGRAPH/Eurographics Symposium on Computer Animation* (pp. 103–112). ACM, New York, NY, USA.
66. Newlove, J. (2007). *Laban for actors and dancers: Putting Laban's movement theory into practice: A step-by-step guide*. UK: Nick Hern Books.
67. Niewiadomski, R., Mancini, M., Baur, T., Varni, G., Griffin, H., & Aung, M. S. (2013). Mmli: Multimodal multiperson corpus of laughter in interaction. In *International Workshop on Human Behavior Understanding* (pp. 184–195). Springer.
68. Niewiadomski, R., Mancini, M., Cera, A., Piana, S., Canepa, C., & Camurri, A. (2018). Does embodied training improve the recognition of mid-level expressive movement qualities sonification? *Journal on Multimodal User Interfaces*, 1–13.
69. Niewiadomski, R., Mancini, M., Volpe, G., & Camurri, A. (2015). Automated detection of impulsive movements in hci. In *Proceedings of the 11th Biannual Conference on Italian SIGCHI Chapter* (pp. 166–169). ACM.
70. Noroozi, F., Kaminska, D., Corneanu, C., Sapinski, T., Escalera, S., & Anbarjafari, G. (2018). Survey on emotional body gesture recognition. *IEEE Transactions on Affective Computing*.
71. O'Reilly, H., Pigat, D., Fridenson, S., Berggren, S., Tal, S., Golan, O., et al. (2016). The eu-emotion stimulus set: A validation study. *Behavior Research Methods, 48*(2), 567–576.
72. Owren, M. J., & Bachorowski, J. A. (2003). Reconsidering the evolution of nonlinguistic communication: The case of laughter. *Journal of Nonverbal Behavior, 27*, 183–200.
73. Paiva, A. (2018). Robots that listen to people's hearts: The role of emotions in the communication between humans and social robots. In *Proceedings of the 26th Conference on User Modeling, Adaptation and Personalization* (pp. 175–175). ACM.
74. Pease, B., & Pease, A. (2008). The definitive book of body language: The hidden meaning behind people's gestures and expressions. Bantam.
75. Piana, S., Mancini, M., Camurri, A., Varni, G., & Volpe, G. (2013). Automated analysis of non-verbal expressive gesture. In *Human Aspects in Ambient Intelligence* (pp. 41–54). Springer.

76. Piana, S., Staglianò, A., Odone, F., & Camurri, A. (2016). Adaptive body gesture representation for automatic emotion recognition. *ACM Transactions on Interactive Intelligent Systems (TiiS)*, *6*(1), 6.
77. Pollick, F. E., Paterson, H. M., Bruderlin, A., & Sanford, A. J. (2001). Perceiving affect from arm movement. *Cognition*, *82*(2), B51–B61.
78. Ramdani, S., Seigle, B., Lagarde, J., Bouchara, F., & Bernard, P. (2009). On the use of sample entropy to analyze human postural sway data. *Medical Engineering & Physics*, *31*(8), 1023–1031.
79. Richman, J., & Moorman, J. (2000). Physiological time-series analysis using approximate entropy and sample entropy. *American Journal of Physiology—Heart and Circulatory Physiology*, *278*(6), H2039.
80. Ruch, W. (1993). Exhilaration and humor. In M. Lewis & J. M. Haviland (Eds.), *The handbook of emotions*. New York: Guilford.
81. Ruch, W., & Ekman, P. (2001). The expressive pattern of laughter. In A. Kaszniak (Ed.), *Emotion, qualia and consciousness* (pp. 426–443). Tokyo: World Scientific Publishers.
82. Russell, J. A. (1980). A circumplex model of affect. *Journal of Personality and Social Psychology*, *39*(6), 1161.
83. Saha, S., Datta, S., Konar, A., & Janarthanan, R. (2014). A study on emotion recognition from body gestures using kinect sensor. In *2014 International Conference on Communication and Signal Processing* (pp. 056–060). IEEE.
84. Savitzky, A., & Golay, M. J. E. (1964). Smoothing and differentiation of data by simplified least squares procedures. *Analytical Chemistry*, *36*(8), 1627–1639.
85. Scherer, K. R. (1982). Emotion as a process: Function, origin and regulation.
86. Scherer, K. R. (1984). On the nature and function of emotion: A component process approach. *Approaches to Emotion*, *2293*, 317.
87. Seely, A., & Macklem, P. (2004). Complex systems and the technology of variability analysis. *Critical Care*, *8*(6), R367–84.
88. Serrano, M., Nigay, L., Lawson, J. Y. L., Ramsay, A., Murray-Smith, R., & Denef, S. (2008). The openinterface framework: A tool for multimodal interaction. In *CHI '08 Extended Abstracts on Human Factors in Computing Systems* (pp. 3501–3506). ACM.
89. Sethares, W. A., & Staley, T. W. (1999). Periodicity transforms. *IEEE Transactions on Signal Processing*, *47*(11), 2953–2964.
90. Shotton, J., Sharp, T., Kipman, A., Fitzgibbon, A., Finocchio, M., Blake, A., et al. (2013). Real-time human pose recognition in parts from single depth images. *Communications of the ACM*, *56*(1), 116–124.
91. Silang Maranan, D., Fdili Alaoui, S., Schiphorst, T., Pasquier, P., Subyen, P., & Bartram, L. (2014). Designing for movement: Evaluating computational models using LMA effort qualities. In *Proceedings of the SIGCHI Conference on Human Factors in Computing Systems* (pp. 991–1000).
92. Todorov, E., & Jordan, M. I. (1998). Smoothness maximization along a predefined path accurately predicts the speed profiles of complex arm movements. *Journal of Neurophysiology*, *80*(2), 696–714.
93. Urbain, J., Niewiadomski, R., Hofmann, J., Bantegnie, E., Baur, T., Berthouze, N., et al. (2013). Laugh machine. In *Proceedings eNTERFACE*, *12*, 13–34.
94. Varni, G., Volpe, G., & Camurri, A. (2010). A system for real-time multimodal analysis of nonverbal affective social interaction in user-centric media. *IEEE Transactions on Multimedia*, *12*(6), 576–590.
95. Wagner, J., Lingenfelser, F., Baur, T., Damian, I., Kistler, F., & André, E. (2011). The social signal interpretation (ssi) framework. In *Proceedings of the 21st ACM International Conference on Multimedia* (pp. 831–834).
96. Walk, R. D., & Homan, C. P. (1984). Emotion and dance in dynamic light displays. *Bulletin of the Psychonomic Society*, *22*(5), 437–440.
97. Wallbott, H. G. (1989). Movement quality changes in psychopathological disorders. In *Normalities and Abnormalities in Human Movement. Medicine and Sport Science*, *29*, 128–146.

98. Wallbott, H. G. (1998). Bodily expression of emotion. *European Journal of Social Psychology, 28*, 879–896.
99. Wallbott, H. G., & Scherer, K. R. (1986). Cues and channels in emotion recognition. *Journal of Personality and Social Psychology, 51*(4), 690–699.
100. Winter, D. (1990). *Biomechanics and motor control of human movement*. Toronto: Wiley Inc.
101. Xiao, B., Georgiou, P., Baucom, B., & Narayanan, S. (2015). Modeling head motion entrainment for prediction of couples' behavioral characteristics. In *2015 International Conference on Affective Computing and Intelligent Interaction (ACII)* (pp. 91–97). IEEE.

Part III
The Robotic Point of View

Chapter 12
The Practice of Animation in Robotics

Tiago Ribeiro and Ana Paiva

Abstract Robot animation is a new form of character animation that extends the traditional process by allowing the animated motion to become more interactive and adaptable during interaction with users in real-world settings. This paper reviews how this new type of character animation has evolved and been shaped from character animation principles and practices. We outline some new paradigms that aim at allowing character animators to become robot animators, and to properly take part in the development of social robots. In particular, we describe the 12 principles of robot animation, which describes general concepts that both animators and robot developers should consider in order to properly understand each other. We conclude with a description of some types of tools that can be used by animators, while taking a part in the development process of social robot applications, and how they fit into the rest of the system.

12.1 Introduction

The art of animation was born more then one hundred years ago in 1896, when Georges Méliès invented the stop-motion technique. Twelve years later, Èmile Cohl became the father of animated *cartoons* with 'Fantasmagorie'. Windsor McCay, however, was coined as the father of animated *movies* for his 1911 work entitled 'Gertie the Dinosaur', in which he created what is considered to be the first animated *character* to actually convey emotions and an appealing personality [4].

Since then these hand-drawn animated characters have been evolving and taking in many different forms and audiences. During the last thirty years, animated characters have become mainly computer-animated, and are being produced by many major animation studios such as Pixar, Walt Disney Animation Studios, Dreamworks or Blue Sky Studios.

T. Ribeiro (✉) · A. Paiva
INESC-ID & Instituto Superior Técnico, University of Lisbon, Lisbon, Portugal
e-mail: me@tiagoribeiro.pt

A. Paiva
e-mail: ana.paiva@inesc-id.pt

N. Noceti et al. (eds.), *Modelling Human Motion*,
https://doi.org/10.1007/978-3-030-46732-6_12

Today we see robots becoming a new form of animated characters. However, this time the characters are jumping out of the big screens, powered by artificial intelligence (AI), and are becoming more interactive, and part of people's daily life. They are being developed in order to be used in social applications, in fields such as education, entertainment or assisted living. Given the technological background required for the creation of such characters, they are being developed by roboticists, software engineers and (AI) scientists, instead of by artists.

While this has been a necessary stage, we believe it is now time for robots and animated characters to reunite, by allowing artists and robot developers to work together, side by side, on the development of such characters. Animation artists have already been providing a contributing voice in the development of expressive, emotional and design traits of robots. However they typically get little to no access to the development of the actual interactive and intelligent behaviors that are performed with humans.

The goal of our work is to establish a solid bridge between these two worlds, which are intrinsically connected, but have been evolving separately, based on different perspectives, fundamental competencies, and end-goals. Such a connection will allow animators to take a new role as artists that are fully part of, and not just accessory, to the development of social robotic products. The same happened upon the emerging of computer animated cartoons and in particular, of 3D animated characters. At that time, animators exploring the new technique also felt the need to look into what had already been done during the last decades, and discover how that knowledge could be adapted for computer animation. On that topic, Lasseter argued that the traditional principles of animation have a similar meaning across different animation medium [18]. Not only were those principles transferred to 3D animated characters, but new tools and methodologies were also created to support the creative and development processes. Establishing robot animation as the new character animation medium will therefore require not only new theories, but also the integration of the technology with new tools and practices.

In this chapter we start by reviewing some character animation theories, along with existing proposals of how to adapt and use them with robots, and present some cases in which the animation process was considered and integrated in the development of socially interactive robots. We then outline a list of principles of animation for robots based on the current state of the art, and on our own previous and diverse experience. These principles are intended to provide thoughts on some general concepts that both animators and robot developers should consider, in order to properly understand each other, and to engage in successful collaborations. We complete the chapter by providing an overview on how the creative and technical tools and workflows may converge into an integrated robot animation pipeline, in which both artists and engineers are able to work together from initial development to the finished product.

12.2 Character Animation

Disney's twelve principles of animation are considered by most to be the commandments of animation. They are a result of more than 60 years of Disney productions, and were compiled into a book called 'The Illusion of Life', by Thomas and Johnston [30], the last two of Disney's Nine Old Men.[1] We summarize these principles further on, in Sect. 12.2.1.

However since The Golden Age of American Animation, Warner Bros. and MGM animators also definitely marked their position as masters of animated cartoons. These animators took exaggeration to another level, by giving special focus on physical exaggeration, in which we can actually identify common sub-types of exaggeration, like extreme distortion or blowing-ups. Most of their animations were largely based on comic plots, which generally included sever physical damage to the characters, thus justifying why they developed so much into blowing-ups and heavy distortion of the characters' body.

Tex Avery, one of the greatest animators of all time, coined the 'Tex Avery Expression', or just a 'Tex Avery', which is a very know eyes-popping-out expression generally used in fear or surprise situations [7].

While we do not want to blow up or physically damage robots while animating them, some of these practices can still provide interesting tips on some specific domains, like robots aimed at entertainment. While entertaining, we want a character to be as much expressive as possible, so entertainment robots will more likely promote the interest for developing and incorporating behaviors and mechanisms inspired by this kind of animation. The EMYS robotic head is an example of how a 'Tex Avery' eyes-popping mechanism can be incorporated into a robot [23].

A common trait in character animation is that each character is made to be very unique and well adapted to its role. Some of the most popular characters created during this time were Bugs Bunny, Daffy Duck, Porky Pig, Elmer Fudd, Yosemite Sam, Tom and Jerry, Scooby Doo and Droopy [3]. They usually carry or use regular props that people end up associating with that character, independently of the plot. Most of them also feature unique catchphrases and often perform secondary action that helps to define the personality of the character they convey. All these features together contribute to the illusion of the character as a being, and to the reinforcement of the connection between viewers and the characters.

Unfortunately, except for Disney-based ones, the practice of these animators is not very well documented. As they were generally jumping around from one studio to another, each animator may have followed different guidelines along his career, there are no compiled guidelines to describe their creative process. However, by viewing their work it is clear that some common traits were followed, just like in the case of extreme exaggeration or the development of characters that we described.

If we are looking at different kinds of animators to draw inspiration from, we must take a look at a genre that actually shares some practical obstacles with robot

[1] A group of nine animators that worked closely with Walt Disney since the debut feature Snow White and the Seven Dwarfs (1937) and onto The Fox and The Hound (1981).

Fig. 12.1 The muppet show's kermit the frog

animation. Puppets are physical characters that are built in order to move and be expressive, and are subject to the laws of physics of our real world. If we replace the word 'Puppets' with 'Social Robots' in this last sentence, it would still be valid.

Puppet animation grew especially popular with Jim Henson's 'The Muppet Show' [8]. Henson's puppets (Fig. 12.1) are generally very simple in movement. Most of them can only open and close their mouth, and wave their arms and body. It was impossible to actually convey human-like expressions with them, and that was not needed. By developing their own non-verbal language, animators were able to portray all kinds of different plots with them. By watching episodes of the series we can find that whenever a *muppet* wants to close its eyes, it will cover them with their hands, as the eyes cannot gaze or shut. This kind of tricks is very inspiring for robot animation.

It is empirically clear that if a character has only a mouth that can open and close, it is impossible to portray emotion by using just its face. That is where animation takes place. Most of the emotional expressions we find in puppets comes from the movement, and not just the poses.

There is no defined happy pose for a *muppet*. Instead, there is a bouncy movement with the arms waving around, that elicits the feeling of excitement and happiness. For fear, the mouth will tremble a lot, and the *muppet* will probably cover its eyes and assume a posture of withdrawal. An angry expression is achieved by leaning the *muppet* against the object or character of hate, closing its mouth, and pulling back its arms.

As in most inspiration from art, the best way to learn the practices of puppet animation is by watching the episodes and using them as reference footage.

In the realm of 3D animated characters, Walt Disney Animation Studios, Pixar, Dreamworks and Blue Sky have become established as the major studios. These studios have been leading teams of some of the best artists in the world to create critically acclaimed animation films such as 'Toy Story', 'Monsters, Inc.', 'Tangled', 'How to Train Your Dragon', 'Ice Age' and many more. In particular, one of Pixar's most popular films is WALL-E, which features a highly expressive animated robot as the main character. Once again John Lasseter, who is a cornerstone in the shift from hand-drawn to 3D animation, has took the time to present some tips for traditional animators to learn how to adapt and animate characters in the 3D world [19].

12.2.1 Disney's Twelve Principles of Animation

For reference, we present a small summary of the original Twelve Principles of Animation defined in 'The Illusion of Life' [30].

Squash and Stretch states that characters should not be solid. The movement and liquidness of an object reflects that the object is alive, because it makes it look more organic. If we make a chair squash and stretch, the chair will seem alive. One rule of thumb is that despite them changing their form, the objects should keep the same volume while squashing and stretching.

Anticipation reveals the intentions of the character, so we know and understand better what they are going to do next.

Staging is the way of directing the viewers attention. It is generally performed by the whole acting process, and also by camera, lights, sound and effects. This principle is related to making sure that the expressive intention is clear to the viewer. The essence of this principle is minimalism, keeping the user focused on what is relevant about the current action and plot.

Follow-Through and Overlapping Action are the way a character, objects or part of them inertially react to the physical world, thus making the movements seem more natural and physically correct. An example of Overlapping action would be hair and clothes that follow the movement of a character. Follow-through action is for example the inertial reaction of a character that throws a ball. After the throw, both the throwing arm and the whole body will slightly swing and tumble along the throwing direction.

Straight Ahead Action and Pose-to-Pose is about the animation process. An animator can make a character go through a sequence of well defined poses connected by smooth in-betweenings (Pose-to-Pose action), or sequentially draw each frame of the animation without necessarily knowing where it is heading (Straight-Ahead action).

Slow In and Slow Out is how the motions are accelerated (or slowed down). Characters and objects do not start or stop abruptly. Instead, each movement has an acceleration phase followed by a slowing down phase, unless it is clearly intended not to. Slow out can be confused with follow-through; however, follow-through extends the action, while the slow-out finishes it smoothly.

Arcs draw the trajectories of natural motions, making them feel less machine-like and more natural and organic. An example is a head that gazes from left to right. A typical robotic movement would make the head rotate only along its vertical axis. A natural movement will make the head slightly lean up or down towards the midpoint of the trajectory while rotating.

Secondary Action is an action that does not contribute directly to the expression of an action, but adds personality and life-likeness. An example would be breathing, blinking the eyes, or holding and scratching different parts of the body.

Timing is a dual principle that focuses especially on two different things. First, it can change how users perceive the emotion of a motion or the physical world in which the character exists. Second, it also relates to the story, and how the story is being told. It is about how the character pauses between the actions, and how it synchronizes to itself and the surroundings.

Exaggeration makes some features more wild and relevant, and is what makes the characters behave as cartoons, as opposite to the dull motion of humans in the real world. An example would be popping out the eyes when startled, or growing a huge red tomato-like head while shouting.

Solid Drawing is about correctly balancing volume and weight of characters and objects. It also warns against symmetric characters and expressions. Characters do not stand stiff and still, unless that is what they are intended to portray.

Appeal of a character is how it expresses and asserts its role, personality and relevance in a story. It is possibly the most subjective principle, as it also relates to how the character can make the viewers believe in its story.

12.2.2 Animation Curves

Animation Curves are tools that are particularly important for animators. An animation curve exists for each Degree of Freedom (DoF) that is being animated in a character, and it shows how that specific DoF varies over time [26].

Figure 12.2 shows the animation curve for the translation DoF of a hypothetical drag race car. In a drag race, the race car only drives forward at full speed. Because this animation curve shows the position changing over time, the speed of the car at some point of the curve is the tangent to the curve on that point (the first derivative). The second derivative (the rate of change of the tangent) thus represents the acceleration of the car.

By analyzing the curve, we see that the car starts by accelerating until about halfway through, when it reaches its maximum speed. We notice this because during the first part of the curve there is an accentuated concavity. Once the curve starts looking straight, the velocity is being kept nearly constant. In the end the car decelerates until it halts.

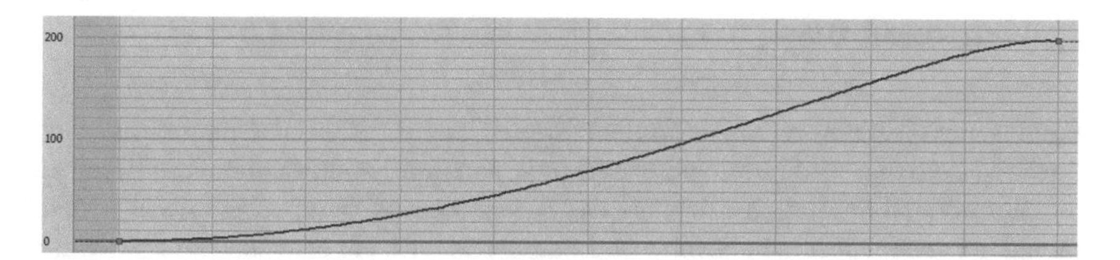

Fig. 12.2 The animation curve of the translation of a drag car accelerating until it reaches a top speed, and then decelerating until it halts. The vertical axis represents distance in generic units

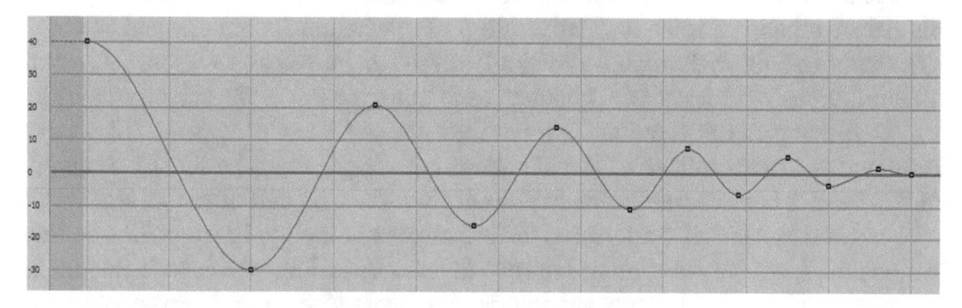

Fig. 12.3 The animation curve of the rotation of a pendulum that is dropped from 40° and balances until it stops. The vertical axis represents the angle in degrees

Animation curves can also be used to represent Rotation. Figure 12.3 shows the animation curve of the rotation of the pivot of a pendulum that is dropped from a height of 40°. It then balances several times while losing momentum due to friction and air resistance, until it stops.

In this curve we see some grey squares where the curve changes. These squares are actually key-frames that were used to design the animation. The curve is a spline interpolation of the movement between these key-frames.

By looking at each key-frame, we see that the angle goes from 40° to −30°, then to about 20°, and so on. Just like in the translation animation curve, the tangent of this curve also represents the velocity of rotation.

If we imagine the pendulum going through the lower-most position of its trajectory (which is the position in which it travels faster), that point would correspond to the 0° line, thus making sense that each spline between two key-frames is steeper at this point, than closer to the key-frames. As the pendulum loses energy and balances less, the steepness becomes lower, which reflects a lower speed, until it comes to a stop.

Animation curves therefore stand as a very important tool for representing, analyzing and adjusting animations. They can also be computationally processed just like a signal, in order to warp the animation and create animation effects. More importantly, the animation curves represent a concept that both animators and engineers can understand, and can use it to connect their thoughts, requirements and obstacles. Furthermore, they provide a technical interface that animators can use, and that can faithfully and mathematically model motion for robots.

12.3 Related Work

Various authors have previously worked towards the idea of robot animation as a well specified field that could even include its own principles of animation. Van Breemen initially defined animation of robots as 'The process of computing how the robot should act such that it is believable and interactive' [6]. He also showed how 'Slow In/Out' could be applied to robots, although he called it Merging Logic.

Wistort has also proposed some principles that should be taken into account when animating robots, which do not accurately follow the ones from Disney [32]. His list of principles refer to 'Delivering on Expectations', 'Squash and Stretch', 'Overlapping/Follow-through animation' (although he refers to it as Secondary Action), 'Eyes', 'Illusion of Thinking' and 'Engagement'. We actually consider that 'Delivering on Expectations' implies the same as Disney's 'Appeal', 'Illusion of Thinking' is closely related to 'Anticipation' and 'Engagement' refers to 'Staging'. Furthermore it is discussable whether or not Eyes must be part of robots at all.

Mead and Mataric also addressed the principles of Staging, Exaggeration, Anticipation and Secondary Action to improve the understanding of a robot's intentions by autistic children [20]. For exaggeration, they were inspired in a process used for the generation of caricatures, by exaggerating the difference from the mean.

Hoffman and Ju have presented some guidelines and techniques, especially based on previous experiences, about designing robots with their expressive movement in mind [15]. They provide useful insights on how the embodiment and expressive motion are tightly connected, and how the design of expressive behaviour may be considered as part of the design of the actual robot, and not just as an after-step.

12.3.1 Use of Animation Concepts and Techniques in Robots

In 2003, Breazeal and colleagues presented the Interactive Theatre [5]. This is one of the first robot animation systems to be developed with interactivity in mind, by blending (AI) and an artistic perspective. Several robotic anemones were animated in collaboration with animators to portray a lifelike quality of motion while reacting to some external stimuli like the approach of a human hand. These animations were driven by parameters which were controlled by a behaviour-based AI system to dynamically change the appearance of its motion depending on events captured by a vision system [13].

The AUR is a robotic desk lamp with 5 DoFs and an LED lamp which can illuminate in a range of the RGB color space [16]. It is mounted on a workbench and controlled through a hybrid control system that allows it to be used for live puppeteering, in order to allow the robot to be expressive while also being responsive. In AUR, the motion is controlled by extensively trained puppeteers, and was composed through several layers. The bottom-most layer moves each DoF based on a pre-designed animation that was made specifically for the scene of the play. If the robot was set to establish eye contact, several specific DoFs would be overridden by an inverse kinematics solution using CCD [31]. A final *animacy* layer added smoothed sinusoidal noise, akin to breathing, to all the DoFs, in order to provide a more lifelike motion to the robot.

Shimon is a gesture based musical improvisation robot created by Hoffman and Weinberg that plays a real marimba [17]. Its behaviour is a mix between its functionality as a musician, for which it plays the instrument in tune and rhythm, and being part of a band, for which it performs expressive behaviour by gazing towards its band mates during the performance.

Travis is a robotic music listening companion also created by Hoffman, that acts as an interactive expressive music dock for smart phones [14]. The system allows a user to dock a smart-phone and request it to play a music from some play-list. The robot plays it through a pair of integrated loudspeakers while autonomously dancing to the rhythm. The music beat is captured by real-time analysis in order to guide the robot's dance movements. Those movements are simple "head banging" and "foot tapping" gestures that are easily programmable.

More recently, Suguitan and Hoffman have created Blossom, a flexible, handcrafted social robot that abides several principles of animation such as squash and stretch, slow in/out and follow-through animation [28]. The robot was built using an innovative compliant tensile structure that allows it to be flexible even in the inside. The exterior has a soft woven cover that can deform and shift freely, thus accentuating its organic movement.

Various interactive social robots have been created at MIT's MediaLab that build on animation concepts and techniques [13]. In particular the AIDA[2] is a friendly driving assistant for the cars of the future. AIDA interestingly delivers an expressive face on top of an articulated neck-like structure to allow to it move and be expressive on a car's dashboard.

Takayama, Dooley and Ju have explored the use of animation principles using the PR-2 robot.[3] This is a large mobile robot with two arms, that can navigate in a human environment. The authors focused on the use of Anticipation, Engagement, Confidence and Timing to enhance the readability of a robot's actions [29]. Once again, the authors refer to 'Engagement', when in fact they follow the 'Staging' principle. Indeed, 'Staging' doesn't sound like a correct term to use in robot animation, because for the first time, we are having animated characters in real settings, and not on a stage. Doug Dooley, a professional animator from Pixar Animation Studios, collaborated on the design of the expressive behaviour so that the robot could exhibit a sense of *thought*, by clearly demonstrating the intention of its actions. *Thought* and *Intention* are two concepts that are in the core of character animation, and in the portrayal of the illusion of life. In this work, the authors also argue for the need of both functional and expressive behaviors, i.e., that some of the robot's behaviours would be related with accomplishing a given task (e.g. picking up an object; opening

[2]http://robotic.media.mit.edu/portfolio/aida (accessed March 02, 2019).

[3]http://www.willowgarage.com/pages/pr2/overview (accessed March 02, 2019).

a door), and that another part would concern its expressiveness in order to convey thought and emotion.

Gielniak et al. have successfully developed an algorithm that creates exaggerated variants of a motion in real-time by contrasting the motion signal, and demonstrated it applied to their SIMON robot [12]. The same authors have also presented techniques to simulate the principles of Secondary Motion [10] and of Anticipation [11] in robot motion.

Walt[4] is a social collaborative robot that that helps factory workers assemble cars. Walt uses a screen to exhibit an expressive face, icons or short animations. Its body is a concealed articulated structure that allows it to gaze around at its co-workers.

Several works by Ribeiro and Paiva aim at creating software technology and tools that allow animators and robot developers to work together. In particular, they have created Nutty Tracks, an animation engine and pipeline, aimed at providing an expressive bridge between an application-specific artificial intelligence, the perception of user and environment, and a physical, animated embodiment [22]. It is able to combine and blend multi-modal expressions such as gazing towards users, while performing pre-designed animations, or overlaying expressive postures over the idle- and gazing- behaviour of a robot.[5] Furthermore, Nutty Tracks can also be used or adapted as a plug-in in animation software such as Autodesk 3ds Max[6] and Maya,[7] SideFX Houdini[8] or even the open-source Blender software.[9] The composing of animation programs in the Nutty Tracks GUI follows a box-flow type of interface greatly inspired by other programming tools commonly used by artists, such as the Unreal Engine,[10] Pure Data[11] or Houdini (see Footnote 8). Figure 12.4 shows the Nutty Tracks GUI. Animation Controllers are connected into a chain of execution that generates and composes animation either procedurally or using animations and postures that were pre-designed using other animation software. The convergence between animation tools and a robot animation engine allows researchers to explore the use of animation principles in such autonomous interactions with humans by focusing, however, on the behaviour selection and management mechanisms, and on pre-designing particular animations that were solely selected and played back on the robots. The development pipeline for Nutty Tracks has also been briefly exemplified with the Keepon robot[12] [25].

More recently the same authors have created ERIK, a new inverse kinematics technique that allows an articulated robot with multiple DoFs (such as a manipulator), to exhibit an expressive posture while aiming towards any given direction [24]. The

[4]http://robovision.be/offer/#airobots (accessed March 02, 2019).

[5]http://vimeo.com/67197221 (accessed March 02, 2019).

[6]https://www.autodesk.com/products/3ds-max/overview (accessed March 02, 2019).

[7]https://www.autodesk.com/products/maya/overview (accessed March 02, 2019).

[8]https://www.sidefx.com/products/houdini (accessed March 02, 2019).

[9]https://www.blender.org (accessed March 02, 2019).

[10]http://www.unrealengine.com (accessed March 02, 2019).

[11]http://puredata.info (accessed March 02, 2019).

[12]https://vimeo.com/155593476 (accessed March 02, 2019).

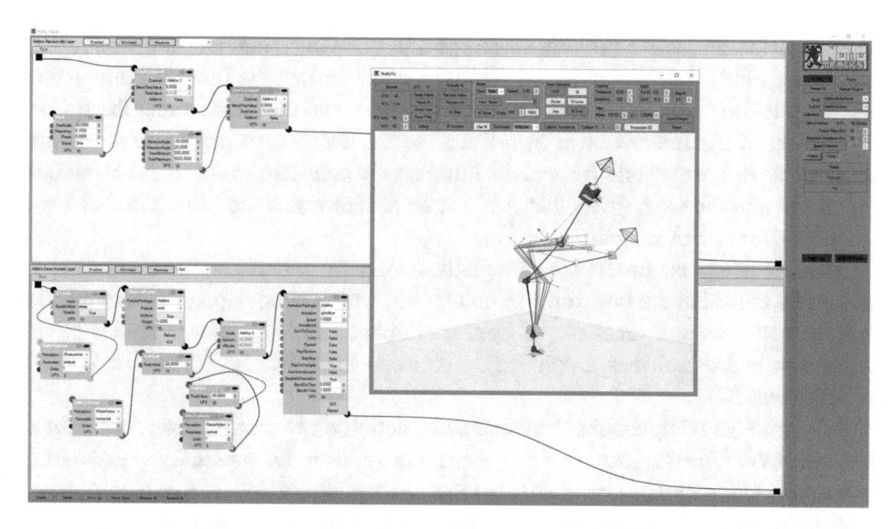

Fig. 12.4 The Nutty Tracks standalone GUI, used for composing animation programs, and to execute them in both a virtual window (for diagnostics) and on the real robot

technique is demonstrated using the custom built, and DIY[13] inspired low-fidelity craft robot Adelino.[14] The purpose of ERIK is to allow complex robots to interact with humans while exhibiting artistically-crafted expressions. By allowing simple, artist-designed expressive postures to be warped in real-time and turned to face any direction, while maintaining continuous movement that complies with the robot's mechanical constraints, the technique brings robot animation a step closer to typical artist-centered character animation pipelines.

12.4 Robot Animation

Before we move on to define our principles of animation for robots, we must first define robot animation. Most animation principles and guidelines report on designing particular motions. In the context of social robotics, our understanding is that robot animation is not just about motion. It is about making the robot seem alive, and to convey thought and motivation while also remaining autonomously and responsive. And because robots are physical characters, users will want to interact with them. Therefore robot animation also becomes a robot's ability to engage in interaction with humans while conveying the illusion of life.

One of the major challenges of bringing concepts of character animation into Human-Robot Interaction (HRI) is at the core of the typical animation process. While

[13]'Do-it-yourself'.

[14]https://vimeo.com/232300140 (accessed March 02, 2019).

in other fields, animation is directed at a specific story-line, timeline, and viewer (e.g. camera), in HRI the animation process must consider that the flow and timeline of the story is driven by the interaction between users and the AI, and that the spacial dimension of the interaction is also linked to the user's own physical motion and placement. Robot animation becomes intrinsically connected with its perception of the world and the user, given that it is not an absent character, blindly following a timeline over and over again.

This challenge is remarkable enough that character animation for robots can and should be considered a new form of animation, which builds upon and extends the current concepts and practices of both traditional and Computer-Graphics (CGI) animation and establishes a connection between these two fields and the field of robotics and AI.

We therefore complement Van Breemen's definition by stating that *robot animation consists of the workflow and processes that give a robot the ability of expressing identity, emotion and intention during autonomous interaction with human users*.

It is important to emphasize the word *autonomous*, as we don't consider robot animation to be solely the design of expressive motion for robots that can be faithfully played back (that would fall into the field of animatronics). Instead it is about creating techniques, systems and interfaces that allow animation artists to design, specify and program ***how*** the motion will be generated, shaped and composed throughout an interaction, based on the behaviour descriptions that are computed by the AI.

One such common and basic behaviour we take as example is face-tracking, which directs a robot's gazing towards the face of the human with whom it is interacting. For a simple robot, e.g., neck with two DoFs, it is easy to implement face-tracking by extracting a vertical and horizontal angle from the system's perception components (e.g. camera, Microsoft Kinect). These two angular components can directly control the two individual motors of the robot's neck. However this is a very limited conception of face-tracking behaviour, and also a very limited form of gaze control in general. Gazing behaviour can also be compound, by featuring not only face-tracking, but also used deictically towards surrounding objects, and in conjunction with other static or motive expressions (e.g. posture of engagement, nodding in agreement). Therefore in the context of robot animation, such gazing behaviour should consider not only an orientation but also the expressivity portrayed through the behaviour in an interactive manner. Furthermore, one must consider that compound gazing behaviour should also be adopted for use with complex embodiments that feature multi-DoF necks, such as industrial manipulators, by considering e.g. the manipulator's endpoint to take on the expressive role of being the character's head, i.e. taking inspiration on an animated snake.

12.4.1 Principles of Robot Animation

A general list of Principles of Robot Animation should also address principles related to human-robot interaction. In this list however, we refrain from deepening such topic

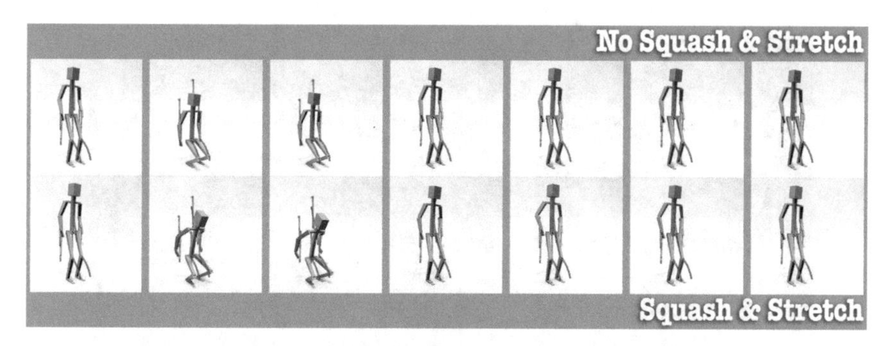

Fig. 12.5 An animation sequence denoting the principle of squash and stretch. The red marks represent the trajectory of the most relevant joints

that is already subject of intensive study [1, 2, 9, 21]. Instead, we have looked into principles and practices of animators throughout several decades, and analysed how the scientific community can and has been trying to merge them into robot animation.

We have noted that not all principles of traditional animation can apply to robots, and that in some cases, robots actually reveal other issues that had not initially existed in traditional animation. Most of these differences are found due to the fact that robots (a) interact with people (b) in the real, physical world.

The following sections reflect our understanding of how the Principles of Robot Animation can be aligned. Although they are stated towards robots, the figures presented show an animated human skeleton, as an easier depiction and explanation of use. Each principle is also demonstrated on the EMYS and the NAO robots in an online video,[15] which can be watched as a complement to provide further clarification. The video first demonstrates each principle using the same humanoid character presented in this section, and then follows with a demonstration of each principle first using the NAO robot, and then using the EMYS robot.

12.4.1.1 Squash and Stretch

For robots to use this principle, it sounds like the design of the robot must include physical squashing and stretching components. However, besides relying on the design [15, 28], we can also create a squash and stretch effect by using poses and body movement.

In Fig. 12.5 we can see how flexing arms and legs while crouching gives a totally different impression on the character. Following the rule of constant volume, if the character is becoming shorter in height, it should become larger in length, and a humanoid robot can perform that by correctly bending its arms and legs. Figure 12.6 presents a snapshot from the video (see Footnote 15) illustrating how this principle looks like on the NAO robot.

[15]https://vimeo.com/49122495 (accessed March 02, 2019).

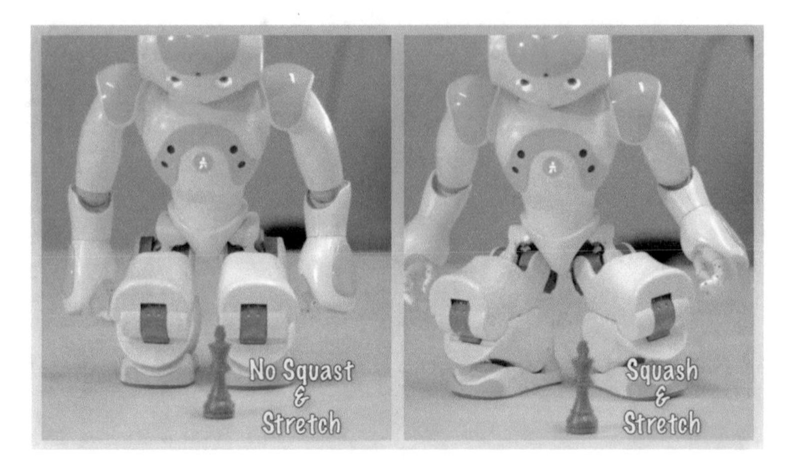

Fig. 12.6 The principle of squash and stretch shown on the NAO robot

12.4.1.2 Anticipation

Anticipating movements and actions helps viewers and users to understand what a character is going to do. That anticipation helps the user to interpret the character or robot in a more natural and pleasing way [29].

It is common for anticipation to be expressed by a shorter movement that reflects the opposite of the action that the character is going to perform. A character that is going to kick a ball, will first pull back the kicking leg; in the same sense, a character that is going to punch another one will first pull back its body and arm. A service robot that shares a domestic or work environment with people can incorporate anticipation to mark, for example, that it is going to start to move, and in which direction, e.g., before picking up an object, or pushing a button.

In Fig. 12.7 we can see how a humanoid character that is going to crouch may first slightly stretch upwards.

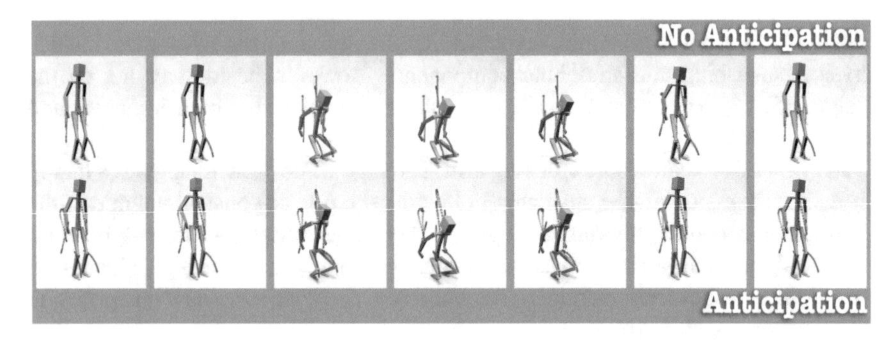

Fig. 12.7 An animation sequence denoting the principle of anticipation. The red marks represent the trajectory of the most relevant joints

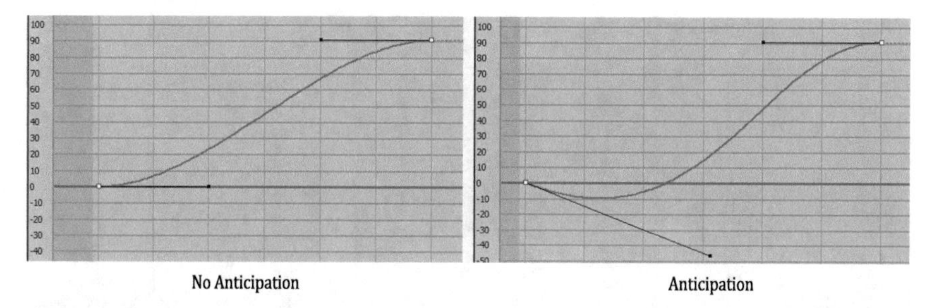

Fig. 12.8 Animation curves demonstrating anticipation. The left curve does not have anticipation; the right curve does

The concept can be better explained by looking at a simple animation curve example. Figure 12.8 shows two animation curves for a 90° rotation of an object. On the left we see a simple animation curve, and at the start and end keyframes we see the tangent of the curve at that point.

On the right we have the same keyframes, but the tangent of the initial keyframe has been changed. Just by adjusting this tangent we have made the object start by slightly rotating 10° backwards before performing the mentioned 90° rotation, thus creating an anticipation effect.

12.4.1.3 Intention

This principle was formerly known as Staging in the traditional principles of animation. In robots, staging results in several things. First, it notes that sound and lights can carefully be used to direct the users' attention to what it is trying to communicate. Second, if a robot is interested in, for example, picking up an object, it can show that immediately by facing such object [29]. In either cases, the key here is showing the intention of the robot.

We can see in Fig. 12.9 a simple idea of a humanoid character that is crouching over a teapot to eventually pick it up. The character immediately looks at the teapot, so users know it is interested in it, and eventually guess that it is going to pick it up, much before the action happens.

That connects Intention with Anticipation; the difference is that while Anticipation should give clues about what the robot is going to do immediately, Intention should tell users about the purpose of all that he is doing, as a pre-action, before the actual action starts. In a crouch-and-pick-up situation, for example, the robot will perform three actions—crouch, pick-up and stand. We should see Anticipation for each of these actions. The Intention, however, should reflect the overall of what the character is *thinking*—it will start looking at the object even before crouching, and will start looking at the destination to where it will take the object even before starting to turn towards that direction.

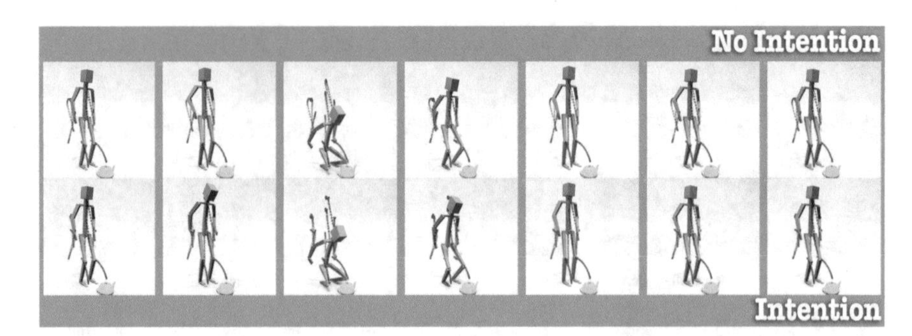

Fig. 12.9 An animation sequence denoting the principle of Intention. The red marks represent the trajectory of the most relevant joints

12.4.1.4 Animated, Procedural and Ad-Hoc Action

This principle was adapted from the Straight-Ahead and Pose-to-Pose action and has strong technical implications on the animation system development. It originally talks about the method used by the animator while developing the animation. Straight-ahead animation is used when the animator knows what he wants to do but has not yet foreseen the full sequence, so he starts on the first frame and goes on sequentially animating until the last one. In pose-to-pose, the animator has pre-planned the animation and timing, so he knows exactly how the character should start and end, and through which poses it should go through.

In robots, this marks in the difference between playing a previously animated sequence, a procedural sequence, or an ad-hoc sequence. As a principle of robot animation, it results in a balance between expressivity, naturalness and responsiveness.

A previously *animated* sequence is self-explanatory. It was carefully crafted by an animator using animation software, and saved to a file in order to be played-back later on. That makes it the most common type of motion to be considered today in robot animation. However it suffers from a lack of interactivity, as the trajectories are played-back faithfully regardless of the state of the interaction. The motion is *procedural* when it is generated and composed from a set of pre-configured motion generators (such as sine-waves). On the other hand, it is *ad-hoc* if it is fully generated in real-time, using a more sophisticated motion-planner to generate the trajectory (e.g. obstacle-avoidance; pick-and-place task). We can say that playing an animation sequence that has previously been designed by an animator is a pose-to-pose kind of animation, while, for example, gaze-tracking a person's face by use of vision, or picking up an arbitrary object would be straight-ahead action.

A pose-to-pose motion can also contain anchor points at specific points of its trajectory (e.g. marking the beat of a gesture), so that the motion may be warped in the time-domain to allow synchronization between multiple motions. Those anchor-points would stand as if they were *poses*, or key-frames in animation terms. The concept of pose-to-pose can also become ambiguous in some case, such as in multi-modal synchronization, where, e.g. an ad-hoc gaze and an animated gesture should

Fig. 12.10 An animation sequence denoting the principles of pre-animated and ad-hoc action. The red marks represent the trajectory of the most relevant joints

meet together at some point in time using anchor-points that define the meeting point for each of them. In that case, the straight-ahead action, planned ad-hoc, can result in an animated sequence generated in real-time, and containing anchors placed by the planner. From there it can be used as if it was a pose-to-pose motion to allow both motions to meet.

It currently sounds certain that the best and most expressive animations we achieve with a robot are still going to be pre-animated. However the message here is that these different types of animation methods imply their own differences in the robotic animation system, and that such system should be developed to support them.

In Fig. 12.10 we can see on top a character performing a pre-animated and carefully designed animation, while in the bottom it is instantaneously reacting to gravity which made the teapot fall, and as such is performing an ad-hoc, straight-ahead animation.

While performing ad-hoc action, like reacting immediately to something, it might not be so important, in some cases, to guarantee principles of animation—if someone drops a cup, it would be preferable to have to robot grab it before it hits the ground, instead of planning on how to do it in a pretty way and then fail to grab it. In another case, if a robot needs to abruptly avoid physical harm to a human, it is always preferable that the robot succeeds in whatever manner it can. An ad-hoc motion planner therefore is likely to not contain many rules about animation principles, but act more towards functional goals (see the "Functional versus Expressive Motion" section in [29]).

12.4.1.5 Slow In and Slow Out

For robot animation, Slow In and Out motion may me implemented within software in two different modalities: interpolation or motion filtering.

The former can be applied when the motion is either pre-animated, or fully planned before execution, so that the system has the full description of the trajectory points.

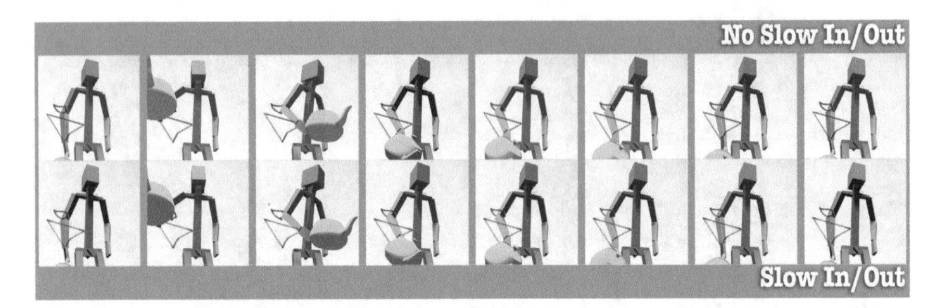

Fig. 12.11 An animation sequence denoting the principle of slow in/out. The red marks represent the trajectory of the most relevant joints. Notice how more frames are placed at the points of the trajectory where the motion changes in direction, in particular within the triangular-shaped portion. More spacing between points, using a fixed time-step, yields a faster motion

By tweaking the tangent type of the interpolation of the animation curve, it is possible to create accelerating and slowing down effects. By using a slow in and slow out tangent, the interpolation rate will slow down when approaching or leaving a key-frame. This means that in order to keep timing unchanged, the rate of interpolation will have to accelerate towards the midpoint between two key-frames. Van Breemen called this Merging Logic and showed how it could be applied to the iCat [6]. In alternative, when the motion is generated ad-hoc, a feed-forward motion filter can be used to saturate the velocity, the acceleration and/or the jerk of the motion.

A careful inspection of the red trajectories in Fig. 12.11 will show us the difference between the top animation and the bottom animation. Each red dot represents an individual frame of the interpolated animation, using a fixed time-step. We can see that in the bottom animation the spacing between the frames changes. It gathers more frames near the key-poses, and less between them. This causes the animation to have more frames on those poses, thus making it slow down while changing direction. Between two key poses the animation accelerates because the interpolation generated less frames there.

This is more noticeable if we look at the animation curves. Figure 12.12 shows a very simple rotation without Slow-In/Out (left) and with (right). In the left image we used linear tangents for the interpolation method, while in the right we used smooth spline tangents.

We can see that with a linear interpolation, the curve looks straight, meaning that the velocity is constant during the whole movement. By using smooth tangents the movement both starts, stops and changes direction with some acceleration, which makes it look smoother.

12.4.1.6 Arcs

Taking as example a character looking to the left and the right. It shouldn't just perform a horizontal movement, but also some vertical movement, so that its head

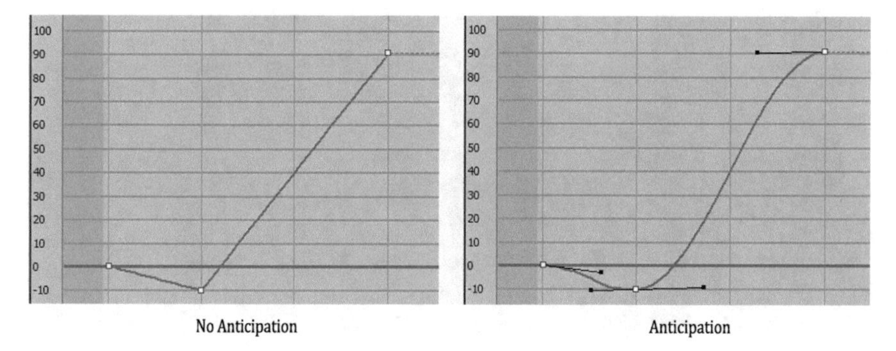

Fig. 12.12 Animation curves demonstrating slow in and slow-out. The left curve does not have slow in/out; the right curve does

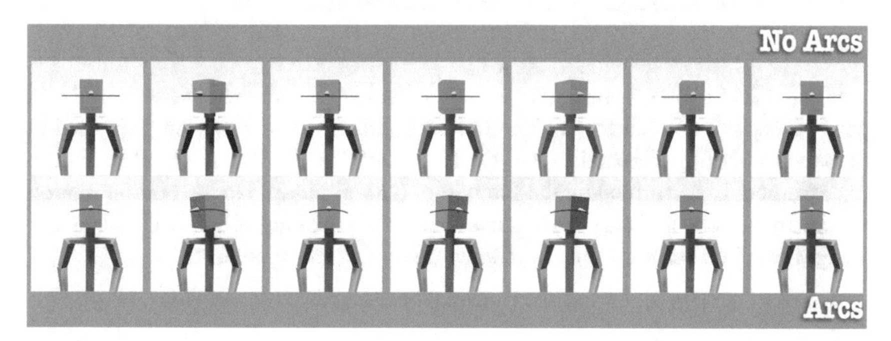

Fig. 12.13 An animation sequence denoting the principle of arcs. The red marks represent the trajectory of the most relevant joints

will be pointing slightly upwards or downwards while facing straight ahead. We can see that illustrated in Fig. 12.13.

This principle is easy to use in pre-animated motion. However, in order to include it in an animation system, we would need to be able to know in which direction the arcs should be computed, and how wide the angle should be. If we have that information, then the interpolation process can be tweaked to slightly bend the trajectory towards that direction, whenever it is too straight.

What actually happens with robots is that depending on the embodiment, it might actually perform the arcs almost automatically. Taking as example a humanoid robot, when we create gestures for the arms, they will most likely contain arcs, due to the fact that the robot's arms are rigid, and as such, in order for the them to move around, the intrinsic mechanics will lead the hands to perform arched trajectories. In traditional animation this principle was extremely relevant as the mechanics of the characters were not rigidly enforced as they are in robots. Arcs still pose as an important principle to be considered in robot animation, both for pre-animated motions and also as a rule in expressive motion planners.

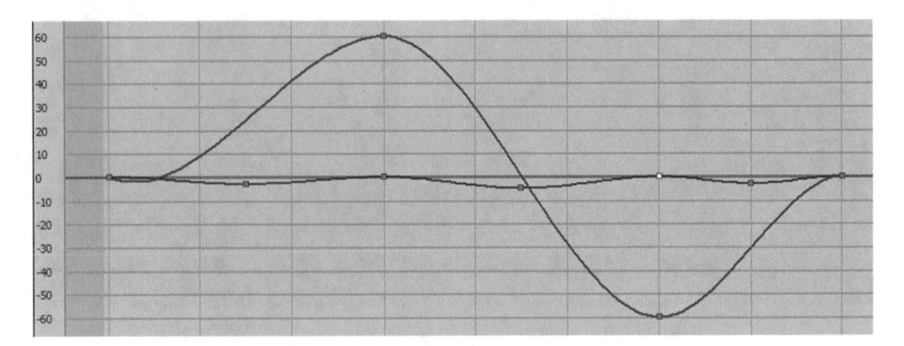

Fig. 12.14 Animation curves demonstrating arcs. The blue curve is the panning DoF, rotating from the rest pose, to its left (60°) and then to its right (−60°), and then back to rest. During this motion, the pitch joint (red curve) slightly waves between those key-frames

Figure 12.14 shows a character gazing sideways. The yellow cone represents the gazing direction at each frame. The red curve illustrates the motion trajectory on the panning DoF (horizontally) and the Pitch DoF (vertically). On the top motion, no movement is performed on the Pitch joint (straight line). On the bottom motion, instead of performing only Yaw movement while looking around, the head also changes its Pitch between each keyframe of the Yaw movement.

12.4.1.7 Exaggeration

Exaggeration can be used to emphasize movements, expressions or actions, making them more noticeable and convincing. As such, it can also make robots seem more like actual characters and not just machines.

Although there are several levels of exaggeration, for robots it is interesting to look at exaggeration of actual movements. It is actually a feature that can be implemented in animation systems by contrasting the motion signal [12].

Figure 12.15 shows not only an amplification of the most relevant features of an animation, but also an added feature—an 'anticipation' backward step. This is meant to show that exaggeration can consist of more then just contrasting the signal, and that by exaggerating the anticipation we can also make the actual action seem more powerful. Because this kind of practice may endanger the robot's surroundings and users if not correctly planned, it is recommended only within pre-animated motion, or for performance and entertainment robots in which the robot's surroundings and mechanical reach are guaranteed to be safe.

Figure 12.16 presents a snapshot from the video[15] illustrating how this principle looks like on the NAO robot, while Fig. 12.17 show the same for the EMYS robot.

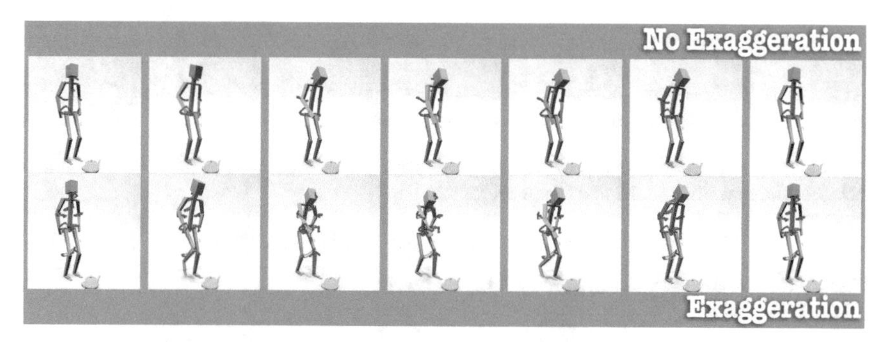

Fig. 12.15 An animation sequence denoting the principle of exaggeration. The red marks represent the trajectory of the most relevant joints

Fig. 12.16 The principle of exaggeration exemplified on the NAO robot

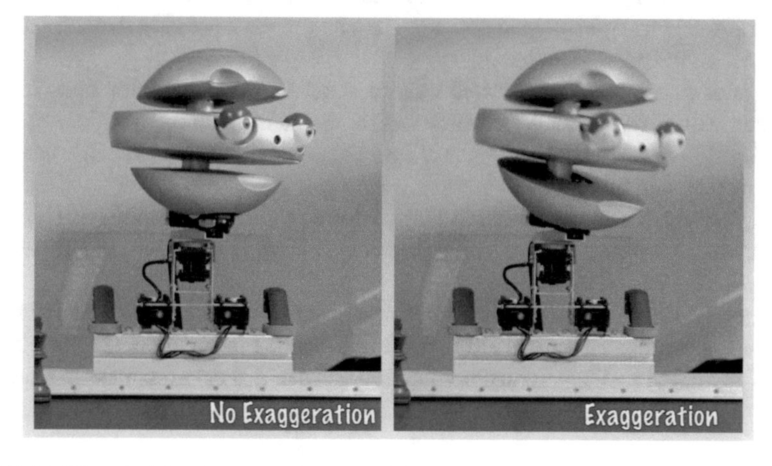

Fig. 12.17 The principle of exaggeration exemplified on the EMYS robot

Fig. 12.18 An animation sequence denoting the principle of secondary action. The red marks represent the trajectory of the most relevant joints

12.4.1.8 Secondary Action and Idle Behavior

During a conversation, people often scratch some part of their bodies, look away or adjust their hair. In Fig. 12.18 we can see a character that is crouching to approach the teapot, and in the meanwhile scratches its gluteus. Using secondary action in robots will help to reinforce their personality, and the illusion of their life.

A character should not stand stiff and still, but should contain some kind of Idle motion, also known as *keep-alive*. Idle motion in robots can be implemented in a very simplistic manner. Making them blink their eyes once and a while, or adding a soft, sinusoidal motion to the body to simulate breathing (lat. *anima*) contribute strongly to the illusion of life.

In the case of facial idle behaviour such as eye-blinking, during a dramatic facial expression these will often go unnoticed or may even disrupt the intended emotion. It is better to perform them at the beginning or end of such expressions, rather than during. Similarly, blinking also works better if performed before and between gaze-shifts.

12.4.1.9 Asymmetry

This principle was derived from the traditional principle of Solid Drawing. Although the traditional principle seemed not to relate with robots, it actually states some rules to follow on the posing of characters.

It states that a character should neither stand stiff and still, nor does it stand symmetrically. We generally put more weight in one leg than on the other, and shift the weight from one leg to the other. It also suggests the need for the idle behavior, and how it should be designed.

The concept of asymmetry stands both for movement, for poses and even for facial expression. The only case in which we want symmetry is when we actually want to convey the feeling of stiffness.

Figure 12.19 shows a character portraying another Principle—Idle Behavior, while also standing asymmetrically. This Idle Behavior is performed by the simulation of breathing and by slightly waving its arms like if they were mere pendulums.

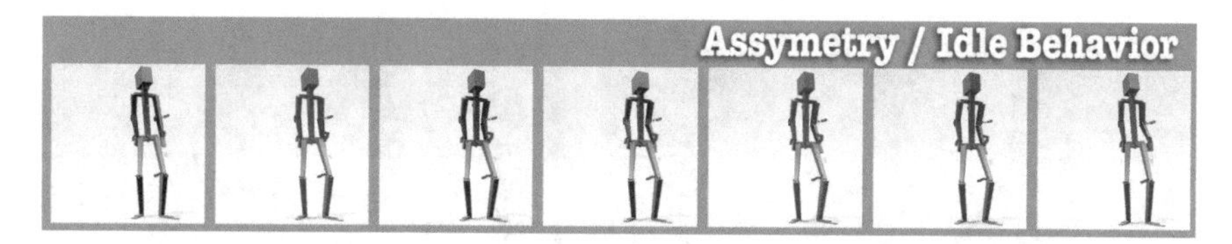

Fig. 12.19 An animation sequence denoting the principles of asymmetry and idle behavior. The red marks represent the trajectory of the most relevant joints

12.4.1.10 Expectation

This principle was adapted from the original Appeal. If we want a viewer or user to love a character, then it should be beautiful and gentle. If we are creating an authoritative robot, it should have more dense and stiff movements. Even if one wants to make viewers and users feel pity for a character (such as an anti-hero), then the character's motion and behaviour should generate that feeling, through clumsy and embarrassing behaviours.

Figure 12.20 shows two characters performing the same kind of behavior, but one of them is performing as a formal character like a butler, while the other is performing as a clumsy character like an anti-hero. In this case the visual appearance of the character was discarded. However, if we had a robotic butler, we would expect him to behave and move formally, and not clumsy.

The expectation of the robot drives a lot of the way users interpret its expression. It relates to making the character understandable, because if users expect the robot to do something that it doesn't (or does something that they are not expecting) they fill fail to understand what they are seeing.

Wistort refers to Appeal as 'Delivering on Expectations' [32], and his arguments have inspired us to agree. He considers that the design and behavior of a robot should

Fig. 12.20 An animation sequence denoting the principle of expectation. The red marks represent the trajectory of the most relevant joints. Notice how the clumsy version balances the teapot around instead of holding it straight, and waves around its left ar instead of holding it closer to its body, delivering a feeling of discourtesy

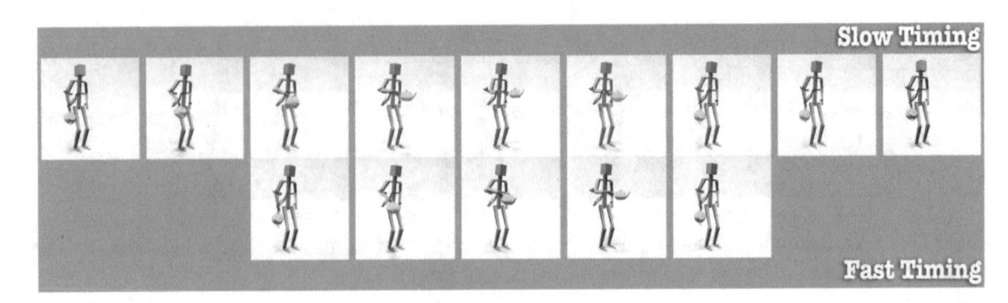

Fig. 12.21 An animation sequence denoting the principle of timing. The red marks represent the trajectory of the most relevant joints

meet, so if it is a robotic dog, then it should bark and wag its tail. But if it is not able to do that, then maybe it should not be a dog. The Pleo robot[16] for example, was designed to be a toy robot for children. So the design of it as a dinosaur works very good, as it does not cause any specific expectation in people—as people do not know any living dinosaurs, and as such, they don't know if Pleo should be able to bark or fetch, so they don't expectation him to be able to do any of that.

12.4.1.11 Timing

Timing can help the users to perceive the physical world to which the robot belongs. If the movement is too slow, the robot will seem like it is walking on the moon.

However, timing can also be used as an expression of engagement. Some studies have revealed a correlation between acceleration and perceived arousal. A fast motion often suggests that a character is active and engaged on what it's doing [27, 29].

Being able to scale the timing is useful to be able to express different things using the same animation, just by making it play slower or faster. In Fig. 12.21 we get a sense that the top character is not engaged as much as the lower character, because we see it taking longer to perform the action. It may even feel like the character is bored with the task. In the fast timing case we are showing less frames of the same animation, to give the impression of it being performed faster. In reality, that would be the result, as a faster paced animation would require less frames to be accomplished using a fixed time-step.

As a principle of robot animation, timing is something that should be carefully addressed when synthesizing motion e.g. using a motion-planner. Such synthesizer will typically solve for a trajectory that meets certain world-space constraints, while also complying with certain time-domain constraints such as the kinematic limits that the robot is allowed to perform. In many cases, a very conservative policy is chosen, i.e., the planner is typically instructed to move the robot very slowly in order to keep as far away as possible from its kinematic limits. However, such a rule may be adding some level of unwanted expressiveness to the motion. We therefore argue that

[16] www.pleoworld.com (accessed March 02, 2019).

when using such planners it is important to consider, within the safety boundaries of the robot's kinematic limits, ways of generating trajectories that can exploit the time-domain in a more expressive way.

12.4.1.12 Follow-Through and Overlapping Action

This principle works like an opposite of anticipation. After an action, there is some kind of reaction—the character should not stop abruptly.

We should start by distinguishing these two concepts here. *Follow-through* animation is generally associated with inertia caused by the character's movement. An example of follow-through is when a character punches another one, and the punching arm doesn't stop immediately, but instead, even after the hit, both body and arm continue to move a bit due to inertia (unless it is punching an 'iron giant'). Overlapping is an indirect reaction caused by the character's action. An example of overlapping is for example the movement of hair and clothes which follow and overlap the movement of the body.

Using follow-through with robots requires some precaution because we do not want the inertial follow-through to hurt a human or damage any other surroundings. Follow-through might also cause a robot to loose balance, so it seems somewhat undesirable. Many robot systems actually will try to defend themselves against the follow-through caused by its own movements, so why would we want it?

In first instance, we consider that follow-through should better not be used in most robots, especially for the first reason we mentioned (human and environment safety). However, when it can be included at a very controlled level, namely on pre-animated motion, it might be useful to help mark the end of an action, and as such, to help distinguish between successive actions. Unlike anticipation, however follow-through is much more likely to be perceived by humans as dangerous, because it can give the impression that the robot slightly lost control over its body and strength. We would therefore imperatively refrain from using it on any application for which the perception of safety is highest, such as in health-care or assistive robotics.

Overlapping animation depends mostly on the robot's embodiment and aesthetics. It might serve as a tip for robot design, by including fur, hair or cloth on some parts of the robot, that can help to emphasize the movement [28]. As such, we find no need to include overlapping animation into the animation process of robots per se, because whatever overlapping parts that the robot might have, should be 'animated' by natural physics. Therefore if one wishes to use it, it should be considered as an animation effect that is drawn by the design of the robot's embodiment, and thus should be developed initially at the robot design stage.

12.5 Animation Tools for Social Robots

When including creative artists such as animators into the development workflow, one of the first question that arises is the tools that the artists can use to author and develop expressive behaviour for the robot. Typically those artists are designated to produce only pre-authored animation files that can be played back by the animation engine. This may be achieved by either developing a custom-build GUI that allows them to directly develop on the system's tools, data types and configurations, or to allow the artists to use their familiar animation tools such as 3dsmax[6], Maya[7], Houdini[8] or Blender[9]. These existing animation packages allow to export animation files using general-purpose formats such as Autodesk FBX.[17] That requires the animation engine to support loading such formats, and to convert them into the internal representation of pre-animated motions. Alternatively, and as most of those software support scriptable plug-ins, one may develop such a plug-in that allows to export the motion data into a format that is designed specifically for the animation engine.

Upon our introduction of the programmable animation engine, and of animation programs, it also becomes necessary to understand how the animators can contribute to such animation programming, alongside with their participation in the motion design.

12.5.1 Animation Design Tools and Plug-Ins

We argue that for simple cases, developing an e.g. FBX import for the actual animation engine run-time environment is a good choice. In this case the learning curve for the animators is almost inexistent, given that they will be working on their own familiar environment. They will only need to adapt to specific technical directions such as maintaining a properly named and specific hierarchy for the joints and animatable elements, so that those can be properly imported later on. When the nature of the project or application does not allow to rely on third-party, or proprietary software, then the only option may be to develop a custom animation GUI, which poses as the most complex and tedious one. However our feeling has been that the creation of plug-ins for existing, third-party animation software provides a good balance between development effort, usability, user-experience and results.

The creation of plug-ins for existing animation software includes the same advantages and requirements as in the first case, of developing an animation-format importer for the engine. Animators will be familiar with the software, but may have to comply with certain technical directions in order for the plug-in to be able to properly fetch and export the motion data. Figure 12.22 shows an example of the Nutty Tracks plug-in for Autodesk 3dsmax. By having the EMYS embodiment already loaded in the Nutty Tracks engine, the plug-in can create an animatable rig for the robot, through the click of a single button, based on the embodiment's hierarchical

[17] https://www.autodesk.com/products/fbx/overview (accessed March 02, 2019).

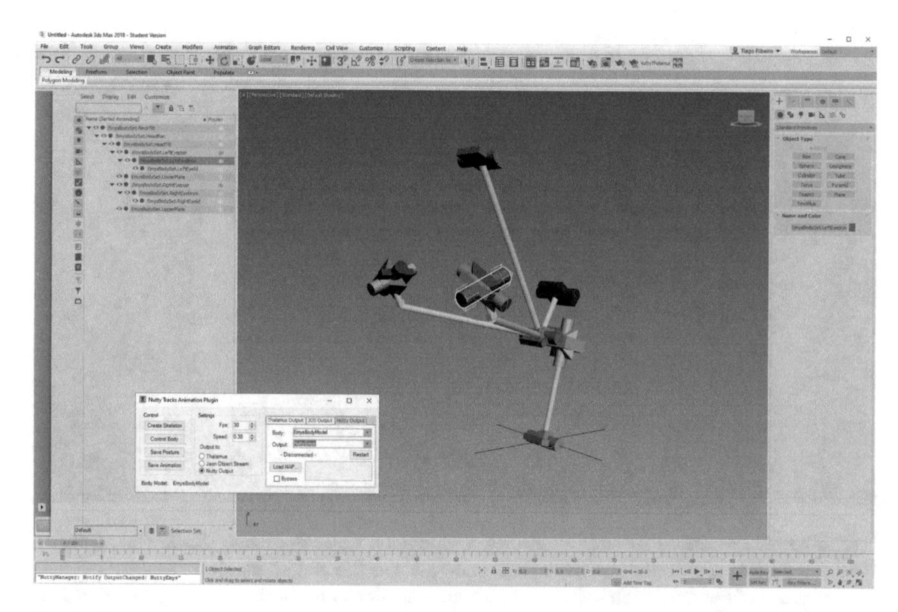

Fig. 12.22 A screenshot of the Nutty Tracks plug-in for Autodesk 3dsmax, illustrating the skeletal animation rig created by the plug-in. An animator can generate this rig through the simple click of a button, and then use the plug-in to export the final animation to a Nutty-compatible animation file

specification including rotation axes, joint limits, etc. Optionally it may even include the actual geometry of the robot for a more appealing experience. From here on an animator may animate each of the gizmos that were created for each of the robot's animatable DoFs, using his or her typical workflow and techniques.

However, the development of such a plug-in also allows to augment the creative development workflow, by adding visual guides directly into the viewports of the animation software, in order to represent technical constraints that are required specifically for robots, such as kinematic ones (e.g. velocity, acceleration, jerk limits). Figure 12.23 shows an example of a plug-in developed for Autodesk Maya, to show the *trajectory-helper* of a given mobile robot platform, which highlights the points in the trajectory that break some of the robot's kinematic constraints. In this case, green means that the trajectory is within the limits, while the other colors each represent a certain limit violation, such as maximum velocity exceeded (orange), or maximum acceleration exceeded (pink) or maximum jerk exceeded (red). Based on this visual guide, the animator knows where the trajectory must be corrected, and is able to readily preview how the fix will look like, while making any further adjustments to the motion in order to ensure the expected intention or expression is properly conveyed without exceeded the physical limits of the robot.

Other useful features may be to perform automatic correction of such constraints, while rendering the result directly within the animation environment, thus allowing the animators to fix the motion that results from enforcing such constraints, in a more interactive way. From what we have gathered however, animators are typically

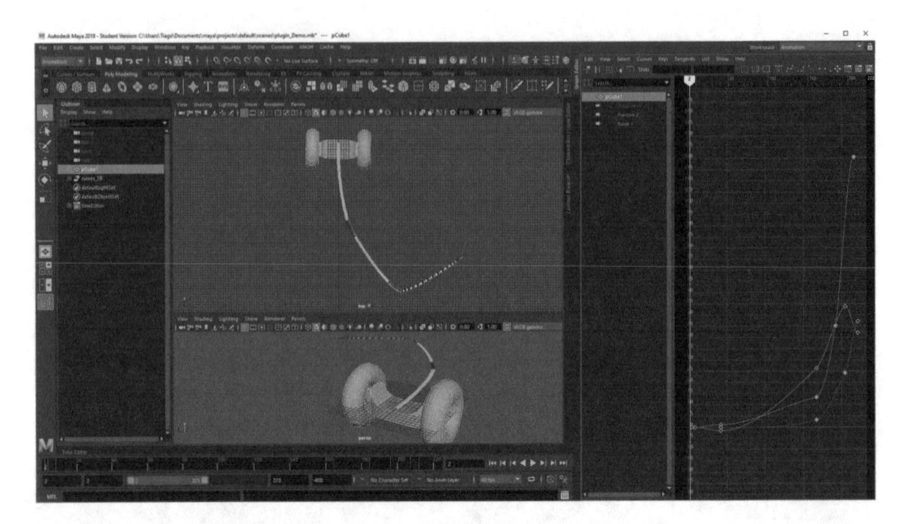

Fig. 12.23 A screenshot illustrating the robot-animation trajectory-helper feature implemented through a plug-in into Autodesk Maya. This feature draws the motion trajectory as a path directly into the scene of the animation software, and highlights the points of the trajectory that break any of the robot's kinematic limits

not happy to have a tool that can change and control their animations. Instead, the preferred option is to keep the artist-animated version of robot untouched by the plug-in, and to create an additional copy of the same robot model. This copy, which we call the *ghost*, will, in turn, not be animatable or even selectable by the animator, but instead, will be fully controlled by the plug-in. Therefore, when the animator is previewing the playback of its animation, the plug-in will take that motion and process it in order to enforce the kinematic limits. The resulting corrected motion is however applied only to the ghost, which therefore moves along with the animated robot. If at any point, the animated motion did exceed the limits, the ghost will be unable to properly follow the animated model due to the signal saturation, which allows the animator to have a glimpse not only of where the motion is failing to comply with the limits, but also how it would look like if the limits were enforced. In some cases the animator might actually feel that the result is acceptable, even if the originally designed motion would report limit violations on a trajectory-helper solution such as the one of Fig. 12.23. Note that in the case of the *ghost-helper* technique, whenever the final animation is exported, it should be exported from the *ghost* robot, which contains the corrected motion, and not the animated robot which does not.

In summary, the two major robot-animation features we have presented, and that can be provided through the use of animation software plug-ins, are the *trajectory-helper*, as presented in Fig. 12.23, and the *ghost-helper*, described in the previous paragraph. Depending on the animator's preferences, and the scripting capabilities of the animation environment, either one or both of the features can be used. The ghost-helper seems to provide a more agile solution, as the animators aren't required to fix all the limit violations. As long as they accept the motion provided through the ghost, the problem is considered to be solved, thus allowing them to complete animations quicker than using the trajectory-helper. The trajectory-helper however allows an animator to better ensure that all the points of the trajectory are smooth and natural, and especially that the automatic correction (achieved e.g. through signal saturation) will not introduce any other unexpected phenomena. This feature is especially important when animating multiple robots,[18] to ensure that each of the individual auto-corrections do not place the robots in risk of colliding.

Without the ability to preview or at least evaluate the animated motion directly within the animation environment, the animators would need to jump between their software, and a custom software that solves and reports on those issues, while providing typically a mediocre or even no visual feedback on what is happening, and what needs to be fixed. Besides making it a more complex workflow, that option also hinders and breaks the animator's own creative process.

Finally, an additional feature that can be developed through plug-ins for existing animation software is the ability to directly play the animations through the robot software or interactive pre-visualisation system. This allows the animators to include testing and debugging into their workflow, by being able to see what will happen with their animations once they become used during interaction with the users and the environment.

12.5.2 Animation Programming Tools

Animators working with social robot application are required to learn some new concepts about how motion works on robots, in order to identify what can or cannot be done with such physical characters, as opposed to what they are used to do in fully virtual 3D characters. Besides having to adapt to certain technical requirements when building their characters and animation rigs, they may also need to learn how to interact with some other pieces of software that will allow them to pre-visualize how the designed motion will look on the robots during actual interactions.

At some point the character animators will acquire so many new competencies and knowledge that they become actual *robot animators*, an evolution of animators that besides being experts on designing expressive motion for robots, may also have learned other technical skills as part of the process. One such skill is what we call animation programming. The difference between a non-robot-programming animator, and a programming-robot animator is akin to the difference between a texture artist and a shader artist (or lighting artist) in the digital media industry. The texture artist is a more traditional digital artist that composes textures that are *statically* used within digital media. A shader artist is able to take such textures, or other pattern-generators, and configure the shaders (i.e., programs) to adapt and change according to the environment parameters and applications. The shaders are, in that sense, *pro-*

[18]https://gagosian.com/exhibitions/2018/urs-fischer-play/ (accessed March 02, 2019).

grammable textures. Similarly, animation programs are *programmable* animations. These can take in certain parameters that are provided throughout the interaction, and using motion sources such as animation files, static poses, or signal generators such as sine-waves and Perlin noise, compose them into a final resulting motion in a way that was both directed by an animator, and managed in real-time interaction by the AI, robotic and perceptual system.

Animation programs can, at a very basic level, be specified by some kind of mark-up code. However, taking inspiration from currently existing tools such as Autodesk's Slate material editor,[19] or the Unreal Material Editor,[20] which provide artist-friendly shader-programming interfaces, we argue for the creation of similar, artist-friendly, animation-programming editors. These new animation programming tools can be built from scratch as standalone GUI application (e.g. Nutty Tracks), or using game development tools such as the Unity Engine,[21] which allows for the scripting of new interface tools. In this case, because a game engine such as Unity3D already provides 3d visualization and animation tools, it could be extended with a robot animation programming tool in order to become a fully-fledged robot animation designing, programming and pre-visualization tool.

Nutty Tracks provides an example of how such an animation-programming editor may be presented.[22] Its programmable animation GUI is also shown in Fig. 12.24. It was conceptualized to allow an animator to load and pre-visualize how animations and expressive postures designed in another software (e.g. 3dsmax) will look like when procedural layers of motion are added, such as ones that generate idle-behaviour, user-face tracking, or inverse kinematics. Such output motion is composed in real-time in Nutty Tracks, while allowing the parameters to be tinkered with, something which could not be properly visualized within the typical animation design software. However the process of composing and tweaking the animation program using animation blocks follows a workflow that is similar to the one found on other artist-friendly applications that inspired us.

Despite such effort, it will still be the case that such an animation program editor will pose as a truly novel tool for the animators, with a steep learning curve. An animator may e.g. be familiar with the concept of an animation layer, which does not match the one used in the visual animation program editor. The idea of composing programmable animations using operator- and generator-blocks may have a parallel with certain motion control nodes found in some animation software, but the way they are used and composed may not seem intuitive or obvious for the traditional 3D animator. As such, it is required that these tools are developed with a user-centered design perspective, in close collaboration with the end-users, who are the actual

[19] https://knowledge.autodesk.com/support/3ds-max/learn-explore/caas/CloudHelp/cloudhelp/2017/ENU/3DSMax/files/GUID-7B51EF9F-E660-4C10-886C-6F6ADE9E8F56-htm.html (accessed March 02, 2019).

[20] https://docs.unrealengine.com/en-us/Engine/Rendering/Materials/Editor/Interface (accessed March 02, 2019).

[21] https://www.unity3d.com (accessed March 02, 2019).

[22] https://vimeo.com/67197221(accessed March 02, 2019).

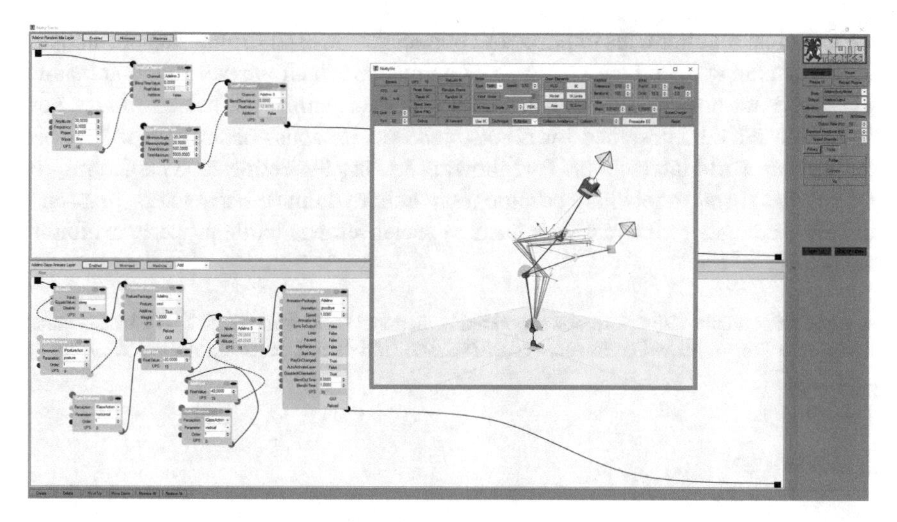

Fig. 12.24 The Nutty Tracks GUI, used for animation programming in a multi-layer, multi-block visual editor. Within the figure, we see several different animation blocks which either generate or operate on motion signals. The integrated 3D visualizer allows and artist to preview the output of the motion based on how he tinkers with the parameters. It additionally includes an inverse kinematics interactive visualizer which allows an animator to tweak the solver, in order to adjust the generated motion to the robot's kinematic capabilities

animators, and to ensure the GUI provides an understandable translation between the animator's mindset, and the underlying mechanics and pipeline of the animation engine.

12.6 Conclusion

Throughout this chapter we have presented our perspective on how robot animation can become an integral process in the development of social robots, based on theories and practices that have been created through the last century, in the fields of both traditional and 3D computer-graphics character animation. We have introduced and described the 12 principles of robot animation, as a foundation that aims at aiding the transfer of the previous character animation practices into the new robot animation ones. In the traditional character animation workflow, characters and their motions are designed to be faithfully played-back on screens. One of the most relevant steps in this transition is the ability to not only design, but also program how animations should be shaped, merged and behave during interaction with human users. We must therefore introduce new techniques and methods that allow such artistically crafted animations to become not only interactive (such as in video-games), but to interact in the real world, with real users. Such new techniques and methods will be provided by new tools and workflows that are designed with artists in mind, and that aim at the technical requirements imposed by robotics. Upon establishing such techniques, such artists may become a new type of animators which we call robot animators. These are not only experts in traditional character animation, but also know how animation must be designed for robots, and how it should be adapted and shaped during real-world interactions. By following and implementing such paradigms, we expect that social robots may become more akin to animated characters, in a sense that they are able to interact with users in social settings while properly exhibiting the illusion of life.

Acknowledgements This work was supported by national funds through FCT—Fundação para a Ciência e a Tecnologia with references UID/CEC/50021/2019 and SFRH/BD/97150/2013.

References

1. Alves-Oliveira, P., Küster, D., Kappas, A., & Paiva, A. (2016). Psychological science in HRI: Striving for a more integrated field of research. In *AAAI Fall Symposium Technical Report, FS-16-01-(PG-2-5)* (pp. 2–5). https://doi.org/10.1016/j.intimp.2016.04.032.
2. Baxter, P., Kennedy, J., Senft, E., Lemaignan, S., & Belpaeme, T. (2016). From characterising three years of HRI to methodology and reporting recommendations. In *ACM/IEEE International Conference on Human-Robot Interaction, 2016-April (December 2017)* (pp. 391–398). https://doi.org/10.1109/HRI.2016.7451777.
3. Beck, J. (2005). *The animated movie guide*. Cappella Bks: Chicago Review Press.
4. Bendazzi, G. (1994). *Cartoons: One hundred years of cinema animation*. Indiana University Press.
5. Breazeal, B. C., Brooks, A., Gray, J., Hancher, M., Mcbean, J., Stiehl, D., et al. (2003). Interactive theatre. *Communications of the ACM*, *46*(7), 76–84.
6. Breemen, A. V. (2004). Animation engine for believable interactive user-interface robots. In *IEEE/RSJ International Conference on Intelligent Robots and Systems—IROS '04* (Vol. 3, pp. 2873–2878). https://doi.org/10.1109/IROS.2004.1389845.
7. Canemaker, J. (1996). *Tex Avery: The MGM years, 1942–1955*. Turner Publishing.
8. Finch, C. (1993). *Jim Henson: The works*. Random House.
9. Fong, T. (2003). A survey of socially interactive robots. *Robotics and Autonomous Systems*, *42*(3–4), 143–166. https://doi.org/10.1016/S0921-8890(02)00372-X. http://linkinghub.elsevier.com/retrieve/pii/S092188900200372X.
10. Gielniak, M. J., Liu, C. K., & Thomaz, A. (2010). Secondary action in robot motion. In *RO-MAN, 2010 IEEE* (pp. 3921–3927). https://doi.org/10.1109/ICRA.2011.5980348.
11. Gielniak, M. J., & Thomaz, A. L. (2011). *Anticipation in Robot Motion. Roman.*
12. Gielniak, M. J., & Thomaz, A. L. (2012). Enhancing interaction through exaggerated motion synthesis. In *ACM/IEEE International Conference on Human-Robot Interaction—HRI '12* (p. 375). https://doi.org/10.1145/2157689.2157813.
13. Gray, J., Hoffman, G., Adalgeirsson, S.O., Berlin, M., & Breazeal, C. (2010). Expressive, interactive robots: Tools, techniques, and insights based on collaborations. In: *ACM/IEEE International Conference on Human-Robot Interaction—HRI '10—Workshop on What do Collaborations with the Arts have to Say About Human-Robot Interaction*.
14. Hoffman, G. (2012). Dumb robots, smart phones: A case study of music listening companionship. In *IEEE International Symposium on Robot and Human Interactive Communication—RO-MAN '12* (pp. 358–363). https://doi.org/10.1109/ROMAN.2012.6343779.
15. Hoffman, G., & Ju, W. (2014). Designing robots with movement in mind. *Journal of Human-Robot Interaction*, *3*(1), 89. https://doi.org/10.5898/JHRI.3.1.Hoffman.

16. Hoffman, G., Kubat, R., & Breazeal, C. (2008). A hybrid control system for puppeteering a live robotic stage actor. In *IEEE International Symposium on Robot and Human Interactive Communication—RO-MAN '08* (pp. 354–359). https://doi.org/10.1109/ROMAN.2008.4600691.
17. Hoffman, G., & Weinberg, G. (2010). Gesture-based human-robot Jazz improvisation. In *IEEE International Conference on Robotics and Automation—ICRA '10* (pp. 582–587). https://doi.org/10.1109/ROBOT.2010.5509182.
18. Lasseter, J. (1987). Principles of traditional animation applied to 3D computer animation. In *ACM International Conference on Computer Graphics and Interactive Techniques—SIGGRAPH '87* (Vol. 21, No. 4, pp. 35–44). https://doi.org/10.1145/37402.37407.
19. Lasseter, J. (2001). Tricks to animating characters with a computer. In *ACM International Conference on Computer Graphics and Interactive Techniques—SIGGRAPH '01* (Vol. 35, No. 2, pp. 45–47). https://doi.org/10.1145/563693.563706. http://dl.acm.org/citation.cfm?id=563706.
20. Mead, R., Mataric, M. J. (2010). Automated caricature of robot expressions in socially assistive human-robot interaction. In *ACM/IEEE International Conference on Human-Robot Interaction—HRI '10—Workshop on What do Collaborations with the Arts Have to Say About Human-Robot Interaction.*
21. Murphy, R., Nomura, T., Billard, A., & Burke, J. (2010). Human-robot interaction. *IEEE Robotics & Automation Magazine*, *17*(2), 85–89. https://doi.org/10.1109/MRA.2010.936953. http://ieeexplore.ieee.org/document/5481144/.
22. Ribeiro, T., Dooley, D., & Paiva, A. (2013). Nutty tracks: Symbolic animation pipeline for expressive robotics. In *ACM International Conference on Computer Graphics and Interactive Techniques Posters—SIGGRAPH '13* (p. 4503).
23. Ribeiro, T., & Paiva, A. (2012). The illusion of robotic life principles and practices of animation for robots. In *ACM/IEEE International Conference on Human-Robot Interaction—HRI '12, 1937* (pp. 383–390).
24. Ribeiro, T., & Paiva, A. (2017). Animating the Adelino robot with ERIK. In *Proceedings of the 19th ACM International Conference on Multimodal Interaction* (pp. 388–396). ACM, Glasgow, UK.
25. Ribeiro, T., Pereira, A., Di Tullio, E., & Paiva, A. (2016). The SERA ecosystem: Socially expressive robotics architecture for autonomous human-robot interaction. In *AAAI Spring Symposium.*
26. Roberts, S. (2004). *Character animation in 3D*. Elsevier.
27. Saerbeck, M., & Bartneck, C. (2010). Perception of affect elicited by robot motion. *Journal of Personality*, 53–60. https://doi.org/10.1145/1734454.1734473. http://portal.acm.org/citation.cfm?doid=1734454.1734473.
28. Suguitan, M., & Hoffman, G. (2019). Blossom: A handcrafted open-source robot. *ACM Transactions on Human-Robot Interaction*, *8*(1).
29. Takayama, L., Dooley, D., & Ju, W. (2011). Expressing thought. In *ACM/IEEE International Conference on Human-Robot Interaction—HRI '11* (p. 69). https://doi.org/10.1145/1957656.1957674.
30. Thomas, F., & Johnston, O. (1995). *The illusion of life: Disney animation*. Hyperion.
31. Wang, L. C., & Chen, C. (1991). A combined optimization method for solving the inverse kinematics problems of mechanical manipulators. *IEEE Transactions on Robotics and Automation*, *7*(4), 489–499. https://doi.org/10.1109/70.86079.
32. Wistort, R. (2010). Only robots on the inside. *Interactions*, *17*(2), 72–74. http://dl.acm.org/citation.cfm?id=1699792.

Chapter 13
Adapting Movements and Behaviour to Favour Communication in Human-Robot Interaction

Katrin Lohan, Muneeb Imtiaz Ahmad, Christian Dondrup, Paola Ardón, Èric Pairet, and Alessandro Vinciarelli

Abstract In this chapter we are presenting an overview on how adaptation of movement and behaviour can favour communication in Human-Robot Interaction (HRI). A model of a communication space based on a action-reaction classification is presented. Past research in HRI is presented for verbal, non-verbal and adaptation of communication. Further, the influence of human aware navigation is discussed and concepts like proxemics, path planing and robot motion are presented. The chapter discusses possible explicated and implicated methods of adaptation as well as it is identifying interruption concepts for communication.

13.1 Introduction

Richmond et al. [112] state that "the importance of communication in human society has been recognized for thousands of years, far longer than we can demonstrate through recorded history". A common example of communication can be of a tourist environment where the tourist manages to interact effectively in diverse scenarios despite being in an exotic culture, where s/he does not share the language, mainly through using non-verbal means including gesture-based communication [133]. This also suggests that humans have an intrinsic capability of adapting their style of communication-based on the situation, or through understanding social cues based on the voice pitch, tone, mood, gestures of the communicating individuals [141]. Con-

K. Lohan (✉) · M. I. Ahmad · C. Dondrup · P. Ardón · È. Pairet
Social Robotics Group, Computer Science Department, Heriot-Watt University,
Edinburgh Center for Robotics, Edinburgh, Scotland
e-mail: katrin.lohan@googlemail.com
URL: http://www.macs.hw.ac.uk/SocialRoboticsGroup

K. Lohan
EMS Institute for Development of Mechatronic Systems, NTB University of Applied
Sciences in Technology, FHO Fachhochschule Ostschweiz, Buchs, Switzerland

A. Vinciarelli
School of Computing Science, University of Glasgow, Glasgow, Scotland
URL: https://www.gla.ac.uk/schools/computing

N. Noceti et al. (eds.), *Modelling Human Motion*,
https://doi.org/10.1007/978-3-030-46732-6_13

sequently, the field of social robotics also envision robots in the future to become a part of society and intends to enable them to perform different kinds of communications. Hence, efforts are made to enable robots to adapt their communication across different social settings [1].

Human communication can be classified as either direct or indirect communication. Direct communication refers to a medium that conveys a clear message and also possesses an intended action. On the contrary, indirect Communication refers to a medium that conveys an acted message rather than directly saying it to the receiver with either an intended on unintended action. Based on the understanding on direct and indirect communication, we, in Fig. 13.1, present a model of the communication space based on actions and reactions particularly in relation to human-robot communication. We define robot's non-verbal communication (gestural, facial expressions based communication) where the action can be indirect and reaction can either be intended or unintended. Robot's Verbal communication (conversational, speech based communication) is defined where the action is direct and reaction can either be intended on unintended. We also define another space of communication referring to mobile robot's navigation where the communication can either be direct (approaching the receiver) and reaction can be intended or indirect (walking beside the receiver) and reaction can be unintended.

We, in this chapter, classify different movements (gestures, facial expressions, eyes, navigation) and behaviours (conversation, dialogue) based on our aforementioned model of communication space. We later present literature on robots adapting their behaviour or communication according to these movement of the receiver (Human). It is important to note that we are presenting communication strategies as dependent on actions and reactions forming a space.

13.2 Communication in HRI

13.2.1 Non-verbal Communication

Nonverbal communication plays a major role in human-human interactions, especially when it comes to conveying socially and psychologically relevant information [67, 111]. Perception and interpretation of nonverbal behavioural cues (facial expressions, vocalisation, etc.) take place, to a large extent, outside conscious awareness [138]. The corresponding cognitive processes are so spontaneous and pervasive that people have been shown to react in the same way to both the cues on displayed by humans and those displayed by machines capable of human-like behaviour (Fig. 13.2).

In other words, machines capable of simulating nonverbal behaviour activate the same cognitive and psychological processes in their users as other humans do. Such a phenomenon is known as *Media Equation* [109] and its main consequence is that

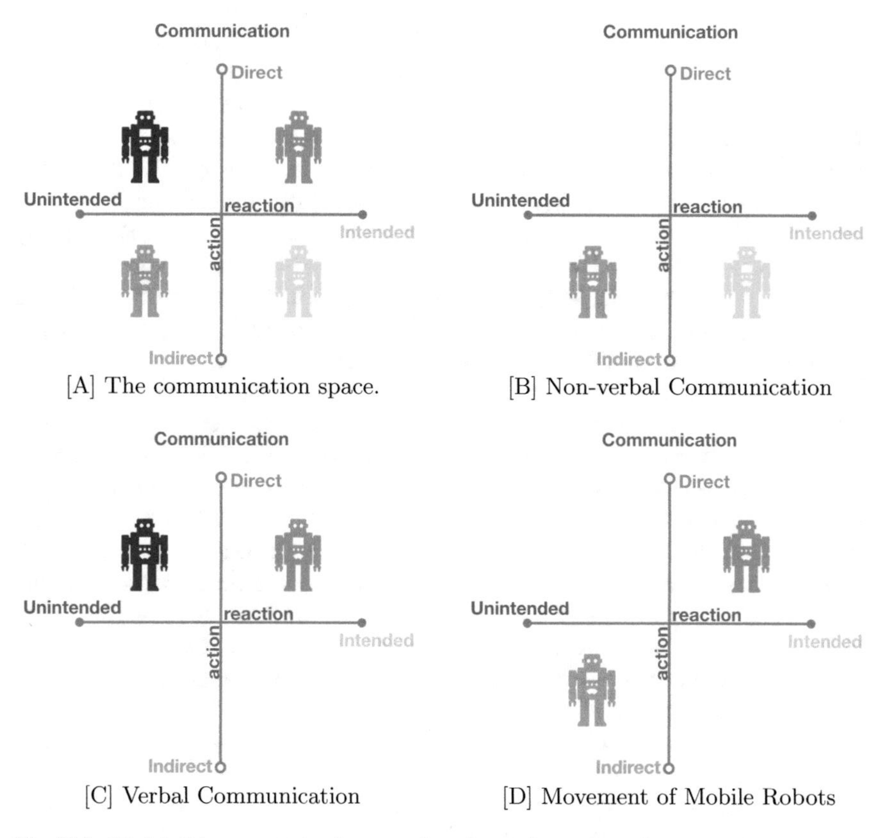

Fig. 13.1 Model of the communication space based on action and reaction

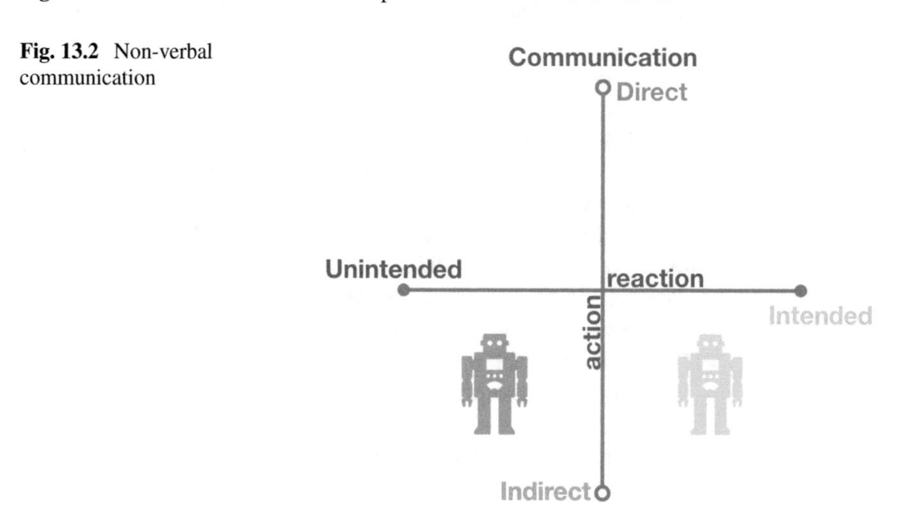

Fig. 13.2 Non-verbal communication

machines displaying human-like behaviour can convey socially and psychologically relevant information in the same way as people do in their interactions.

It is probably for the reasons above that the *International Journal of Social Robotics* has published no less than 39 articles on the artificial generation of nonverbal cues in its first 10 years of life (roughly the 10% of the 425 works that the journal has published in this period). According to the terminology introduced in [139], nonverbal behavioural cues can be grouped into five major classes called *codes*, namely *face and head behaviour* (facial expressions, gaze, head pose, etc.), *gestures and posture* (hand movements, body orientation, etc.), *vocal behaviour* (everything in speech except words), *appearance* (shape of the body, attractiveness, etc.), and *use of space and environment* (proxemics, interpersonal distance, etc.).

The majority of the works focuses on the behaviour of face and head (22 works out of the 39 under examination). In 9 cases, the experiments show that there is a relationship between the use of gaze, in particular eye contact, and the perception that human users develop of the robot, especially when it comes to trust [23, 124]. Some of these works have further shown that making the perception of the robot more positive has an effect on the outcome of the interaction with the users, including, e.g., improved message retention [93] and recall of short stories in children [64]. For what concerns facial expressions (6 works out of the 22), the problem most commonly addressed is their effective representation, i.e., how to convey all the messages that a human face can convey while having at disposition only a few degrees of freedom [17, 30]. Finally, the synthesis of the head pose (6 of the 22 articles dedicated to the face and head behaviour code) is used mostly to investigate the expression of emotions (see, e.g., [85, 106]).

The second most commonly explored code is gestures (12 works) and posture (4 works). In both cases, the problem that tends to be addressed most frequently is the expression of emotions. Particular attention has been paid to the use of deictic gestures (pointing to objects or places in the environment where the interaction between robots and their users takes place) to ensure that the rapport between robots and humans is reinforced, whether in terms of higher engagement (see, e.g., [7]) or improved immediacy (see, e.g., [64]).

A significant number of articles (7 out of the 39 examined in this section) are dedicated to vocal behaviour and, in particular, to the use of synthetic speech in Human-Robot Interaction (HRI). Like in the case of other codes, the problem most frequently addressed is the expression of emotions, in particular through the use of prosody [37] and non-linguistic utterances [108]. However, in line with the psychological literature showing that nonverbal speech properties interplay significantly with the impression people convey, several articles have addressed the problem of improving the perception people develop about a robot through the use of synthetic vocal cues (see, e.g., [23, 108]).

The rest of the 39 works examined in this section rely on the synthesis of the last two codes, namely appearance (4 works) and use of space and environments (3 works). In the first case, the main attempt is to elicit the perception of typically human characteristics such as animacy, intelligence and gender (see, e.g., [16, 28]). In the second case, the accent is on the use of the physical distance between robots and users as a social cue, especially when it comes to conveying social roles [68] and immediacy [64]. In addition to journal publications, non-verbal communication in navigation has been addressed by a number of conference articles. These describe methods such as prompting [105], i.e. small movements to communicate ones intention such as inching forwards at an intersection or hesitation [42], or legible movement [76] to communicate the intention and goal of the robot while driving/walking.

Overall, the analysis presented in this section confirms that the role of nonverbal communication is as important in HRI as it is in Human-Human interaction. The attempts to synthesise nonverbal behavioural cues appears to cover all the *codes* that where identified in [139] (see beginning of this section) and to address the main goals that nonverbal communication appears to address in the case of humans, namely expressing emotions, conveying impressions, regulating interaction, etc. The main difference with respect to the psychological literature, is the relatively low number of works trying to use multimodal stimuli (only 8 out of 39). However, this might depend on technical difficulties and on the wide spectrum of possible embodiments that can make it difficult, if not impossible, to combine multiple cues.

In addition to the above, the social robotics community addresses two nonverbal communication channels that the psychological literature has not considered for different reasons. The first is the use of touch, a form of communication that has been recognised as a possible code to be added to the five considered so far. However, at least in the Western culture, the use of touch is limited only to private settings that tend to be less accessible to scientific research which is why analysis of touch has not been studied in great depth. The second is the use of non-human nonverbal cues such as lights, e.g. [84], or acoustic signals that a robot can use, but a human cannot. In both cases, the publication of works on the International Journal of Social Robotics can be considered as a further confirmation of how significant nonverbal communication is in HRI.

13.2.2 Verbal Communication

Tomasello [133], highlighted that humans communicate to request assistance's, to transmit information to others, and to share attitudes as a way of connecting with each other. Communication is a joint activity which largely depends on the ability to keep common attention, to share the relevant background knowledge and joint experience in order to get the content across and make sense in the exchanges [33]. Both [132] and [56] support the theory that language originated when early hominids started gradually changing their primate communication systems, acquiring the ability to form a theory of other minds and a shared intentionality.

Language structure can be based on systems of sounds (speech), gestures (sign languages), or graphic or tactile symbols (writing). Here we are interested in sound systems for language to construct meaning. Lohan et al. [79] have shown that there is a strong relation between spoken words and the semantic of the word for action description represented by a different behaviour profile of the receiver of the action

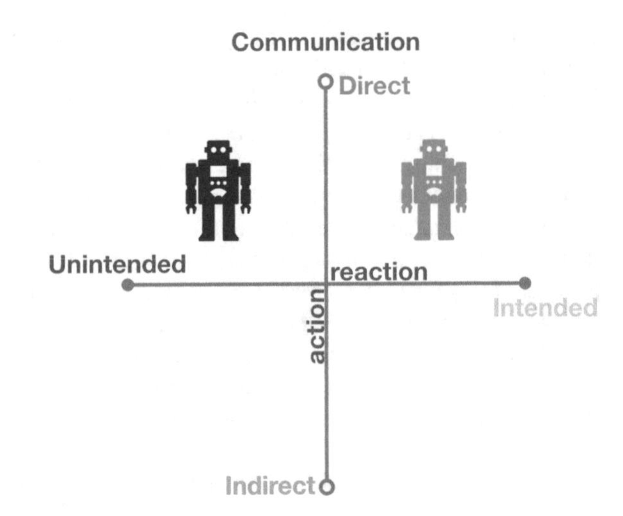

Fig. 13.3 Verbal communication

description. Their work suggest that this behaviour change also indicates concept understanding. Furthermore, they describe an influence on humans' movement when explaining different actions [79]. Verbal communication in form of words is thus not a one way street, but lies in a social communication of at least two partners and changes the recipients behaviour. Furthermore, the sound signal in our communication helps humans to structure the conversation into a time dependent system that can convey further information, e.g. in which order to complete sub-tasks as suggested by Bilac et al. [19] and Theofilis et al. [130]. When putting yourself into a robot's shoes these features become vital to follow an ongoing communication with a human.

By using concepts like contingency [50] and acoustic packaging [87], robots can not only appear to engage with humans in an interaction, but can tune their input towards their perceptional needs [46, 77, 78, 118] (Fig. 13.3).

13.2.3 Adapting Robot Behaviours to User's Social Cues

As highlighted above, humans communicate both verbally and non-verbally in various forms to complete a social communication/interaction, hence, social roboticists have also attempted to apply both principles during Human-Robot Communication. Based on the theoretical and empirical foundation of the impact of verbal and non-verbal cues during communication, researchers have implemented novel means for social robots to adapt their communication to these social cues such as: affective states (emotions), gestures, voice tone, and several others [1]. A recent systematic review on the adaptivity during HRI across health-care, education, public spaces and in-home domains revealed that most of the robot's communication adaptation has been based on the user's emotions or facial expressions and gestures during the interaction [1]. Different methods to adapt robots' communication applied in various Human-Robot Communication contexts have resulted in several positive findings in terms of users' attitudinal preferences as well as improving their level of social engagement [4, 74] and task performances [3]. The review also suggested that most adaptation has been applied in the context of games and it remains an open question to understand when should a robot interrupt a user or communication during the interaction.

We understand that Levinger's model of human-human relationship development explains the reason for the positive findings with respect to integration adaptation in human-robot communication [75]. Levinger [75] presented a model highlighting five stages of human relationships: (1) *acquaintance*, (2) *buildup*, (3) *continuation*, (4) *deterioration* and (5) *termination*. We are particularly interested in the first three stages to describe the theoretical relevance of the finding related in literature on empirical evaluation of adaptivity in HRI. There exists a number of factors that involve acquainting with someone (human) such as first impressions, physical appearance, behaviour, attitude and personality [44]. According to one of the attitude similarity theories, the similarity of attitudes, individual preferences, previous relational history is among the reinforcing factors towards creating an element of attraction between the two individuals [26]. Other factors include common circumstance between the two individuals [95].

The second and third stage of Levinger's model deals with the maintenance of the human relationship. We understand that a number of behaviours are performed by humans to maintain a relationship. These behaviours have been categorised into two types (routine and strategic behaviours) [123]. Routine behaviours are defined as "those behaviours where people engage in for other reasons which serve to maintain a relationship as a side effect (such as performing daily tasks together)" [18]. On the other hand, strategic behaviours are those "which individuals enact with the conscious intent of preserving or improving the relationship" [123]. Particularly, we are interested in the strategic behaviours such as: have a social dialogue, recalling past events, providing support, giving advise or increasing trust [43].

Keeping the theoretical perspective of human-human relationship in mind, researchers in HRI have also highlighted various human-robot relationship maintenance strategies. Researchers believe that different strategic behaviours could be applied to the robots during long-term interaction in different social settings [18]. These strategic behaviours include adapting to the dialogue, recalling user' past events, understanding and reacting to user emotions, and several similar behaviours [47]. Similarly, existing methods to adapt robot's communication according to user verbal and non-verbal behaviours enable the robot to generate similar strategic behaviours, consequently, there exists a relevance between the existing findings and the aforementioned human-human relationship theories. In essence, it can be inferred from most findings that humans create relationships with robots in a similar fashion. When a robot adapts its behaviour through understanding human emotions or through understanding their gestures, it creates an element of attraction and it, as a result, generates an increase of interest during the Human-Robot Communication.

13.3 Adapting the Movements of Mobile Robots to Favour Communication

One of the most common examples of adapting the movements of a robot to favour direct and indirect communication is navigation. It can easily be seen that being outside of the field of view of your interaction partner, e.g. [22], or being too far away to be perceived in enough detail to make out gestures or hear sounds, e.g. [86], is vastly detrimental to communication. For this reason, navigation of mobile robots is an important factor to improve communication with a human interaction partner. This ranges from finding the optimal distance to interact with someone verbally or via gestures to planning paths that maintain a certain formation with a walking human which allows you to still be perceived and heard. This section introduces some of the most used and interesting approaches to adapt the movements of a mobile robot on a 2D-plane, i.e. navigation, to favour explicit communication. Implicit communication also plays a major role in robot navigation but will not be addressed here (Fig. 13.4).

13.3.1 *Distance*

The most commonly used principle in human-aware navigation (adaptive path planning in the presence of humans) is the so-called *proxemics* which is a term describing interpersonal distances and was coined by Hall [54]. This theory divides the space around a human into four distinct zones, i.e. *Intimate Space*, *Personal Space*, *Social Space*, and *Public Space*, which are themselves divided into a close and far phase. Figure 13.5 shows these zones and distances with the public space having no defined

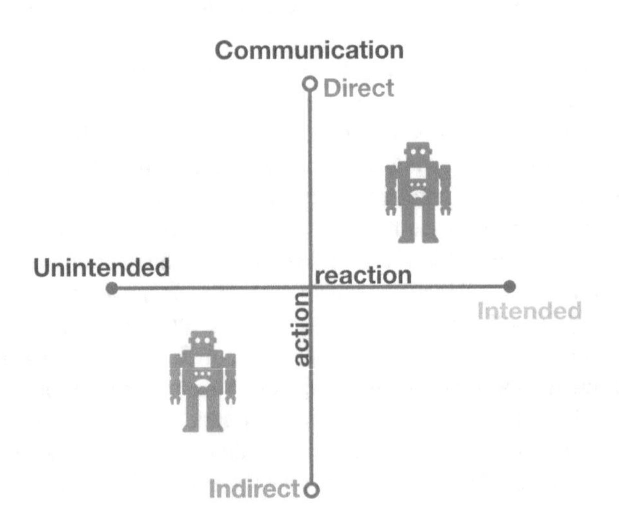

Fig. 13.4 Movement of mobile robots

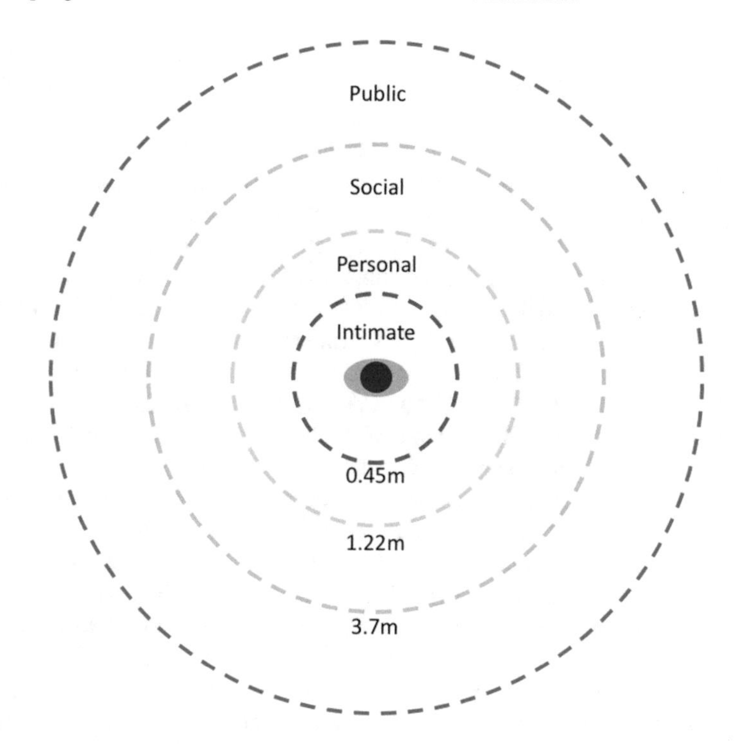

Fig. 13.5 Proxemics zones according to Hall [54]

outer limit. In general interaction among strangers happens in the Social Space or beyond (>1.22 m). Intrusions into the Personal or Intimate Space without consent are perceived as rude or even threatening and therefore create annoyance and stress as pointed out by Hall [54]. Similarly, in human-aware navigation, robots aim to avoid these zones when circumventing humans as investigated by, e.g. [97] and [98]. According to Hall, at this distance conversation is conducted at a normal voice level. The visual focus extends to the nose and parts of both eyes or nose, mouth, and one eye. Which parts of the face are visible at a certain distance play a vital role in communication because gaze has been identified as an important tool in HRI, e.g. [46]. According to Hall [54], at distances of >7.6 m voice, facial expressions, and movement must be exaggerated. Hence finding the best distance for communication is an active field of study in HRI. A discussion on the different shapes of these zones can be found in the work by Rios-Martinez [113].

One out of many examples of research that investigates distances for communication is the work by Torta et al. [135] where the optimal approach distance and angle between a small humanoid robot[1] and a sitting person is investigated. Torta et al. [135] present an attractor based navigation framework that includes the definition for a *Region of Approach* which is optimal to communicate between the two agents.

[1]NAO—https://www.softbankrobotics.com/emea/en/nao.

In the conducted experiment, a NAO robot is approaching a sitting person from different angles, with the purpose of starting a conversation. The approach is stopped when that person presses a button at a distance perceived as suitable to achieve the task. Torta et al. [135] show that an approach from the front is preferable over an approach from the side and found that the distance at which the participants stopped the robot to have a conversation loosely correlates with the close phase of the social space as defined by Hall [54]. One of the very few examples of research on the long-term habituation effects of approach distances is the work by Walters et al. [140]. They use a standing participant and a mobile service robot instead of a NAO in a confined space in an otherwise similar experimental setting as [135], i.e. the robot approaches the participant from the front and is stopped via a button when it is close enough to have a conversation, and inspect the long-term effect on this most suitable approach distance. Over the first couple of weeks, this distance seemed to decrease and then remain stable for the remainder of the experiment.

Looking at the changing nature of people's preferences when it comes to distance, is not only influenced by habituation as described above but also by the resulting robot performance. Mead and Mataric [86] conducted an experiment in which the human participant explained certain objects to a mobile robot via speech and gestures from a fixed location. The robot, however, altered its position during the trials. After an object had been explained, the robot would change its distance to the human before the next object explanation phase started. After each explanation, the robot signalled success or failure of understanding the explanation to the participant where the success rate depended on the distance to the human and was modelled as a normal distribution with its peak at 2.25 m distance to the human and a standard deviation of 1.0 m. Before and after the experiment, to evaluate if the proxemics preferences of the participants changed, Mead and Mataric [86] had the robot approach the participant until they said "stop" when they thought that the robot would be at an appropriate distance for the task. Comparing the measurements from before and after the experiment, they found that humans indeed adapt their proxemics preferences to the area of peak performance of the robot whereas in the control condition where the success rate was modelled uniformly this effect did not appear.

All of the above, highlights the importance of finding the correct distance for a robot to communicate with its interaction partner. Research is still ongoing but the commonly accepted opinion seems to be that while it varies for each individual based on personal preferences but also on the performance of the robot (due to, e.g. environmental factors), the social space as defined by Hall [54] seems to be a good approximation. This is partly owed to the resulting simplicity stemming from using a small set of fixed thresholds which facilitates easy decision making under uncertainty.

13.3.2 Path Planning

When it comes to navigation in human-populated environments, a great body of work is dedicated to avoiding humans and enabling the robot to fulfil a given task without being interfered with. At the same time, the robot should treat the humans around it in a manner that makes them feel safe and adheres to factors such as comfort, sociability, and naturalness as defined by Kruse et al. [72]. A few examples of this being [41, 66, 81, 82, 116, 120, 127, 136]. On the other hand, there is also a body of research that focuses not on avoiding humans but on moving to interact with them in a more explicit way such as verbal communication or jointly executing a given task. These approaches are described in the following.

13.3.2.1 Robot Motion

When communicating, humans tend to assume formations where they place themselves in a spatial arrangement that faces inward around a space to which everyone has immediate access. As noted by Ciolek and Kendon [32], this creates conditions in which each participant can effectively exchange glances, gestures, and words. An example for such a formation of two people is the so-called *f-formation* [63]. Maintaining formation for approaching a group is, therefore, an active field of research one example being the work by Althaus et al. [6] where the robot assumes a position in an existing group of people which allows it to effectively communicate with everybody. Others attempt to create an f-formation between human and robot such as [73]. Apart from assuming the correct position to interact with humans, the way in which a robot approaches a single person or a group is of great importance for its acceptance as an interaction partner, e.g. [12, 25, 40, 65, 69, 92, 129], where the consensus seems to be that approaching someone from the front or from an area that is visible to the human is of great benefit and that appearing out of hidden zones should be avoided [119]. This has even been adopted by research focused circumvention of humans that explicitly seek to avoid the area behind a person to not cause discomfort such as [104, 116, 120] or avoiding to pass behind obstacles that obstruct the field of view [31].

Interaction with groups is not just about how to explicitly interact with them, i.e. joining and participating in the conversation, but also how to implicitly interact with them, i.e. avoiding interference. Several research groups have investigated approaches to avoid passing through the centre of a group (≥ 2 participants) to disturb their communication by blocking their vision. Some examples of this are [49, 104, 116]. To the best of the authors' knowledge, there is no work on finding the right point in time when to pass through the centre of a group given there is no way to avoid them. The general consensus is to circumvent them if possible and to pass through if not. This is especially true when using simple approaches such as Gaussian cost functions based on proxemics.

13.3.2.2 Joint Motion

Some tasks require that both human and robot move in unison to a common goal. This could be the case for a museum tour guide robot where the human or the group of humans is supposed to follow the robot guide. This scenario has been one of the first that was adopted by the community and some of the most commonly known examples for these kind of robots are Rhino by Burgard et al. [24], Robox by Arras et al. [11], Minerva by Thrun et al. [131], Rackham by Clodic et al. [34], Mobot by Nourbakhsh et al. [94], and Cice by Macaluso et al. [83]. A similar system has also been used for therapeutic purposes by Hebesberger et al. [57]. These systems rely on the robot being in front of the group or person so they can be seen at all times to make it easy to follow. Moreover, the navigation systems should aim to produce goal directed non-jittery motion to allow the human to follow the robot easily as stated by Kruse et al. [72]. There are other factors that play a role when guiding a person or a group such as monitoring and adapting to their speed, reacting to path alterations, finding a path that is not only comfortable to take for the humans that are guided but also for the humans that might be encountered on the way, and more direct interaction like reengaging someone if they suspend the tour [103].

A more complex task when it comes to joint motion is walking side-by-side. This formation is mainly adopted to allow both the robot and the human to see each other and, therefore, favours communication. Morales Saiki and Morales [88, 89] present an approach for side-by-side motion that is based on the observation of people and created a model of an autonomous robot which emulates this behaviour. As they phrase it, this increases the shared utility. A recent approach by Ferrer et al. [45] looks at how to accomplish walking side-bye-side in crowded urban environments. This is particularly difficult as it imposes spatial constraints which might result in the side-by-side movement not being possible. According to Costa [35], this is when humans assume different types of formations depending on the space and the number of people in the group. While there are differences between male and female groups typical formations of 2–3 people are walking abreast so side-by-side, or in a V shape with the walking direction being from bottom to top. These kind of formations can be achieved by using so called social forces that pull the robot towards a shared goal and the people in its group while repelling it from obstacles and other people. An example of using atractor forces to the centre of the group the robot is with was provided by Moussaïd et al. [90], or similarly a force that attracts the robot to the other people in its group by Xu et al. [142].

13.3.3 Adaptive Robot Navigation Summary

In summary, the distance between robot and human is of paramount importance for communication between the two. Distance is also one of the easiest variables to change using a mobile robot. The difficulty is given by the task of finding the "correct" distance and the "correct" way of approaching someone. Meaning the distance chosen has to be one at which the interaction partner feels comfortable interacting with the robot and is able to perceive all its movements and hear all its utterances. The approach has to come from a direction where the human is not surprised and can gauge the intention of the robot. When more than one person is to be approached, the correct formation has to be assumed or maintained. In the case of both human and robot moving and not just aiming to avoid each other, joint path planning can be achieved in a way that optimises the navigation task, improves the communication between the robot and the human(s), or both. If the robot does not only act as a simple guide but should also be available for communication while navigating, a side-by-side formation is assumed to allow both the human and the robot to perceive each other.

To date, no holistic approach that would be able to solve all these tasks has been developed. There are solutions for parts of these problems using different techniques but their combination is non-trivial. Moreover, the problem hinted at earlier of knowing when to interrupt someone by approaching them or passing through the “we-space” [70] of a group of people if no other path can be found is a non-trivial and unsolved problem as well. There are certain techniques, however, that can be employed to detect the opportune moments to interrupt someone which will be detailed in the following section.

13.4 When to Interrupt: “Understanding of Cognitive Load”

13.4.1 Interruption During HRI

Speier et al. [121] defined interruption as “externally generated, randomly occurring, discrete event that breaks the continuity of cognitive focus on a primary task”. Similarly, interruptibility is defined in terms of the responsiveness of the individual; either a person or a robot at a certain point in time [122]. Interruptibility is particularly crucial in relation to Human-Robot Communication as it is vital for a robot to understand when to interrupt a human or for a robot to understand when it has been interrupted during a communication. The process of interruptibility in the context of HRI requires an understanding of social cues based on both verbal and non-verbal communication. Past research on interruptibility in the field of HRI has considered both dimensions of interruption. We find a number of methods to understand ways for the robot to interrupt humans during a communication. Satake et al. [115] proposed a model for a mobile robot to interrupt and approach a person at the shopping mall. They used people’s positions and walking speed as an indicator to decide on approaching an individual in a safe and polite manner. Other researchers have used estimation of person’s engagement in different environment including in shopping mall as helping/advertising assistants [62], in hotels as receptionist [20] and in bars as bartenders [48] to enable the robot understand interruptibility in a safe and polite manner. Most recently, [15] presented an interruptibility scale or a framework for a robot to decide when to interrupt the user. In essence, the scale had five levels ranging from highly interruptible suggesting that the individual is not busy and is conscious of robot's presence to interruptibility unknown suggesting that the individual is present but cannot recognize if the person should be interrupted. To understand the availability of the person, they used two sources of information—*person state* and *interruption context*. Person's state was calculated through understanding their head orientation, gaze direction, audible signals and body postures where as context was understood through objects such as mobile phone, laptop in the scene. This aforementioned information was used to classify the scales of interruptibility and later was used by the robot to interrupt or don't interrupt the person. Palinko et al. [102] also presented an interruptibility estimator and applied it during the robot communication with multiple humans. They used signals based on head pose and eye gaze on top of individual silence to recognize interruptibility and found in an experimental study that the robot can barge-in in the conversation in a more efficient manner while considering the non-verbal signals. However, they also recognized that linguistic (verbal) modelling of the conversation may result in an efficient model of interruptibility during a human-robot group conversation setting.

We see a reasonable amount of work towards creating robots that can understand when to interrupt the human during HRI. On the contrary, there is limited research on the methods that enables the human to interrupt the robot using non-verbal communication. Prior literature has shown the use of tactile sensors or limited gestures to interrupt the robot. For instance; [55] investigated the use of palm gesture as a sign of interruption to interrupt the conversation with the robot. Other researchers have further used tactile sensor, specifically in the case of the NAO robot [38] but otherwise limited research has been witnessed in the past. We understand that the use of gestures is particularly relevant. However, it remains a challenge to interrupt a conversational robot in social settings such as in the malls or hotels in a non-verbal manner. The fields of Human-Computer interaction [51] and Ubiquitous computing [137] have highlighted on computing *workload* or cognitive load as an estimator of interruption. We recognize that cognitive load could also be used as an indicator for the robot to stop or reduce the communication. More specifically, the situation that requires the robot to provide situational awareness in a sensitive environment. For instance, the robot giving information on what is happening on an oil rig where different robots are deployed to carry out various tasks [80]. We understand that such as situation may enhance humans' mental load. In such a scenario, human's cognitive or mental load can be used as an indicator for a conversational robot to learn about reducing or adjusting the communication and this can result in effective communication.

13.4.2 Cognitive Load

Cognitive load points to the load placed on human's working (short-term) memory during a task. It is defined as a construct that can be measured in three dimensions: (1) mental load, (2) mental effort and (3) performance [128]. We also understand that the working memory differs among different individuals consequently it calls for estimating cognitive load in real-time [13]. As highlighted, in a conversational robot scenario, one of the interruption strategies can be based on cognitive load as a robot must adapt according to the mental load of the user.

It remains a challenge to measure the individual's mental load in a non-intrusive and robust manner in real-time [29, 96]. Prior work has measured mental load in three different ways: through subjective evaluations such as NASA TLX (NASA Task Load Index[2]), through understanding physiological behaviours, or through performance-based objective measures (Mathematical Equations). However, subjective rating and performance-based objective measures are not continuous and cannot be used in real-time. On the contrary, physiological behaviours based on Pupil Diameter (PD), Blinking Rate (BR), Heart Rate (HR), Heart Rate Variability (HRV), Electroencephalography (EEG) and Galvanic Skin Response (GSR) are continuous and can be used to estimate mental load in real-time [96]. Empirical studies conducted with all of these aforementioned behaviours have observed that changes in one's behaviour can be attributed to a higher level of cognitive load. For instance, a low HRV and a higher HR is associated with a high cognitive load [36, 91]. In addition, an increase in the amount of PD, and decrease in the number of eye blinks reflects on a higher mental load [59, 110, 114]. We also understand that data collected on physiological behaviours through the various state of the art sensors is not only continuous but is also a robust and accurate representation of the particular behaviour [60, 134]. Recently, [2] have proposed to develop a system as shown in Fig. 13.6 to estimate users' mental load in real-time. Their aim is to collect data on these behaviours including PD, BR, HR and HRV using various state of the art sensors [60, 134] and later use this data to understand correlations between them and finally use it in a linear mixed-effect regression model to estimate of cognitive load in real-time.

In summary, we believe that cognitive or mental load could be utilised as an indicator for interruption for conversational robots and that an empirical evaluation is needed in the future.

[2]NASA Task Load Index—https://humansystems.arc.nasa.gov/groups/TLX/downloads/TLXScale.pdf

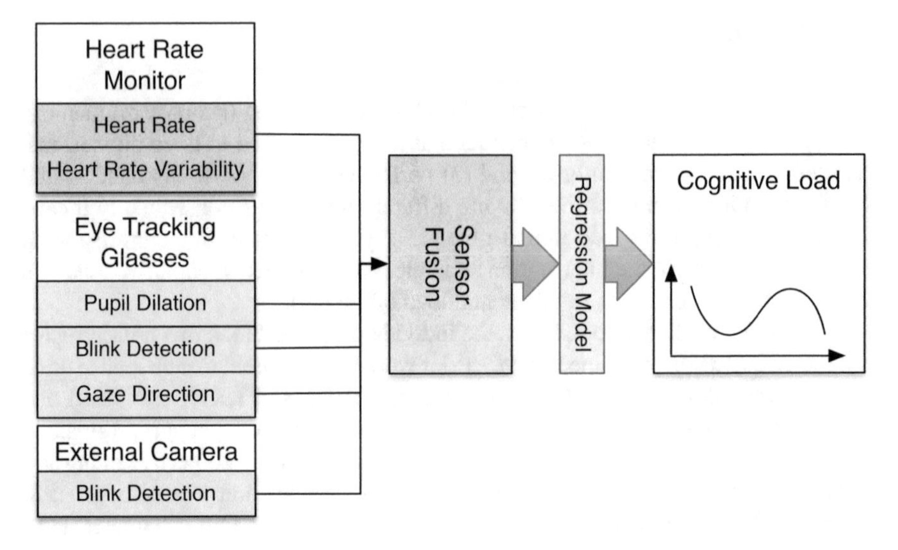

Fig. 13.6 Concept overview: for a system non-intrusive measurement for mental load [2]

13.5 Implicitly Learning Behaviours and Movements Favouring Communication

In the previous section we have shown examples of explicitly changing the behaviour and movement of a robotic agent to favour communication and when to interrupt a person or persons while interacting. This section focuses on an example scenario, i.e. single and dual arm manipulation of objects, where explicitly changing the behaviour and movement of a robotic agent to favour communication arguably presents a greater challenge than in, e.g. mobile navigation (see Sect. 13.3. In order to achieve the desired behaviours and movements, relying on human expert knowledge when learning from demonstration [107] presents a way of implicitly generating movements and behaviours that favour communication. This approach, therefore, relies on the humans' subconsciously demonstrating movements which are goal directed, i.e. achieve the manipulation task, but at the same time also follow social norms. Hence, when using learning from demonstration, one can assume that the resulting behaviour emulates human behaviour which naturally favours communication.

In order to achieve an autonomous manipulation, it is important to develop a reasoning technique that is able to hierarchically learn a fluid object interaction given the eminent dynamic nature of indoor environments. There are extensive studies on manipulation [52, 53, 145], grasping [5, 21, 39], and learning [27, 101, 125, 143]. However, due to their complexity, little attention has been paid to their interaction and joint integration in robotic systems so far. This complexity also makes this problem a prime candidate for approaches relying on implicitly learning movements that favour communication. For these reasons, one of the most common techniques used for manipulation tasks is learning from demonstration combined with hierarchical learning techniques to endow a robot with self-learning capabilities and interaction understanding. The advantages of using learning from demonstration have been shown in single-arm manipulator systems and, to a smaller extent, for dual-arm manipulator set-ups. Specifically, a robot understanding human demonstrations involving dual-arm manipulation will acquire the required knowledge to imitate the task and behaviour and will improve its model through trial and error experiments. Nonetheless, to learn human-like behaviour with a dual-arm system does not suffice for robust manipulation, but the integration of the understanding of the objects and environment is also essential. Information such as the affordance, grasping point, the object's fragility, and its manoeuvrability can be extracted with the help of vision, force, tactile, and pressure sensors. The following presents these two learning problems in more detail.

13.5.1 Learning Manipulation Tasks

As mentioned above, a popular technique of learning different manipulation tasks is learning from demonstration in combination with hierarchical learning (see Fig. 13.7). As done in [99, 100], a combination of absolute and relative skills is carried out to achieve complex dual-arm tasks. An absolute skill implies motions such as move or turn an object in a particular manner. Instead, a relative skill describes the synchronisation requirements between manipulators, such as opening a bottle's screw cap, or holding a parcel employing force contact. A primitive skill is represented by its coupling term [58] and a frame of reference. Learning coupling terms only requires a human demonstrator teaching the characteristic skill. The different coupling terms might be better formulated with different mathematical representa-

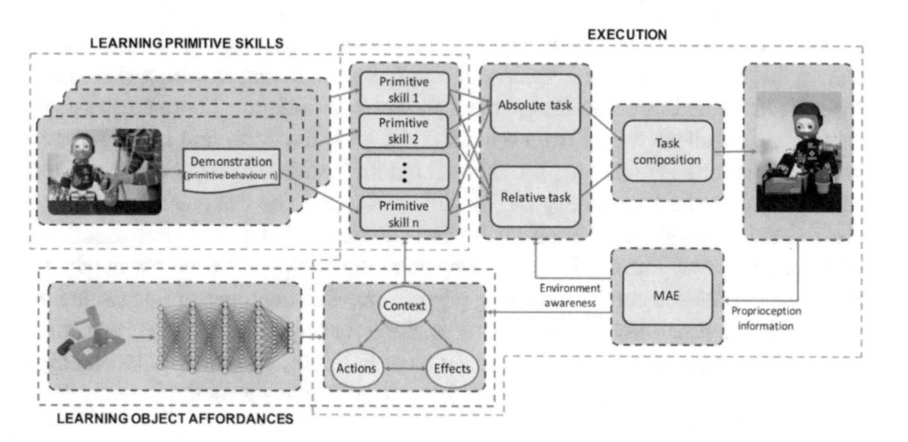

Fig. 13.7 The structure of a hierarchical machine learning framework

tions, e.g. a weighted combination of non-linear radial basis functions to encode the dynamics of a task, an analytical obstacle avoidance expression, or among others, a force profile to control the environmental interaction. Dynamic Movement Primitives (DMP)-based formulation [117] offers the needed modularity. This means that instead of learning a task as a whole, the framework harvests a collection of primitive skills. Creating a repertoire of skills referred to as a library, allowing the demonstrator to teach in a one-at-a-time fashion, i.e. to focus on one feature of the demonstration at a time [14]. This modular library can be employed for movement recognition purposes, where a demonstrated skill can be compared against the existing ones in the library. If the observed behaviour does not match any existing primitive, it is identified as a new skill and can be added to the framework's library [61]. This feature allows incremental learning by exploration or further human demonstrations.

13.5.2 Learning Object Grasp Affordances

Humans are very good at communicating to each other and coordinate on different tasks, being handovers a clear example. When it comes to grasping, most of the state-of-the-art literature explores methodologies by focusing exclusively on attributes of the target object and grasp stability metrics. However, as humans we also take the environment in which this task is executed into account and are able to adapt in the presence of other agents. Therefore, in order to achieve performance that emulates human behaviour, favours communication, and is able to collaboratively achieve manipulation tasks the system should be able to grasp the object considering the physical qualities of the world [8]. These qualities cannot only be inferred from the object, but we also have to consider the characteristics of the surroundings. In [9] this approach is used in a system where the grasping action affordance towards an object is the result of the association of different semantic features that describe the object and the surrounding environment. Ardón et al. [10] propose a method that includes environmental context to reason about object affordance to then deduce its grasping regions. This affordance is the result of a ranked association of visual semantic attributes harvested in a knowledge base graph representation. These attributes are the result of a collected data from human input, thus they represent social rules for grasping. These rules inherently provide affordance features for collaborative manipulation that ease the communication for different tasks. The designed framework is assessed using standard learning evaluation metrics and the zero-shot affordance prediction scenario. The resulting grasping areas are compared with unseen labelled data to assess their accuracy matching percentage. The outcome of this evaluation indicates the applicability of the proposed method for object interaction applications in indoor environments. Other examples such as [71, 126, 144] also focus on object affordances to improve the robot-object interaction which shows the importance of this aspect of manipulation.

Past research has extensively investigated approaches to autonomous collaborative manipulation. Nonetheless, grasping is still an open challenge due to the large variety of object shapes and robotic platforms as well as interaction variants among agents and humans that differ in the communication schemes. The current state of the art methods is limited to specific robot manipulator, grasping scenarios, and objects. Further, the current approaches need a large amount of data to train the learning model without being able to successfully generalise among object instances. Due to all these complications, we argue that implicitly learning movements that favour communication via learning from demonstration is the only feasible method to date.

13.6 Discussion and Conclusion

In this chapter concepts have been presented introducing three different forms of communication from human-human to human-robot interaction (see Sect. 13.2). It delineates how the dynamical process of communication can be represented through building a subspace spanned between action and reaction (see Sect. 13.1). The influence of adaptation of movements and behaviours in navigation is introduced and the consequences of this adaptations are conceptualized (see Sect. 13.3). The time dimension is crucial in communication, consequently, understanding a pattern on intercepting turns in communication are vital and are described in the section on interruption. It is discussed that building a theory of mind of the conversation partner might be the best way to identify a possible interruption window (see Sect. 13.4). The impact of implicitly learning about "social norm" in collaborative movements is presented in our Sect. 13.5, which leads to the question on the impact of bias on communication. Human-Human communication is undoubtedly fluent and dynamic as well as influenced by social norms. Past research has presented somewhat stable mechanisms and concepts to favour communication, but there are still open questions of the influence of social norms. Positively, this has been identified in the community as a way forward represented through the focus on gender-stereotypes and cross-culture comparisons of communication. Furthermore, ethical concerns as to what a robotic system should elicit, that it is capable of, is another positive trend in the current research, e.g. the research on trust and explainabilty of AI.

References

1. Ahmad, M., Mubin, O., & Orlando, J. (2017a). A systematic review of adaptivity in human-robot interaction. *Multimodal Technologies and Interaction*, *1*(3), 14.
2. Ahmad, M., Keller, I., & Lohan, K. (2019). Integrated real-time, non-intrusive measurements for mental load (in print). In *Workshop on AutomationXp at the ACM CHI Conference on Human Factors in Computing Systems*, Glasgow, UK.
3. Ahmad, M. I., Mubin, O. (2018). Emotion and memory model to promote mathematics learning-an exploratory long-term study. In *Proceedings of the 6th International Conference on Human-Agent Interaction* (pp. 214–221). ACM.

4. Ahmad, M. I., Mubin, O., & Orlando, J. (2017b). Adaptive social robot for sustaining social engagement during long-term children-robot interaction. *International Journal of Human-Computer Interaction, 33*(12), 943–962.
5. Ala, R., Kim, D. H., Shin, S. Y., Kim, C., & Park, S. K. (2015). A 3d-grasp synthesis algorithm to grasp unknown objects based on graspable boundary and convex segments. *Information Sciences, 295*, 91–106.
6. Althaus, P., Ishiguro, H., Kanda, T., Miyashita, T., Christensen, H. I. (2004). Navigation for human-robot interaction tasks. In *2004 IEEE International Conference on Robotics and Automation, 2004. Proceedings. ICRA '04* (Vol. 2, pp. 1894–1900). IEEE.
7. Anzalone, S., Boucenna, S., Ivaldi, S., & Chetouani, M. (2015). Evaluating the engagement with social robots. *International Journal of Social Robotics, 7*(4), 465–478.
8. Ardón, P., Pairet, È., Ramamoorthy, S., & Lohan, K. S. (2018) Towards robust grasps: Using the environment semantics for robotic object affordances. In *Proceedings of the AAAI Fall Symposium on Reasoning and Learning in Real-World Systems for Long-Term Autonomy* (pp. 5–12). AAAI Press.
9. Ardón, P., Pairet, È., Petrick, R., Ramamoorthy, S., & Lohan, K. (2019a). Reasoning on grasp-action affordances. In *Annual Conference Towards Autonomous Robotic Systems* (pp. 3–15). Springer.
10. Ardón, P., Pairet, È., Petrick, R. P., Ramamoorthy, S., & Lohan, K. S. (2019b). Learning grasp affordance reasoning through semantic relations. *IEEE Robotics and Automation Letters, 4*(4), 4571–4578.
11. Arras, K. O., Tomatis, N., & Siegwart, R. (2003). Robox, a remarkable mobile robot for the real world. In *Experimental robotics VIII* (pp. 178–187). Springer.
12. Avrunin, E., & Simmons, R. (2014). Socially-appropriate approach paths using human data. In *The 23rd IEEE International Symposium on Robot and Human Interactive Communication, 2014 RO-MAN* (pp. 1037–1042). IEEE.
13. Baddeley, A. (1992). Working memory. *Science, 255*(5044), 556–559.
14. Bajcsy, A., Losey, D. P., O'Malley, M. K., & Dragan, A. D. (2018). Learning from physical human corrections, one feature at a time. In *Proceedings of the 2018 ACM/IEEE International Conference on Human-Robot Interaction* (pp. 141–149). ACM.
15. Banerjee, S., & Chernova, S. (2017). Temporal models for robot classification of human interruptibility. In *Proceedings of the 16th Conference on Autonomous Agents and Multiagent Systems, International Foundation for Autonomous Agents and Multiagent Systems* (pp. 1350–1359).
16. Bartneck, C., Kanda, T., Mubin, O., & Al Mahmud, A. (2009). Does the design of a robot influence its animacy and perceived intelligence? *International Journal of Social Robotics, 1*(2), 195–204.
17. Bennett, C., & Šabanovic, S. (2014). Deriving minimal features for human-like facial expressions in robotic faces. *International Journal of Social Robotics, 6*(3), 367–381.
18. Bickmore, T. W., & Picard, R. W. (2005). Establishing and maintaining long-term human-computer relationships. *ACM Transactions on Computer-Human Interaction (TOCHI), 12*(2), 293–327.
19. Bilac, M., Chamoux, M., & Lim, A. (2017). Gaze and filled pause detection for smooth human-robot conversations. In *2017 IEEE-RAS 17th International Conference on Humanoid Robotics (Humanoids)* (pp. 297–304). IEEE.
20. Bohus, D., & Horvitz, E. (2009). Dialog in the open world: Platform and applications. In *Proceedings of the 2009 international conference on multimodal interfaces* (pp. 31–38). ACM.
21. Bonaiuto, J., & Arbib, M. A. (2015). Learning to grasp and extract affordances: The integrated learning of grasps and affordances (ilga) model. *Biological Cybernetics, 109*(6), 639–669.
22. Boyle, E. A., Anderson, A. H., & Newlands, A. (1994). The effects of visibility on dialogue and performance in a cooperative problem solving task. *Language and Speech, 37*(1), 1–20.
23. van den Brule, R., Dotsch, R., Bijlstra, G., Wigboldus, D., & Haselager, P. (2014). Do robot performance and behavioral style affect human trust? *International Journal of Social Robotics, 6*(4), 519–531.

24. Burgard, W., Cremers, A. B., Fox, D., Hähnel, D., Lakemeyer, G., Schulz, D., et al. (1999). Experiences with an interactive museum tour-guide robot. *Artificial Intelligence*, *114*(1), 3–55.
25. Butler, J. T., & Agah, A. (2001). Psychological effects of behavior patterns of a mobile personal robot. *Autonomous Robots*, *10*(2), 185–202. https://doi.org/10.1023/A:1008986004181.
26. Byrne, D. (1997). An overview (and underview) of research and theory within the attraction paradigm. *Journal of Social and Personal Relationships*, *14*(3), 417–431.
27. Calinon, S., D'halluin, F., Sauser, E. L., Caldwell, D. G., & Billard, A. G. (2010). Learning and reproduction of gestures by imitation. *IEEE Robotics & Automation Magazine*, *17*(2), 44–54.
28. Carpenter, J., Davis, J., Erwin-Stewart, N., Lee, T., Bransford, J., & Vye, N. (2009). Gender representation and humanoid robots designed for domestic use. *International Journal of Social Robotics*, *1*(3), 261.
29. Chen, F., Ruiz, N., Choi, E., Epps, J., Khawaja, M. A., Taib, R., et al. (2012). Multimodal behavior and interaction as indicators of cognitive load. *ACM Transactions on Interactive Intelligent Systems (TiiS)*, *2*(4), 22.
30. Cheng, L. C., Lin, C. Y., & Huang, C. (2013). Visualization of facial expression deformation applied to the mechanism improvement of face robot. *International Journal of Social Robotics*, *5*(4), 423–439.
31. Chung, W., Kim, S., Choi, M., Choi, J., Kim, H., Cb, Moon, et al. (2009). Safe navigation of a mobile robot considering visibility of environment. *IEEE Transactions on Industrial Electronics*, *56*(10), 3941–3950.
32. Ciolek, T. M., & Kendon, A. (1980). Environment and the spatial arrangement of conversational encounters. *Sociological Inquiry*, *50*(3–4), 237–271.
33. Clark, H. H. (1996). *Using language*. Cambridge University Press.
34. Clodic, A., Fleury, S., Alami, R., Chatila, R., Bailly, G., Brethes, L., et al. (2006). Rackham: An interactive robot-guide. In *The 15th IEEE International Symposium on Robot and Human Interactive Communication, 2006. ROMAN 2006* (pp. 502–509). IEEE.
35. Costa, M. (2010). Interpersonal distances in group walking. *Journal of Nonverbal Behavior*, *34*(1), 15–26.
36. Cranford, K. N., Tiettmeyer, J. M., Chuprinko, B. C., Jordan, S., & Grove, N. P. (2014). Measuring load on working memory: The use of heart rate as a means of measuring chemistry students' cognitive load. *Journal of Chemical Education*, *91*(5), 641–647.
37. Crumpton, J., & Bethel, C. (2016). A survey of using vocal prosody to convey emotion in robot speech. *International Journal of Social Robotics*, *8*(2), 271–285.
38. Csapo, A., Gilmartin, E., Grizou, J., Han, J., Meena, R., Anastasiou, D., Jokinen, K., & Wilcock, G. (2012). Multimodal conversational interaction with a humanoid robot. In *2012 IEEE 3rd International Conference on Cognitive Infocommunications (CogInfoCom)* (pp. 667–672). IEEE.
39. Dang, H., & Allen, P. K. (2010). Robot learning of everyday object manipulations via human demonstration. In *2010 IEEE/RSJ International Conference on Intelligent Robots and Systems* (pp. 1284–1289). IEEE.
40. Dautenhahn, K., Walters, M., Woods, S., Koay, K. L., Nehaniv, C. L., Sisbot, A., Alami, R., Siméon, T. (2006). How may I serve you?: A robot companion approaching a seated person in a helping context. In *Proceedings of the 1st ACM SIGCHI/SIGART Conference on Human-Robot Interaction* (pp. 172–179). ACM.
41. Dondrup, C., & Hanheide, M. (2016). Qualitative constraints for human-aware robot navigation using velocity costmaps. In *IEEE International Symposium on Robot and Human Interactive Communication (RO-MAN)*.
42. Dondrup, C., Lichtenthäler, C., & Hanheide, M. (2014). Hesitation signals in human-robot head-on encounters: a pilot study. In *Proceedings of the 2014 ACM/IEEE international conference on Human-robot interaction* (pp. 154–155). ACM.
43. Duck, S. (2007). *Human relationships*. Sage.

44. Feingold, A. (1992). Good-looking people are not what we think. *Psychological Bulletin, 111*(2), 304.
45. Ferrer, G., Zulueta, A. G., Cotarelo, F. H., & Sanfeliu, A. (2017). Robot social-aware navigation framework to accompany people walking side-by-side. *Autonomous Robots, 41*(4), 775–793. https://doi.org/10.1007/s10514-016-9584-y.
46. Fischer, K., Lohan, K., Saunders, J., Nehaniv, C., Wrede, B., & Rohlfing, K. (2013). The impact of the contingency of robot feedback on HRI. In *2013 International Conference on Collaboration Technologies and Systems (CTS)* (pp. 210–217). IEEE.
47. Fong, T., Nourbakhsh, I., & Dautenhahn, K. (2003). A survey of socially interactive robots. *Robotics and Autonomous Systems, 42*(3), 143–166.
48. Foster, M. E., Gaschler, A., & Giuliani, M. (2013). How can I help you': Comparing engagement classification strategies for a robot bartender. In *Proceedings of the 15th ACM on International Conference on Multimodal Interaction* (pp. 255–262). ACM.
49. Garrell, A., & Sanfeliu, A. (2010). Local optimization of cooperative robot movements for guiding and regrouping people in a guiding mission. In *2010 IEEE/RSJ International Conference on Intelligent Robots and Systems* (pp. 3294–3299). IEEE.
50. Gergely, G., & Watson, J. S. (1999). Early socio-emotional development: Contingency perception and the social-biofeedback model. *Early Social Cognition: Understanding Others in the First Months of Life, 60*, 101–136.
51. Grandhi, S., & Jones, Q. (2010). Technology-mediated interruption management. *International Journal of Human-Computer Studies, 68*(5), 288–306.
52. Gribovskaya, E., & Billard, A. (2008). Combining dynamical systems control and programming by demonstration for teaching discrete bimanual coordination tasks to a humanoid robot. In *2008 3rd ACM/IEEE International Conference on Human-Robot Interaction (HRI)* (pp. 33–40). IEEE.
53. Guenter, F., Hersch, M., Calinon, S., & Billard, A. (2007). Reinforcement learning for imitating constrained reaching movements. *Advanced Robotics, 21*(13), 1521–1544.
54. Hall, E. T. (1969). *The hidden dimension*. New York: Anchor Books.
55. Han, J., Campbell, N., Jokinen, K., & Wilcock, G. (2012). Investigating the use of non-verbal cues in human-robot interaction with a nao robot. In *2012 IEEE 3rd International Conference on Cognitive Infocommunications (CogInfoCom)* (pp. 679–683).
56. Hauser, M. D., Chomsky, N., & Fitch, W. T. (2002). The faculty of language: What is it, who has it, and how did it evolve? *Science, 298*(5598), 1569–1579.
57. Hebesberger, D., Dondrup, C., Körtner, T., Gisinger, C., & Pripfl, J. (2016). Lessons learned from the deployment of a long-term autonomous robot as companion in physical therapy for older adults with dementia—A mixed methods study. In *11th ACM/IEEE International Conference on Human-Robot Interaction (HRI)*.
58. Hogan, N. (1985). Impedance control: An approach to manipulation: Part ii-implementation. *Journal of Dynamic Systems, Measurement, and Control, 107*(1), 8–16.
59. Holland, M. K., & Tarlow, G. (1972). Blinking and mental load. *Psychological Reports, 31*(1), 119–127.
60. HRV E. (2018). How do you calculate the hrv score? Webpage, https://help.elitehrv.com/article/54-how-do-you-calculate-the-hrv-score.
61. Ijspeert, A. J., Nakanishi, J., & Schaal, S. (2002). Movement imitation with nonlinear dynamical systems in humanoid robots. In *Proceedings 2002 IEEE International Conference on Robotics and Automation (Cat. No. 02CH37292)* (Vol. 2, pp. 1398–1403). IEEE.
62. Kato, Y., Kanda, T., & Ishiguro, H. (2015). May I help you?: Design of human-like polite approaching behavior. In *Proceedings of the Tenth Annual ACM/IEEE International Conference on Human-Robot Interaction* (pp. 35–42). ACM.
63. Kendon, A. (1976). The f-formation system: The spatial organization of social encounters. *Man-Environment Systems, 6*, 291–296.
64. Kennedy, J., Baxter, P., & Belpaeme, T. (2017). Nonverbal immediacy as a characterisation of social behaviour for human-robot interaction. *International Journal of Social Robotics, 9*(1), 109–128.

65. Kessler, J., Schroeter, C., Gross, H. M. (2011). Approaching a person in a socially acceptable manner using a fast marching planner. In *Intelligent robotics and applications* (pp. 368–377). Springer.
66. Kirby, R., Simmons, R., & Forlizzi, J. (2009). COMPANION: A constraint-optimizing method for person-acceptable navigation. In *RO-MAN 2009—The 18th IEEE International Symposium on Robot and Human Interactive Communication* (pp. 607–612). https://doi.org/10.1109/ROMAN.2009.5326271.
67. Knapp, M., & Hall, J. (1972). *Nonverbal communication in human interaction*. Harcourt Brace College Publishers.
68. Koay, K., Syrdal, D., Ashgari-Oskoei, M., Walters, M., & Dautenhahn, K. (2014). Social roles and baseline proxemic preferences for a domestic service robot. *International Journal of Social Robotics*, *6*(4), 469–488.
69. Koay, K. L., Sisbot, E. A., Syrdal, D. S., Walters, M. L,. Dautenhahn, K., & Alami, R. (2007). Exploratory study of a robot approaching a person in the context of handing over an object. In *AAAI Spring Symposium: Multidisciplinary Collaboration for Socially Assistive Robotics* (pp. 18–24).
70. Krueger, J. (2011). Extended cognition and the space of social interaction. *Consciousness and Cognition*, *20*(3), 643–657.
71. Krüger, N., Geib, C., Piater, J., Petrick, R., Steedman, M., Wörgötter, F., et al. (2011). Object-action complexes: Grounded abstractions of sensory-motor processes. *Robotics and Autonomous Systems*, *59*(10), 740–757.
72. Kruse, T., Pandey, A. K., Alami, R., & Kirsch, A. (2013). Human-aware robot navigation: A survey. *Robotics and Autonomous Systems*, *61*(12), 1726–1743. https://doi.org/10.1016/j.robot.2013.05.007, http://www.sciencedirect.com/science/article/pii/S0921889013001048.
73. Kuzuoka, H., Suzuki, Y., Yamashita, J., & Yamazaki, K. (2010). Reconfiguring spatial formation arrangement by robot body orientation. In *Proceedings of the 5th ACM/IEEE International Conference on Human-Robot Interaction, HRI '10* (pp. 285–292). IEEE Press, Piscataway, NJ, USA. http://dl.acm.org/citation.cfm?id=1734454.1734557.
74. Leite, I., Pereira, A., Castellano, G., Mascarenhas, S., Martinho, C., & Paiva, A. (2011). Modelling empathy in social robotic companions. In *International Conference on User Modeling, Adaptation, and Personalization* (pp. 135–147). Springer.
75. Levinger, G. (1983). Development and change. Close relationships (pp. 315–359).
76. Lichtenthäler, C., Lorenz, T., & Kirsch, A. (2012). Influence of legibility on perceived safety in a virtual human-robot path crossing task. In *RO-MAN* (pp. 676–681). IEEE.
77. Lohan, K. S., Pitsch, K., Rohlfing, K., Fischer, K., Saunders, J., Lehmann, et al. (2011). Contingency allows the robot to spot the tutor and to learn from interaction. In *2011 First Joint IEEE International Conference on Development and Learning and on Epigenetic Robotics*.
78. Lohan, K. S., Rohlfing, K., Pitsch, K., Saunders, J., Lehmann, H., Nehaniv, C., et al. (2012). Tutor spotter: Proposing a feature set and evaluating it in a robotic system. *International Journal of Social Robotics*, *4*, 131–146.
79. Lohan, K. S., Griffiths, S., Sciutti, A., Partmann, T., & Rohlfing, K. (2014). Co-development of manner and path concepts in language, action and eye-gaze behavior (Topics in Cognitive Science Accepted).
80. Lopes, J., Robb, D. A., Ahmad, M., Liu, X., Lohan, K., & Hastie, H. (2019). Towards a conversational agent for remote robot-human teaming. In *2019 14th ACM/IEEE International Conference on Human-Robot Interaction (HRI)* (pp. 548–549). IEEE.
81. Lu, D. V., Allan, D. B., & Smart, W. D. (2013). Tuning cost functions for social navigation. In *Social robotics* (pp. 442–451). Springer.
82. Lu, D. V., Hershberger, D., & Smart, W. D. (2014) Layered costmaps for context-sensitive navigation. In *2014 IEEE/RSJ International Conference on Intelligent Robots and Systems (IROS 2014)* (pp 709–715). IEEE.
83. Macaluso, I., Ardizzone, E., Chella, A., Cossentino, M., Gentile, A., Gradino, R., et al. (2005). Experiences with cicerobot, a museum guide cognitive robot. In *AI* IA 2005: Advances in artificial intelligence* (pp. 474–482). Springer.

84. May, A. D., Dondrup, C., & Hanheide, M. (2015). Show me your moves! Conveying navigation intention of a mobile robot to humans. In *2015 European Conference on Mobile Robots (ECMR)* (pp. 1–6). IEEE. https://doi.org/10.1109/ECMR.2015.7324049.
85. McColl, D., & Nejat, G. (2014). Recognizing emotional body language displayed by a human-like social robot. *International Journal of Social Robotics*, *6*(2), 261–280.
86. Mead, R., & Mataric, M. J. (2015). Robots have needs too: People adapt their proxemic preferences to improve autonomous robot recognition of human social signals. *New Frontiers in Human-Robot Interaction*, 100.
87. Meyer, M., Hard, B., Brand, R. J., McGarvey, M., & Baldwin, D. A. (2011). Acoustic packaging: Maternal speech and action synchrony. *IEEE Transactions on Autonomous Mental Development*, *3*(2), 154–162.
88. Morales, Y., Kanda, T., & Hagita, N. (2014). Walking together: Side-by-side walking model for an interacting robot. *Journal of Human-Robot Interaction*, *3*(2), 50–73. https://doi.org/10.5898/JHRI.3.2.Morales.
89. Morales Saiki, L. Y., Satake, S., Huq, R., Glas, D., Kanda, T., & Hagita, N. (2012). How do people walk side-by-side?: Using a computational model of human behavior for a social robot. In *Proceedings of the Seventh Annual ACM/IEEE International Conference on Human-Robot Interaction, HRI '12* (pp. 301–308). ACM, New York, NY, USA. http://doi.acm.org/10.1145/2157689.2157799.
90. Moussaïd, M., Perozo, N., Garnier, S., Helbing, D., & Theraulaz, G. (2010). The walking behaviour of pedestrian social groups and its impact on crowd dynamics. *PloS One*, *5*(4), e10,047.
91. Mukherjee, S., Yadav, R., Yung, I., Zajdel, D. P., & Oken, B. S. (2011). Sensitivity to mental effort and test-retest reliability of heart rate variability measures in healthy seniors. *Clinical Neurophysiology*, *122*(10), 2059–2066.
92. Mumm, J., & Mutlu, B. (2011). Human-robot proxemics: Physical and psychological distancing in human-robot interaction. In *International Conference on Human-Robot Interaction*, Lausanne, Switzerland.
93. Niculescu, A., van Dijk, B., Nijholt, A., Li, H., & See, S. (2013). Making social robots more attractive: The effects of voice pitch, humor and empathy. *International Journal of Social Robotics*, *5*(2), 171–191.
94. Nourbakhsh, I. R., Kunz, C., & Willeke, T. (2003). The mobot museum robot installations: A five year experiment. In *2003 IEEE/RSJ International Conference on Intelligent Robots and Systems, 2003 (IROS 2003). Proceedings* (Vol. 4, pp. 3636–3641). IEEE.
95. Orbuch, T., & Sprecher, S. (2006). Attraction and interpersonal relationships. In John Delamater (Ed.), *Handbook of social psychology*. Springer: Handbooks of sociology and social research.
96. Paas, F., Tuovinen, J. E., Tabbers, H., & Van Gerven, P. W. (2003). Cognitive load measurement as a means to advance cognitive load theory. *Educational Psychologist*, *38*(1), 63–71.
97. Pacchierotti, E., Christensen, H. I., & Jensfelt, P. (2005). Human-robot embodied interaction in hallway settings: a pilot user study. In *IEEE International Workshop on Robot and Human Interactive Communication, 2005. ROMAN 2005* (pp. 164–171). https://doi.org/10.1109/ROMAN.2005.1513774.
98. Pacchierotti, E., Christensen, H. I., & Jensfelt, P. (2006). Evaluation of passing distance for social robots. In *The 15th IEEE International Symposium on Robot and Human Interactive Communication, RO-MAN 2006* (pp 315–320). https://doi.org/10.1109/ROMAN.2006.314436.
99. Pairet, È., Ardón, P., Broz, F., Mistry, M., & Petillot, Y. (2018). Learning and generalisation of primitives skills towards robust dual-arm manipulation. In *Proceedings of the AAAI Fall Symposium on Reasoning and Learning in Real-World Systems for Long-Term Autonomy* (pp. 62–69). AAAI Press.
100. Pairet, È., Ardón, P., Mistry, M., & Petillot, Y. (2019a). Learning and composing primitive skills for dual-arm manipulation. In *Annual Conference Towards Autonomous Robotic Systems* (pp. 65–77). Springer.

101. Pairet, È., Ardón, P., Mistry, M., & Petillot, Y. (2019b). Learning generalizable coupling terms for obstacle avoidance via low-dimensional geometric descriptors. *IEEE Robotics and Automation Letters*, *4*(4), 3979–3986.
102. Palinko, O., Ogawa, K., Yoshikawa, Y., & Ishiguro, H. (2018). How should a robot interrupt a conversation between multiple humans. In *International Conference on Social Robotics* (pp. 149–159). Springer.
103. Pandey, A. K., & Alami, R. (2009). A step towards a sociable robot guide which monitors and adapts to the person's activities. In *2009 International Conference on Advanced Robotics* (pp. 1–8). IEEE.
104. Pandey, A. K., & Alami, R. (2010). A framework towards a socially aware mobile robot motion in human-centered dynamic environment. In *2010 IEEE/RSJ International Conference on Intelligent Robots and Systems (IROS)* (pp. 5855–5860). IEEE.
105. Peters, A. (2011). Small movements as communicational cues in HRI. In T. Kollar & A. Weiss (Eds.), *HRI 2011—Workshop on human-robot interaction pioneers* (pp. 72–73).
106. Rosenthal-von der Pütten, A. M., Krämer, N., & Herrmann, J. (2018). The effects of human-like and robot-specific affective nonverbal behavior on perception, emotion, and behavior. *International Journal of Social Robotics*, 1–14.
107. Ravichandar, H., Polydoros, A. S., Chernova, S., & Billard, A. (2020). Recent advances in robot learning from demonstration. *Annual Review of Control, Robotics, and Autonomous Systems*, *3*.
108. Read, R., & Belpaeme, T. (2016). People interpret robotic non-linguistic utterances categorically. *International Journal of Social Robotics*, *8*(1), 31–50.
109. Reeves, B., & Nass, C. (1996). *How people treat computers, television, and new media like real people and places*. Cambridge University Press.
110. Reilly, J., Kelly, A., Kim, S. H., Jett, S., & Zuckerman, B. (2018). The human task-evoked pupillary response function is linear: Implications for baseline response scaling in pupillometry. *Behavior Research Methods*, 1–14.
111. Richmond, V., & McCroskey, J. (1995). *Nonverbal Behavior in interpersonal relations*. Allyn and Bacon.
112. Richmond, V. P., McCroskey, J. C., & McCroskey, L. (2005). Organizational communication for survival: Making work. *Work*, *4*.
113. Rios-Martinez, J., Spalanzani, A., & Laugier, C. (2015). From proxemics theory to socially-aware navigation: A survey. *International Journal of Social Robotics*, *7*(2), 137–153. https://doi.org/10.1007/s12369-014-0251-1.
114. Sabyruly, Y., Broz, F., Keller, I., & Lohan, K. S. (2015). Gaze and attention during an hri storytelling task. In *2015 AAAI Fall Symposium Series*.
115. Satake, S., Kanda, T., Glas, D. F., Imai, M., Ishiguro, H., & Hagita, N. (2009). How to approach humans? Strategies for social robots to initiate interaction. In *Proceedings of the 4th ACM/IEEE International Conference on Human Robot Interaction* (pp. 109–116). ACM.
116. Scandolo, L.,& Fraichard, T. (2011). An anthropomorphic navigation scheme for dynamic scenarios. In *2011 IEEE International Conference on Robotics and Automation (ICRA)* (pp. 809–814). IEEE.
117. Schaal, S. (2006). Dynamic movement primitives—A framework for motor control in humans and humanoid robotics. In *Adaptive motion of animals and machines* (pp. 261–280). Springer.
118. Schillingmann, L., Wrede, B., & Rohlfing, K. J. (2009). A computational model of acoustic packaging. *IEEE Transactions on Autonomous Mental Development*, *1*(4), 226–237.
119. Sisbot, A. (2008). Towards human-aware robot motions. Ph.D. thesis, Université Paul Sabatier-Toulouse III.
120. Sisbot, E., Marin-Urias, L., Alami, R., & Simeon, T. (2007). A human aware mobile robot motion planner. *IEEE Transactions on Robotics*, *23*(5), 874–883. https://doi.org/10.1109/TRO.2007.904911.
121. Speier, C., Valacich, J., & Vessey, I. (1997). The effects of task interruption and information presentation on individual decision making. In *ICIS 1997 Proceedings* (p. 2).

122. Spivey, M. (2007). *The Continuity of Mind.* https://doi.org/10.1093/acprof:oso/9780195170788.001.0001.
123. Stafford, L., Dainton, M., & Haas, S. (2000). Measuring routine and strategic relational maintenance: Scale revision, sex versus gender roles, and the prediction of relational characteristics. *Communications Monographs*, *67*(3), 306–323.
124. Stanton, C., & Stevens, C. (2017). Don't stare at me: The impact of a humanoid robot's gaze upon trust during a cooperative human-robot visual task. *International Journal of Social Robotics*, *9*(5), 745–753.
125. Steil, J. J., Röthling, F., Haschke, R., & Ritter, H. (2004). Situated robot learning for multi-modal instruction and imitation of grasping. *Robotics and Autonomous Systems*, *47*(2–3), 129–141.
126. Sung, J., Lenz, I., & Saxena, A. (2017). Deep multimodal embedding: Manipulating novel objects with point-clouds, language and trajectories. In *2017 IEEE International Conference on Robotics and Automation (ICRA)* (pp. 2794–2801). IEEE.
127. Svenstrup, M., Bak, T., & Andersen, H. J. (2010). Trajectory planning for robots in dynamic human environments. In *2010 IEEE/RSJ International Conference on Intelligent Robots and Systems (IROS)* (pp. 4293–4298). IEEE.
128. Sweller, J., Van Merrienboer, J. J., & Paas, F. G. (1998). Cognitive architecture and instructional design. *Educational Psychology Review*, *10*(3), 251–296.
129. Takayama, L., & Pantofaru, C. (2009). Influences on proxemic behaviors in human-robot interaction. In *2009 IEEE/RSJ International Conference on Intelligent Robots and Systems (IROS)* (pp. 5495–5502). IEEE. https://doi.org/10.1109/IROS.2009.5354145.
130. Theofilis, K., Lohan, K. S., Nehaniv, C. L., Dautenhahn, K., & Werde, B. (2013). Temporal emphasis for goal extraction in task demonstration to a humanoid robot by Naive users. In *2013 IEEE Third Joint International Conference on Development and Learning and Epigenetic Robotics (ICDL)* (pp. 1–6). IEEE.
131. Thrun, S., Beetz, M., Bennewitz, M., Burgard, W., Cremers, A. B., Dellaert, F., et al. (2000). Probabilistic algorithms and the interactive museum tour-guide robot minerva. *The International Journal of Robotics Research*, *19*(11), 972–999.
132. Tomasello, M. (1996). Communicating meaning: The evolution and development of language. In *The cultural roots of language* (pp. 275–307).
133. Tomasello, M. (2010). Origins of human communication. MIT press.
134. Tommaso, D. D. (2018). Tobii pro glasses 2 python controller. Webpage, https://github.com/ddetommaso/TobiiProGlasses2_PyCtrl.
135. Torta, E., Cuijpers, R. H., Juola, J. F., van der Pol, D. (2011). Design of robust robotic proxemic behaviour. In *Social robotics* (pp. 21–30). Springer.
136. Tranberg Hansen, S., Svenstrup, M., Andersen, H. J., & Bak, T. (2009). Adaptive human aware navigation based on motion pattern analysis. In *The 18th IEEE International Symposium on Robot and Human Interactive Communication, 2009. RO-MAN 2009* (pp. 927–932). IEEE.
137. Turner, L. D., Allen, S. M., & Whitaker, R. M. (2015). Interruptibility prediction for ubiquitous systems: Conventions and new directions from a growing field. In *Proceedings of the 2015 ACM International Joint Conference on Pervasive and Ubiquitous Computing* (pp. 801–812). ACM.
138. Uleman, J., Adil Saribay, S., & Gonzalez, C. (2008). Spontaneous inferences, implicit impressions, and implicit theories. *Annual Reviews of Psychology*, *59*, 329–360.
139. Vinciarelli, A., Pantic, M., Bourlard, H., & Pentland, A. (2008). Social signals, their function, and automatic analysis: A survey. In *Proceedings of the International Conference on Multimodal Interactions* (pp. 61–68).
140. Walters, M. L., Oskoei, M. A., Syrdal, D. S., & Dautenhahn, K. (2011). A long-term human-robot proxemic study. In *RO-MAN 2011—The 20th IEEE International Symposium on Robot and Human Interactive Communication* (pp. 137–142).
141. Watson, J. S. (1985). Contingency perception in early social development. In *Social perception in infants* (pp. 157–176).

142. Xu, S., & Duh, H. B. L. (2010). A simulation of bonding effects and their impacts on pedestrian dynamics. *IEEE Transactions on Intelligent Transportation Systems*, *11*(1), 153–161.
143. Ye, G., & Alterovitz, R. (2017). Guided motion planning. In *Robotics research* (pp. 291–307). Springer.
144. Zhu, Y., Fathi, A., & Fei-Fei, L. (2014). Reasoning about object affordances in a knowledge base representation. In *European Conference on Computer Vision* (pp. 408–424). Springer.
145. Zollner, R., Asfour, T., & Dillmann, R. (2004). Programming by demonstration: Dual-arm manipulation tasks for humanoid robots. In *2004 IEEE/RSJ International Conference on Intelligent Robots and Systems (IROS) (IEEE Cat. No. 04CH37566)* (Vol. 1, pp. 479–484). IEEE.

Chapter 14
Expressivity Comes First, Movement Follows: Embodied Interaction as Intrinsically Expressive Driver of Robot Behaviour

Carlos Herrera Perez and Emilia I. Barakova

Abstract Social robotics is concerned with the development of embodied agents that can interact naturally with humans in social contexts. Such agents need to gather information about the interaction in a way similar to that of humans—that is, relying not only on verbal communication but taking into account the expressivity and intentionality of movement and the intonation of speech. It is commonly accepted that expressivity derives from a set of specialized behaviours, which often function as expressions of emotions. *In this paper, we advocate for an embodied dynamic interaction approach, arguing that not just certain specialized behaviours are expressive, but rather all embodied interaction, insofar as it creates a relationship with the world, is intrinsically expressive and provides important contextual cues.* This non-reductionist approach highlights the importance of movement understanding for emotion and cognition generally. Drawing from emotion theory, we present an interdisciplinary approach that uses dance as an empirical and experiential domain of research naturally concerned with the issue of expressivity beyond paradigmatic expressions. In particular, the Laban system that captures expressivity in dance serves as the foundation for an interaction design of embodied objects, robots in particular, capable of embedding (i.e. performing and understanding) movement expressivity in social interaction. In conclusion, we argue that there are grounds for more research in social robots that base their interactions on dynamical principles, going beyond occasional expressivity.

C. Herrera Perez · E. I. Barakova (✉)
Eindhoven University of Technology, Eindhoven, The Netherlands
e-mail: e.i.barakova@tue.nl

N. Noceti et al. (eds.), *Modelling Human Motion*,
https://doi.org/10.1007/978-3-030-46732-6_14

14.1 Introduction

Social, cognitive and brain-inspired robotics is increasingly concerned with the qualitative aspects of human–robot interaction, such as the expressivity of human motion. Modelling expressivity is extremely relevant for social robotics, as it is widely accepted that body language and nonverbal communication are rich sources of meaning [1] and can account for the majority of information transmitted during interpersonal interactions [2]. There is ample scope in real-world applications for robots with the ability to engage in complex and sound interactions with humans, which go from assistive technologies and the use of robots in education, therapy and entertainment, to any industrial application in collaborative robotics. Furthermore, and even though some robotic systems may be conceived to operate detached from any human activity, there may be aspects of intelligence that can only be understood in the context of social agents. In particular, expressivity may be a phenomenon key to understanding embodied intelligence and adaptivity.

Humans have evolved not only in response to environmental pressures that require instrumental abilities but also within a social context marked by complex embodied interactions with fellow agents. Sociocultural evolution can be a determinant of traits in humans as biological evolution [3]. Furthermore, cultural and genetic evolution can interact with one another and influence both transmission and selection [4]. Thus, cognitive and adaptive functions that lack a social dimension may at their core be socially determined. The capacity to ascribe, interpret and embed meaning in movement may thus not be a complementary skill for cognitive systems, but instead, it may be integrated with its primary cognitive functions.

Psychology has traditionally made a distinction between instrumental and expressive behaviour, which makes these conceptually independent. In short, instrumental behaviour is about getting something done, while expressive behaviour is about sending out social signals (cf. [5]). In socio-cognitive robotics, the challenge is to bridge the traditionally separated areas of industrial robots (which perform complex goal-oriented behaviours) and social robots (able to interact with humans, often with little functionality beyond). Yet, most approaches tend to consider expression as the outcome of specialized "expressive" behaviours, whose goal is to reflect an internal state often correlated to "an emotion". This approach has the advantage of producing straightforward results, and yet more sophisticated models may be required to fulfil the potential of social robotics.

This paper aims to look beyond the instrumental versus expressive distinction to explore a dynamic interaction approach. Not only social robots should be goal-oriented—there may be functions we regard as solely instrumental or cognitive that demand attention to the social dimension and the capacity for expressive interaction as well.

The resulting view advocates for a dynamic interaction approach, arguing that not only certain specialized behaviours are expressive, but also that all embodied interaction, insofar as it creates a relationship with the world, is intrinsically expressive and provides important contextual cues. In order to further our understanding of what such an approach would entail, we will borrow insights from two apparently independent areas: emotion and dance research.

This chapter is organized as follows. Section 14.2 explores the potential relationship between expressivity and higher cognitive functions, including language. In particular, it explores the hypothesis that movement understanding in social contexts may be a fundamental process in the evolution of language and compositional semantics.

Section 14.3 examines how the underlying concept of emotion determines approaches to motion perception and understanding. In particular, the idea that only specialized behaviours are expressive is related to discrete emotion theories, while the complex system approach to emotion motivates the study of expressivity beyond the occurrence of emotional episodes.

In Sect. 14.4, we introduce the idea that dance can help the development of movement understanding in social robotics. In particular, we discuss the Laban approach to motion modelling, and we present ongoing work that exploits the Laban approach for the development of social robotics.

In Sect. 14.5, we point at present limitations, future developments and the implication of complex unscripted motion understanding for social robotics.

14.2 Grounding Expressivity in Movement

The relationship between expressivity and movement is very much related to the symbol grounding problem. The latter deals with agents capable of processing symbols and questions how such symbols acquire their meaning. A common hypothesis since the 1950s is that the processes that facilitate the behaviour of an adaptive agent are information-processing mechanisms; thus, the capacity of managing abstract symbols is due to the very nature of the cognitive machinery. This approach has nevertheless undergone great criticism, especially from the area of embodied robotics [6]. The difficulties in grounding symbols are one of the foundational arguments of the embodied paradigm in cognitive science, which argues that "the peculiar nature of bodies shapes our very possibilities for conceptualization and categorization" [7].

The embodied cognition paradigm has successfully applied this approach to studying a variety of problems. Work in embodied robotics has demonstrated repeatedly that bodily configurations are determinant of the information flows between an agent and its world, and therefore, cognitive challenges cannot be abstracted without taking into account the concrete type of embodiment at issue. For instance, it is argued that the world needs not to be represented, and depending on the embodiment, an agent can extract the relevant information from the world itself [6].

Another important finding in embodied cognition is the discovery of mirror neurons and their role in understanding action and for learning through imitation. Such neurons are activated whenever the action is performed or observed in another. The human mirror system, which comprises multiple cortical regions and shows evidence of significant activation when the subject is either observing actions or executing actions of a certain class, behaves similarly to individual mirror neurons found within non-human primate brains [8]. Mirror neuron theory thus suggests that expressivity does not require the occurrence of specialized expressive behaviours—embodied behaviour is a source of information about intentions and internal states, even if an expression is unintended by the agent.

Furthermore, this basic mechanism for motion understanding may be the basis for higher cognitive functions such as language. "Mirror neurons might be at the heart of language parity - the hearer can often get the meaning of the speaker via a system that has a mirror mechanism for gestures at its core" [9]. Thus, "a clear challenge is to go beyond models of speech comprehension to include sign language and models of production and to link language to visuomotor interaction with the physical and social world" [9].

An interesting hypothesis suggests fundamental similarities between the way we process grammatical structures and the way we perceive movement. In neuroscience, Rizzolatti and Arbib [10] have shown that some of the neural structures in charge of action recognition form the basis for communication. "A plausible hypothesis is that the transition from the australopithecines to the first forms of 'Homo' coincided with the transition from a mirror system, enlarged, but used only for action recognition, to a human-like mirror system used for intentional communication". They argue that language capacities have evolved from the ability to understand movement in others. Additional evidence in support of this hypothesis is that the language centres in the brain did not evolve from early forms of voice communication in animals, which are connected more with emotional centres than with semantic abilities Rizzolatti and Arbib [10].

Other researchers accept that the faculty of language has a sensory-motor component [11]. There is a large body of psychological and neuroimaging experiments that have interpreted their findings in favour of functional equivalence between action generation, action simulation, action verbalization and action perception [12]. Iacoboni et al. [13] have argued that motor imitation may underlie aspects of language acquisition, and Binkofski et al. [14] have argued that Broca's region subserves mental imagery of motion. Broca's area retains a function that is not directly related to language processing; to be exact, the neurons in this area have response characteristics that may give rise to an imitation of complex motor behaviours, including language.

The idea that networks used for action recognition could be exploited in compositional semantics is also present in robotics. Jun Tani has investigated hierarchical structures for actions and motor imagery [15] showing that multiple timescales recurrent neural networks can realize predictions of sensory streams and abstract compositional information [16]. Robots are guided by hand through sequences of movements, such as grasping a ball and lifting it. The neural network is trained to make predictions in proprioception and vision as the robot goes through a series of predetermined movements. Following training, the neural network can not just replicate the movements, but also "learn to extract compositional semantic rules with generalization in the higher cognitive level" [17]. MTRNN can also "acquire the capabilities of recognizing and generating sentences by self-organizing a hierarchical linguistic structure" [18]. Olier et al. [19, 20] combined deep recurrent networks with probabilistic methods to show that robots can create concepts by coupling sensing and actions towards objects, accentuating the shortcoming of most robotics approaches, where reasoning is based solely on observations.

These experiments show that continuous recurrent neural networks can support "the compositionality that enables combinatorial manipulations of images, thoughts and actions" [15], grounding compositionality naturally on sensory-motor interaction. This touches on the classic AI problem of symbol grounding, which has seen many attempts to ground language in action and perception (see [21]). An influential work is the perceptual symbol system hypothesis [22], which claims that perceptual experience captures bottom-up patterns of activation in sensorimotor areas, through the association in the brain of multimodal sensory information.

Motion expressivity can, therefore, be significant not just for the exchange of information between interacting agents, but for the evolution and development of social and cognitive skills, such as categorization and language. This may be so regardless of whether an expression is intended or unintended, or whether dealing with specialized expressive behaviour or embodied behaviour in general.

14.3 Expressivity Beyond the Emotions

14.3.1 Emotions as Basic Phenomena

The scientific study of embodied expression is intrinsically connected with that of emotion. When researchers ask the question of how to make robots more expressive in order to have a natural interaction with humans, the first answer that comes to mind is by developing their ability to recognize and express emotions. Since understanding the affective dimension of interaction is obviously crucial, we cannot leave aside that the concept of emotion we adopt has great implications for the way we approach the problem of the expressivity of movement.

A common assumption is that expressivity is best studied by focusing on specialized behaviours whose primary function is to communicate emotions. This assumption has its roots in the history of emotion theory. Darwin was the first to theorize about "the expression of emotion in humans and animals", in the framework of developing the evolutionary explanation of behaviour.

> Certain complex actions are of direct or indirect service under certain states of the mind, in order to relieve or gratify certain sensations, desires, etc.; and whenever the same state of mind is induced, however feebly, there is a tendency through the force of habit and association for the same movements to be performed, though they may not then be of the least use. ([23], [1965, p. 28])

Darwin was concerned with a particular set of behaviours, *serviceable habits*, conceived as reflex-like/involuntary behaviours, which seem to lack functional character in the adult, beyond a potential communicative role insofar as they reveal the emotional state that caused them (that is relieved by the expression).

The idea that the behavioural patterns we call expressive mainly serve the function of communicating emotional states is closely linked to theories of basic emotions (e.g. [3, 5, 10, 16, 20, 24, 25]) [26]. Basic emotion theories state that some neurophysiological pathways (e.g. those involved in the fight-or-flight response) have evolved to provide adaptive responses or action programs. All components of a basic emotional response get triggered together during the occurrence of the emotion. Physiological components of the response (arousal) will play an adaptive role (preparing the neuromuscular system for a certain type of interaction), while expressive components will serve mainly a communicative role.

The basic emotion conjecture is used in research on human expression [27]. Basic emotions are hypothesized to determine how certain facial expressions (typically static snapshots) can communicate the basic emotion that triggered such a response. Complex emotions are conceived as built upon the basic ones, whether as a combination of them or specifically through other cognitive or behavioural components. Yet, there is no agreement on what the set of basic emotions are. Ekman's original proposal for a set of basic emotions included anger, disgust, fear, happiness, sadness, and surprise [28], while Plutchik [29] advocated for a set of eight emotions grouped into four pairs of polar opposites (joy–sadness, anger–fear, trust–distrust, surprise–anticipation).

Emotions as Complex Adaptive Phenomena

Much research in social robotics follows the basic emotion approach. A set of emotions is defined, and robots are endowed with a pattern recognition system to match expressions with one item in the emotion set (e.g. Liu et al. 2017), or with the capacity to emulate facial expressions with an emotion in the emotion set (e.g. [30, 31]). The advantages of the basic emotion approach for modelling are evident: having a predefined set of basic emotions facilitates modelling efforts, producing tangible results that nevertheless require quasi-structured interactions.

Basic emotions are nevertheless far from being universally accepted, and not just about the actual set of basic emotions, as their existence has for long been a subject of controversy in psychology. A large body of work advocates for an approach to emotions as complex adaptive phenomena. Recent dynamical system approaches explain emotions as emergent from the dynamic interaction of a multitude of neurophysiological and cognitive components with the real world [32]. Emotions refer to global properties of dynamic behaviour—thus, complexity does not arise from basic emotions, but the other way around. In this framework, we should question the nature of the expressivity of movement beyond the idea of specialized behaviours that somehow codify messages.

Appraisal theories consider that emotions necessarily involve an evaluation (whether conscious/rational or unconscious/automatic) of the relationship between the agent and the environment [33, 34]. Yet this evaluation is not just a cognitive event, but in contrast, it also involves both relational aspect and a motivational aspect [34]. Primary appraisal is conceived as an embodied process that cannot be abstracted from environmental dynamics, nor from the social embodiment. For Frijda, "the most general characteristic of expressive behaviour" is that it "establishes or modifies a relationship between the subject and some object or the environment at large" [33]. Expressive behaviour is relational activity concerned with the relationship between the agents and operates "mainly not by modifying the environment, but by modifying the location, accessibility, and sensory and locomotor readiness" [33].

The notion of action readiness is fundamental to understand what happens during emotion. Embodied agents always present a state of action readiness, a disposition to interact with a specific part of the environment, and thus, the main function of expressive behaviour is to establish, maintain or disrupt a certain relationship with the environment, while emotional episodes are marked changes in action readiness [33]. The adaptive processes that underlie emotion are therefore not restricted to the occurrence of paradigmatic emotional episodes, but are ongoing and shape behaviour at all levels.

During interaction, emotion expression is not perceived from a third-person perspective, as if perceived by a detached observer of a communicative act. Expression, insofar as it is relational, is relevant for dynamics of interaction for both the expressing agent and the receiving agent. A second-person perspective is therefore at stake [35]. The real challenge is thus not just to model agents who express, but agents who engage in interactions where those expressions play a role in configuring the dynamic relationship between agents that interact.

14.4 The Dance Approach to Understanding the Expressivity of Movement

Our concept of emotion, therefore, shapes our approach to expression in robotics. For social robotics, the challenge of considering emotions as complex phenomena is multiple: firstly, specific expressions are not the only expressive behaviours; any embodied behaviour can be a source of important information, however subtle. Expressivity concerns not only particular expressive patterns or the occurrence of emotional episodes but also the continuum of embodied interaction. As a consequence, humans have evolved an extraordinary sensitivity to expressivity in movement.

The second challenge is to conceptualize this sort of embodied information. The argued connection between movement, expressivity and language may be the key to developing social robots capable of rich interactions with humans. The preliminary linking hypothesis is that humans perceive movement as meaningful, engaging neural structures shared by other forms of communication and compositional meaning. These result in principles (laws, regularities, structures) of movement that prescribe how an observer extracts meaning from movement, which in turn could be investigated and formalized into some grammar of movement.

This is not a simple question as this process occurs mainly unconsciously in humans. Its formalization requires empirical work that examines how we understand motion. Rather than starting from scratch, a possible approach is to consider the field of dance as an empirical ground for studying meaning in movement. Dance is inherently focused on movement production and compositional meaning, and the sort of experiences watching another person move can produce in the spectator. Theories about the functional use of the body and the meaning that originate from dance are continually tested and refined in teaching and performance.

For instance, one of the key questions in dance is phrasing, or how expressive movement is structured into movement primitives so that the composition of (meaning in) movement can be better understood. In robotics too, the question of motion primitives is essential for robot programming. It is worthwhile to "assess the potential of dance notations for decomposing complex robot actions into sequences of elementary motions" [36].

While dance is not, strictly speaking, a science, it has gone through an intense period of research and experimentation, especially during the last century. Several systems have developed over the decades that have attempted to explain in a systematic way the physical production of movement (e.g. techniques such as Limon's, Graham's, etc.), the embodied cognitive states that facilitate such movement (prominent in the dance and Somatics approach, [37]), as well as the significance and experience of dance as something to be watched, known as dance aesthetics.

The idea that dance is related to language dates back to Antiquity. Plutarch (46–120 AD) called dancing "mute poetry", and poetry "speaking dance" (cf. [38]). Researchers in dance science have claimed that dance draws on the same cognitive infrastructure as does the capacity for language [38]. Earlier, Collinwood had claimed that every kind of language is "a specialised form of bodily gesture, in this sense, it may be said that the dance is the mother of all languages" [39].

Orgs et al. [40] distinguish between "the processing of syntactic information of postures, movements and movement sequences on the one hand, and processing of semantics of movement intentions on the other hand". Successful message passing between performer and spectator [40] provides cues about what they are thinking, feeling, sensing or intuiting (the four mental factors in Laban's theory). "The idea of 'engagement' allows that a dance might be termed 'successful' on the basis of its ability to create a clear embodiment in movement of a choreographer's intentions and its development of a clear choreographic structure or syntax, rather than on the basis of the aesthetic or genre preferences of the viewer" [35].

Thus, in order to understand expressivity in open-ended movement, robotics may exploit the wealth of experiential knowledge that dance has developed. Particularly useful are approaches that shed light on how movement is perceived, how meaning is composed in movement and how movement patterns can be formalized.

The Laban System

One established theory that has found applications in robotics is Laban's movement analysis. Rudolf Von Laban was a choreographer who created a method for describing, visualizing, interpreting and documenting human movement. His method became known as written dance [41, 42], and the Laban notation system [43] became known as Laban Kinetography [44, 45] and Labanotation [46]. Bartenieff and Lewis [47] are responsible for a significant and unique elevation of this notation method, transforming it into a tool for the qualitative analysis of movement known as Laban Movement Analysis (LMA).

LMA emphasizes the embodied processes underlying motor actions rather than the resultant motor action or trajectory [48]. In this respect, it is able to capture the intentionality and the emotional expressivity of a movement. This is achieved by introducing four movement components "Body, Effort, Shape and Space (BESS)", and the qualitative analysis results from the degree to which these components are integrated throughout the development of a movement primitive.

Especially interesting for robotics is the *effort* component, sometimes called by Laban the *dynamics* of the movement, which points to the subtle characteristics of the way a movement is performed with respect to its intentionality and its emotional load [24, 49, 50]. Attention to strength, control and timing of the movement is of particular interest. The *time* subcategory of *effort* is expressed through deceleration or acceleration within a movement phrase, ranging between two opposites, sustained time and quick time. Further interpretation of movement is obtained through the analysis of *flow,* which can range between the *free* or *bound* flow. The flow category of effort is responsible for the continuity of motions, varying from uncontrolled to more controlled use of flow within the movement, thus relating to the unfolding dynamics of movement.

Such components in the analysis are relevant not only to describe and choreograph dance movements but also to discern qualitative differences in everyday movement. For instance, the difference between punching and reaching for an object is small in terms of body organization, since both rely on the extension of an arm. The strength and the control of the movement differ, as well as its timing, and this difference reflects the emotional load (e.g. anger in the case of punch and neutrality in the second) and the intentionality of the movement (to hit and cause harm, versus taking an object for a purpose).

In our previous work [48, 51], we proposed a framework to facilitate robots' expression and interpretation of movements based on acceleration patterns, as an important dynamic characteristic that can facilitate nonverbal expressive interaction between humans and robots (Fig. 14.1). A robot can, for instance, distinguish between a simply instrumental movement from an emotionally expressive movement. In the research presented by Barakova and Lourens [51], simple car-like robots were used to isolate movements, so that intentionality or effect may not be inferred through the embodiment, i.e. the shape of the robot. In this case, only the movement is subject to interpretation, since the shape is very simple and not anthropomorphic. Further research has explored this topic as well [52, 53].

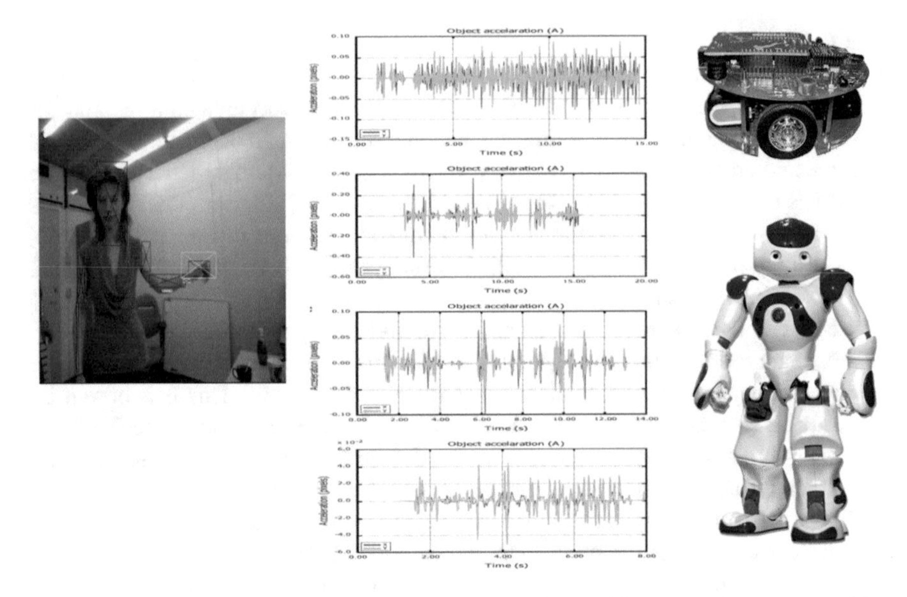

Fig. 14.1 Analysis of experiments with humans showed that different emotions expressed by movement may have the same movement trajectory, but the acceleration patterns of these movements differ. The acceleration patterns of the movement (the plots in the middle column denoting correspondingly happy, angry, sad and polite waving) make it possible for a robot to understand movement in humans by analysing the detected acceleration patterns, and the robot can express emotions, by performing movements with these acceleration profiles

Several attempts to show the expressivity of movement using these approaches were successful on simple robots [48, 51, 54, 55, 56, 57]. However, the application of LMA on humanoid robots has so far shown a limited degree of success because these robots (such as NAO) have motors that are not fast enough, and the acceleration patterns, as proposed by Lourens et al. [48] and Barakova and Lourens [51] are impossible to achieve by such robots. If the speed limitations of humanoid robots are to be overcome, LMA can be a useful framework, because it can help the robot understand human movement by analysing its acceleration patterns and allow the robot to express dynamic emotions and thus convey the intentionality of movement.

These limitations may be overcome in simulated robots. Masuda et al. [58] developed a method, based on LMA, to modify the qualitative aspects of a given (simulated) robot motion to give it an affective character. The method works on arbitrary whole-body movements, which points to the idea that motion expressivity is not exclusive of a certain number of specialized behaviours, or in particular facial expressions, but any movement can contain valuable information regarding affectivity.

Another way to use LMA is proposed by Perugia et al. [59]. In this work, LMA is used to analyse the behaviours of elderly persons interacting with a robot that responds to touch and expresses emotion-provoking behaviours. Although the robot in these studies does not attempt to recognize the movement by itself, the dynamics of the movement is captured by the E4 band and the evolution of the movement can be traced. This study aimed to make a detailed evaluation of the engagement levels and the quality of the interaction of elderly persons with social robots.

14.5 Conclusion

In this paper, we have tried to widen the perspective on movement expressivity, from a reductionist view that considers specialized expressive behaviours that communicate emotional episodes, to a dynamic-interactive view that considers all movement to be both emotional and expressive. The fact that the movement expressivity is essentially dynamic and is inherent in the interaction not only provided humans with an overwhelming drive to attribute meaning movement observed in others as a result of sociocultural evolution, but also has facilitated the evolution of higher cognitive function.

Despite recent advances, social robotics is still in its early days, and it has only dealt with salient aspects of movement expressivity. The long-term challenge to reach maturity has to do with the subtlety and elusiveness of meaning in movement beyond paradigmatic expressions. Expressivity in human movement is a fundamental aspect of communication and interaction, but this normally happens unconsciously and is embedded in cultural forms of embodied interaction, which complicates making universal claims about its underlying mechanisms.

Some expressive phenomena rise to the surface of consciousness—these are gestures and paradigmatic expressions. They play important roles in interactions with children, in storytelling and social representations, among other things. Their importance cannot be underestimated, and thus, social robotics is dedicating important efforts to understand and exploit them in human–robot interaction. Yet, they may be just the tip of the iceberg, and they will never be fully understood unless we look below the surface.

In this paper, we have offered two interdisciplinary connections that may help to shed light on the long-term challenges of the field. First, we have drawn attention to the connections between social robotics and emotion as embodied and dynamic phenomena, and how fundamental questions in emotion theory can be used in a more holistic way to make progress in social robotics. Emotion research has a long history, and some controversies lie at the core of the social robotics challenge. One of them, discussed in this paper, is whether affectivity derives from a set of basic emotions, or whether there is an affective continuum from where discrete emotions emerge. In any case, when working in the field of social robotics, researchers should be aware that they are taking a stance that is relevant beyond their field and that points to fundamental issues in emotion and cognition.

The second contribution of this paper is to draw attention to dance as a field that attempts to unveil the unconscious structure of movement understanding and the attribution of meaning. In order to understand a natural phenomenon, it is necessary to experiment, try things out, test them, combine and separate and fail. Science is to synthesize and create knowledge, but there are many realms of human activity that can facilitate the experiences necessary to create scientific knowledge and for technology to exploit it later on.

Human movement expressivity has been the subject of intense experimentation in dance, which has produced an overwhelming body of experiential knowledge that social robotics can exploit. Choreographers have made hypotheses about what people see and what they look for when they watch someone dance; they have used such hypotheses to create pieces and they have presented them in front of audiences, whose feedback was used to validate and refine their understanding.

The Laban system is an example of how dance research can produce systematic hypotheses that can be used and exploited by social robotics. The discussed experiments exemplify a simple possible use of dance methods and LMA in particular, for creating embodied, context-dependent and dynamic interpretations of movement-based interactions [60]. We have seen how by analysing LMA categories such as effort or time, robots can extract or produce information. Yet, despite the idea that movement contains affective information that should not be structured into basic emotions, to assess the validity of the models, most experimental work relies on basic emotion categories such as fear, anger or sadness.

This framework thus imposes human movement expressivity and its interpretation in the field of robotics. Human expressivity has emerged through coevolution and the constant interaction of humans operating in social environments. For human–robot expressive interaction to be grounded and naturally occur, we propose that it be the result of interaction and evolution. The development of Social robotics should thus go hand in hand with the development of new spheres of interaction that, in principle, would go beyond the regularities of human–human interaction. Further development of social robotics hence requires exploiting dynamic emotion models, which, in order to be assessed, do not rely on the specific categories of basic emotions, but instead focus on the relational and interactive qualities of new robot–human interactions.

References

1. Argyle, M. (2013). *Bodily communication*. Routledge.
2. Onsager, M. (2014). Understanding the importance of non-verbal communication. *Body Language Dictionary*.
3. Klüver J. (2008). The socio-cultural evolution of our species. The history and possible future of human societies and civilizations. *EMBO Reports*, *9*(Suppl. 1), S55–S58. https://doi.org/10.1038/embor.2008.35.
4. Creanza, N., Kolodny, O., & Feldman, M. W. (2017). Cultural evolutionary theory: How culture evolves and why it matters. *Proceedings of the National Academy of Sciences, 114*(30), 7782–7789.
5. Rugg, G. (2013). *Blind spot: Why we fail to see the solution right in front of us*. Harper Collins.
6. Brooks, R. A. (1991). Intelligence without representation. *Artificial Intelligence, 47*(1–3), 139–159.
7. Lakoff, G., & Johnson, M. (1999). *Philosophy in the flesh: The embodied mind and its challenge to western thought* (Vol. 28). New York: Basic books.

8. Molenberghs, P., Cunnington, R., & Mattingley, J. B. (2012). Brain regions with mirror properties: A meta-analysis of 125 human fMRI studies. *Neuroscience & Biobehavioral Reviews, 36*(1), 341–349.
9. Arbib, M. A. (2016). Toward the language-ready brain: Biological evolution and primate comparisons, contribution to a special issue on "Language Evolution" (T. Fitch, Ed.). *Psychonomic Bulletin and Review*. https://doi.org/10.3758/s13423-016-1098-2.
10. Rizzolatti, G., & Arbib, M. A. (1998). Language within our grasp. *Trends in Neurosciences, 21,* 188–194.
11. Hauser, M. D., Chomsky, N., & Fitch, W. T. (2002). The faculty of language: What is it, who has it, and how did it evolve? *Science, 298*(5598), 1569–1579.
12. Grezes, J., & Decety, J. (2001). Functional anatomy of execution, mental simulation, observation, and verb generation of actions: A meta-analysis. *Hum. Brain Mapp, 12*(1), 1–19.
13. Iacoboni, M., Woods, R., Brass, M., Bekkering, H., Mazziotta, J., & Rizzolatte, G. (1999). Cortical mechanisms of human imitations. *Science, 286*, 2526–2528.
14. Binkofski, F., Amunts, K., Stephan, K., Posse, S., Schormann, T., et al. (2000). Broca's region subserves imagery of motion: A combined cytoarchitectonic and fMRI study. *Hum. Brain Mapp, 11*, 273–285.
15. Nishimoto, R., & Tani, J. (2009). Development of hierarchical structures for actions and motor imagery: A constructivist view from synthetic neuro-robotics study. *Psychological Research PRPF, 73*(4), 545–558.
16. Yamashita, Y., & Tani, J. (2008). Emergence of functional hierarchy in a multiple timescale neural network model: A humanoid robot experiment. *PLoS Computational Biology, 4*(11), e1000220.
17. Park, G., & Tani, J. (2015). Development of compositional and contextual communicable congruence in robots by using dynamic neural network models. *Neural Networks, 72,* 109–122.
18. Hinoshita, W., Arie, H., Tani, J., Ogata, T., & Okuno, H. G. (2010). Recognition and generation of sentences through self-organizing linguistic hierarchy using MTRNN. In *Trends in Applied Intelligent Systems* (pp. 42–51). Springer Berlin Heidelberg.
19. Olier, J. S., Campo, D. A., Marcenaro, L., Barakova, E., Rauterberg, M., & Regazzoni, C. (2017a). Active estimation of motivational spots for modeling dynamic interactions. In *2017 14th IEEE International Conference on Advanced Video and Signal Based Surveillance (AVSS)* (pp. 1–6). IEEE.
20. Olier, J. S., Barakova, E., Regazzoni, C., & Rauterberg, M. (2017). Re-framing the characteristics of concepts and their relation to learning and cognition in artificial agents. *Cognitive Systems Research, 44,* 50–68.
21. Cangelosi, A. (2010). Grounding language in action and perception: From cognitive agents to humanoid robots. *Physics of Life Reviews, 7*(2), 139–151.
22. Barsalou, L. W. (1999). Perceptual symbol systems. *Behavioral and Brain Sciences, 22*(4), 577–660.
23. Darwin, C. (1872/1965). *The expression of the emotions in man and animals*. London, UK: John Marry.
24. Eskiizmirliler, S., Forestier, N., Tondu, B., & Darlot, C. (2002). A model of the cerebellar pathways applied to the control of a single-joint robot arm actuated by McKibben artificial muscles. *Biological Cybernetics, 86*(5), 379–394.
25. Tomkins, S. S. (1984). Affect theory. *Approaches to emotion, 163*, 163–195.
26. Plutchik, R. (1962). *The emotions: Facts, theories and a new model*. New York, NY, US.
27. Ekman, P. (1972). Universals and cultural differences in facial expressions of emotions., chapter nebraska symposium on motivation. *J. Cole, lincoln, neb.: university of nebraska press edition,* 207–422.
28. Ekman, P. (1992). An argument for basic emotions. *Cognition and Emotion, 6,* 169–200.
29. Plutchik, R. (1980). A general psychoevolutionary theory of emotion. In R. Plutchik & H. Kellerman (Eds.), *Emotion: Theory, research, and experience* (Vol. 1, pp. 3–33)., Theories of emotion New York, NY: Academic Press.

30. Kirby, R., Forlizzi, J., & Simmons, R. (2010). Affective social robots. *Robotics and Autonomous Systems, 58*(3), 322–332.
31. Saldien, J., Goris, K., Vanderborght, B., Vanderfaeillie, J., & Lefeber, D. (2010). Expressing emotions with the social robot probo. *International Journal of Social Robotics, 2*(4), 377–389.
32. Scherer, K. R. (2009). Emotions are emergent processes: They require a dynamic computational architecture. *Philosophical Transactions of the Royal Society B: Biological Sciences, 364*(1535), 3459–3474.
33. Frijda, N. H. (1986). *The emotions*. Cambridge University Press.
34. Lazarus, R. S. (1991). *Emotion and adaptation*. Oxford University Press on Demand.
35. Vincs, K., Schubert, E., & Stevens, C. (2009, January). Measuring responses to dance: Is there a 'grammar' of dance? In *WDA 2008: Proceedings of the 2008 World Dance Alliance Global Summit* (pp. 1–8). QUT Creative Industries and Ausdance.
36. Salaris, P., Abe, N., & Laumond, J. P. (2017). Robot choreography: The use of the kinetography Laban system to notate robot action and motion. *IEEE Robotics and Automation Magazine, 24*(3), 30–40.
37. Eddy, M. (2009). A brief history of somatic practices and dance: Historical development of the field of somatic education and its relationship to dance. *Journal of Dance & Somatic Practices, 1*(1), 5–27.
38. Hagendoorn, I. (2010). Dance, language and the brain. *International Journal of Arts and Technology, 3*(2–3), 221–234.
39. Collingwood, R. G. (1958). *The principles of art* (Vol. 11). Oxford University Press.
40. Orgs, G., Caspersen, D., & Haggard, P. (2016). You move, I watch, it matters: Aesthetic communication in dance. *Shared representations: Sensorimotor foundations of social life* (pp. 627–654).
41. Laban, R. (1928). *Schrifttanz*. Wien: Universal Edition.
42. Laban, R. (1930). *Schrifttanz—Kleine Tänze mit Vorübungen*. Wien: Universal Edition.
43. Laban, R. (1956). *Principles of dance and movement notation*. New York: Macdonald & Evans.
44. Knust, A. (1953). *Kinetographie Laban*. Parts A–O, Eight volume, unpublished manuscript (in German).
45. Knust, A. (1958). *Handbook of Kinetography Laban*. Hamburg: Das Tanzarchiv.
46. Hutchinson, A. (1956). Labanotation a tool for the exploration and understanding of movement. *Physical Education, 18,* 144.
47. Bartenieff, I., & Lewis, D. (1980). *Body movement: Coping with the environment*. Gordon and Breach Science Publishers.
48. Lourens, T., Van Berkel, R., & Barakova, E. (2010). Communicating emotions and mental states to robots in a real time parallel framework using Laban movement analysis. *Robotics and Autonomous Systems, 58*(12), 1256–1265.
49. Porr, B., & Wörgötter, F. (2003). Isotropic sequence order learning. *Neural Computation, 15*(4), 831–864.
50. Laban, R., & Lawrence, F. C. (1947). Effort: A system analysis, time motion study. *London: MacDonald & Evans*.
51. Barakova, E. I., & Lourens, T. (2010). Expressing and interpreting emotional movements in social games with robots. *Personal and Ubiquitous Computing, 14*(5), 457–467.
52. Heimerdinger, M., & LaViers, A. (2019). Modeling the interactions of context and style on affect in motion perception: Stylized gaits across multiple environmental contexts. *International Journal of Social Robotics, 11,* 495–513.
53. Knight, H, & Simmons, R. (2015). Layering Laban effort features on robot task motions. In *Proceedings of the Tenth Annual ACM/IEEE International Conference on Human–Robot Interaction Extended Abstracts* (pp. 135–136). ACM.
54. Hoffman, G., & Weinberg, G. (2010). Shimon: An interactive improvisational robotic marimba player. In *CHI'10 Extended Abstracts on Human Factors in Computing Systems* (pp. 3097–3102). ACM.
55. Arbib, M. A. (2005). From monkey-like action recognition to human language: An evolutionary framework for neurolinguistics. *Behavioral and Brain Sciences, 28,* 105–167.

56. Barakova, E. I., & Chonnaparamutt, W. (2009). Timing sensory integration. *IEEE Robotics and Automation Magazine, 16*(3), 51–58.
57. Barakova, E. I., & Vanderelst, D. (2011). From spreading of behavior to dyadic interaction—A robot learns what to imitate. *International Journal of Intelligent Systems, 26*(3), 228–245.
58. Masuda, M., Kato, S., & Itoh, H. (2010). Laban-based motion rendering for emotional expression of human form robots. In B. H. Kang & D. Richards (Eds.), Knowledge management and acquisition for smart systems and services. PKAW 2010. *Lecture Notes in Computer Science* (Vol. 6232). Berlin, Heidelberg: Springer.
59. Perugia, G., van Berkel, R., Díaz-Boladeras, M., Català-Mallofré, A., Rauterberg, M., & Barakova, E. I. (2018). Understanding engagement in dementia through behavior. The Ethographic and Laban-Inspired Coding System of Engagement (ELICSE) and the Evidence-Based Model of Engagement-related Behavior (EMODEB). *Frontiers in Psychology, 9*, 690.
60. Schafir, T., Vincs, K., Schubert, E., & Stevens, C. (2009, January). Measuring responses to dance: Is there a 'grammar' of dance? In *WDA 2008: Proceedings of the 2008 World Dance Alliance Global Summit* (pp. 1–8). QUT Creative Industries and Ausdance.
61. Barakova, E. I., Gorbunov, R., & Rauterberg, M. (2015). Automatic interpretation of affective facial expressions in the context of interpersonal interaction. *IEEE Transactions on Human-Machine Systems, 45*(4), 409–418.
62. Laban, R., & Lawrence, F. C. (1947). *Effort*. Macdonald & Evans.
63. Plutchik, R. (2001). The nature of emotions: Human emotions have deep evolutionary roots, a fact that may explain their complexity and provide tools for clinical practice. *American Scientist, 89*(4), 344–350.

Chapter 15
Gestures in Educational Behavior Coordination. Grounding an Enactive Robot-Assisted Approach to Didactics

Hagen Lehmann and Pier Giuseppe Rossi

Abstract The ability to coordinate behaviors at an interindividual level has shaped human social evolution by enabling the formation and maintaining the cohesion of large social groups. Nonverbal communication always played a central role in this process—a role that now can be expanded. With the introduction of social robots, capable of emulating human appearance and movements to communicate with us through social signals, the mechanisms of human–human nonverbal communication offer us a way to improve human social communication with robots in a variety of fields—from information to assistance to people with special needs. In this article, we explore this possibility with reference to educational robotics, and, more precisely, to robot-supported didactics. In the first part of this chapter, we discuss the concepts of behavior coordination and structural coupling as evolutionary mechanism underlying human social structures and illustrate the importance of nonverbal communication in social interactions. We will give examples of different nonverbal communication channels and illustrate, with recent paradigmatic studies, how they can be used for social robotics in different cultural settings. In the second part of the chapter, focusing on educational robotics, we illustrate how nonverbal communication between humans and robots can be used as feedback channel between teachers and students in order to reinforce the structural coupling in an enactive robot-assisted approach in didactics.

15.1 Behavior Coordination in Human Evolution

Human social evolution is, to a large extend, driven by the human capability to communicate about past experience, and in this way to pass on and to accumulate cultural techniques [1, 2]. Humans transmit information to each other via a plethora of different signals. These signals can roughly be categorized into verbal and nonverbal. Verbal signals include language and utterances, like shouts and laughter. Nonverbal signals include touch, facial expressions, body posture, and gestures.

H. Lehmann (✉) · P. G. Rossi
Università degli Studi di Macerata, Macerata, Italy
e-mail: hagen.lehmann@unimc.it

N. Noceti et al. (eds.), *Modelling Human Motion*,
https://doi.org/10.1007/978-3-030-46732-6_15

While communicating, humans exhibit a multiplicity of these nonverbal behaviors at the same time, and many of them are displayed subconsciously. The expression of these behaviors, as well as their recognition, involves almost the entire body [3]. Humans are able to use the posture of conspecifics, the way they move in terms of speed and expressivity, their tone of voice and general appearance to deduce or even understand internal states like emotions or level of arousal. This understanding enables us to feel empathy for one another [4], which plays an important role in the formation and maintenance of social cohesion in large groups of individuals [5], like human societies. Since most of the cues used to "understand and feel for" the other are nonverbal, the importance of nonverbal communication for human social evolution cannot be overestimated [6]. Face, eyes, and hands play a central role in this process [7]. Crucial for the interaction with others are subconscious eye movements like gaze and pupil dilation and hand and arm gestures [8]. Most of these nonverbal signals have facilitating, regulating, and illustrating functions [9], and are as such part of the embodied information exchange that makes coordinated communication between two or more people possible.

15.1.1 Embodiment and Structural Coupling

Humans can be represented as complex self-organizing systems dynamically embedded in complex self-organizing environment(s) [10]. In this theoretical perspective, the process of adaptation is often thematized in terms of "coevolution." The general idea is that of an dense interaction, made of exchanges of energy and matter, between two operatively independent self-organizing systems. Typically coevolution is characterized as a symmetrical relation of reciprocal perturbations and endogenous processes of self-regulation that coordinates the dynamics of a system with the dynamics of its environment. Until both these two systems maintain their organization, the dynamical evolution of each of them consists of a series of endogenously generated states of activity that are compatible with the self-organizing states of the other system. Humberto Maturana and Francisco Varela, within the theory of autopoiesis, offered a particularly well-defined notion of coevolution in terms of "structural coupling" [11]. Introduced by Maturana and Varela to conceptualize the adaptive coupling as a cognitive coupling, this notion indicates the capability, typical of biological systems, to effectively act within their domain of existence to maintain and develop their organization and their mode of existence. According to the theory of autopoiesis, at the level of the dense interactions between conspecifics characterizing social environments, structural coupling becomes "behavioral coupling": a symmetrical relation of reciprocal perturbation and endogenous self-regulations that generates the interdependence of the behavioral conducts of the interacting systems. In humans, behavioral coupling is the basic structure of social interaction based on communication [11].

When developing enaction in the 1990s (e.g., [12]), Varela put this notion of structural coupling at the center of his theory. The loop shown in Fig. 15.1 illustrates

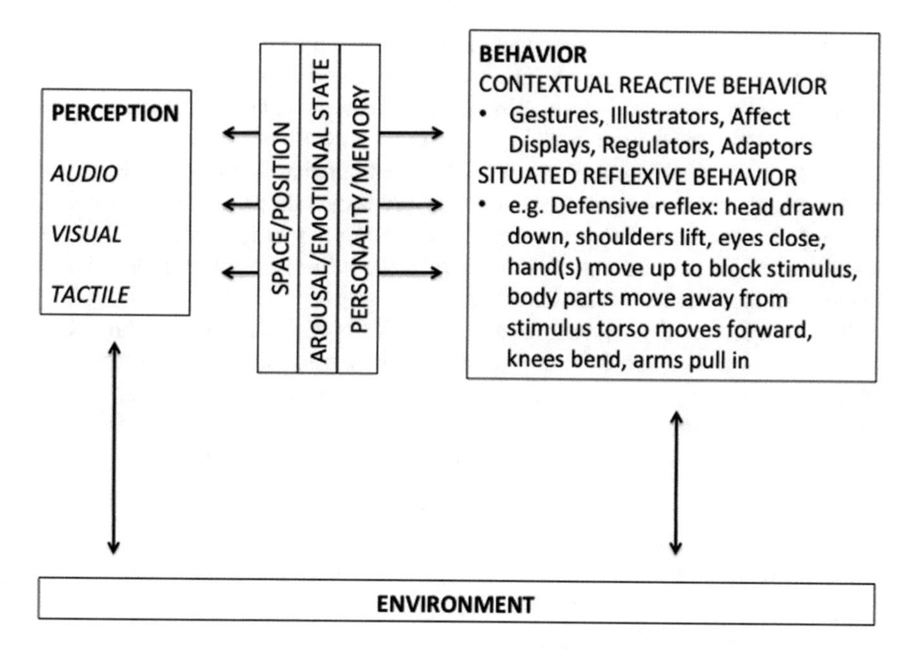

Fig. 15.1 Structural coupling between environment and human perception and behavior

the structural coupling between an individual and its environment. The changes in the dynamics of the environment generate perturbations in the dynamic of the system, which reacts on these changes via different self-regulative behaviors to compensate them. These behaviors generate in turn perturbations in the environment, and so on. In case of social interactions between two or more humans, the internal equilibria can be represented also by the individuals personality, which depends on the individuals phylo- and ontogenetic history, and the perceptible changes can be represented by the different verbal and nonverbal communication signals.

In order for a social exchange to be successful, i.e., to achieve a common goal, which in its simplest form could mean to have a conversation, the behaviors of the individual and its environment need to be coordinated [13]. This type of coordination can be found on all levels of embodied behavior, from eye movements [14] to coordinated neuronal patterns [15].

In order for robots to be accepted into mixed human–robot ecologies [16], it is important that not only their verbal, but also their nonverbal behavior is aligned with the expectations of the users. As pointed out above, human nonverbal behavior incorporates a multitude of signals. Specifically for robots that are operating in close physical and even social proximity to humans, the same should be true. For example, it has been shown that different robot blinking patterns can influence how the robot is perceived [17]. This is even more true for contextual reactive behaviors like gestures.

Research has shown that with an increasing level of autonomy and human likeness in appearance of robots, their human users have the tendency to anthropomorphize them [18, 19, 20] (Damiano and Dumouchel; Eyssel and Kuchenbrandt). Since the goal of social robotics is to enable intuitive and comfortable interaction with between robots and humans, robots should be enabled to become part of the structural coupling of humans and their environment by endowing them with capabilities of behavior coordination. In other words, if we understand both human–social interactions, and human–robot interactions as coevolutionary processes, or processes of structural coupling, we can apply the principles of enaction in the design process of robotic behaviors. In the second part of this chapter, we will discuss the implication of enaction further from an educational perspective.

Research on social coordination shifted into focus of evolutionary anthropology in the middle of the 1960s. One important task was to find a categorization for nonverbal behaviors that explained many of the observed phenomena and allowed for predictions of group dynamics. Ekman and Friesen [21], for example, separated nonverbal behaviors into contextual reactive and situated reflexive.

15.1.2 Reflexes

According to Ekman and Friesen's definition, the latter included the orientation reaction and the startle reflex, if something or someone touches us or appears quickly and unexpectedly in the personal zone of a person [22]. In this case, the person unwillingly draws the head in and lifts the shoulders to protect the neck, closes the eyes to protect them, draws the arms in and moves the hands up to protect the body, bends the knees slightly, and moves the body away from the stimulus [23]. Another reflex in this category would be the orientation reaction, which is exhibited when an unexpected event occurs around a person not fast enough to initiate the startle reflex. In this case, the person's body will stiffen, and the person will orient herself toward the stimulus and exhibit a general outward alertness [24]. On the other hand, there are reactive contextual behaviors, which are usually used to influence conversation dynamics. They can have an illustrative function emphasizing what is currently said, a regulatory function facilitating turn-taking during conversations, or they can specific linguistic meaning like most hand gestures.

15.1.3 Facial Cues

For humans, the highest concentration of different sensors is located in the face, harboring the mouth, the nose, the eyes and to a certain extent the ears as sensory input channels is also the focal point when communicating with conspecifics. As highly visual species humans automatically "face" their counterpart when they want to start a social exchange or when they are addressed by someone else, in order to see her/his intentions. Since the hairless human face allows for the visibility of very small muscle movements, it is not surprising that facial expressions are one of the most efficient channels for the transmission of information about the emotional states of the other, and that a lack of facial expressivity creates in humans a sense of eeriness. Social eye movements like gaze following, change of pupil size, and blinking have been shown to be among the most powerful signals humans use to create, maintain, or disturb group cohesion or peer-to-peer interaction [25, 26]. The specific visibility of the human eye [27] turns it to a communication channel that is unique in nature.

15.1.4 Gestures

Despite the importance of the above-mentioned communication channels, the importance of the hand and arm gestures for nonverbal communication is central. When engaged in social exchanges, in which one is not required to have "ones hands full," the hands are usually used to illustrate and emphasize what is currently said and even thought, as well as to regulate the conversational dynamics of an interaction. This is usually done via a set of cultural depending gestures. These gestures are essential for ensuring comfortable and intuitive social exchanges. In contrast to other subconscious nonverbal communication signals, gestures are population dependent [28, 29, 30].

Communicative gestures have evolved in different parts of the world, which were isolated from each other for long periods of time. This, in combination with the physical constraints of the human body, led to the effect that the same gesture can have very different meanings in different cultures. However, it is important to point out that despite these differences it is possible, albeit on a very basic level, to establish communication via gestures between members of very different cultural backgrounds. This hints at the long evolutionary history and importance of gestures as communication channel in human evolution. In some cases, the differences can be quite striking. For example, going from Europe to Japan and seeing a Japanese person waving her hand in front of her face with the face turned toward you could lead to quite a severe misunderstanding. This gesture, in Europe commonly understood as an insult with the meaning "Are you crazy?" is meant as an apologetic negation in Japan (Fig. 15.2).

But even within Europe, the differences are very noticeable. In southern Europe, namely in Italy, gestures are used much more frequently during conversations when compared to countries of northern Europe. Comparing the frequency and expressivity in the use of hand gestures during a discussion among Scandinavians or among Italians would illustrate the point (Fig. 15.3).

These examples show that gestures, which have played a crucial role during the early social evolution of our species, remain very much alive in human social communication. Research exploring different aspects of human cognition has demonstrated the universal importance of gestures for enhanced information transfer [32] and lexical retrieval [33]. It has even been shown that using gestures helps to reduce the cognitive load when explaining complex problems to others [34]. In this way, gestures not only reflect our cognitive state, but also shape it.

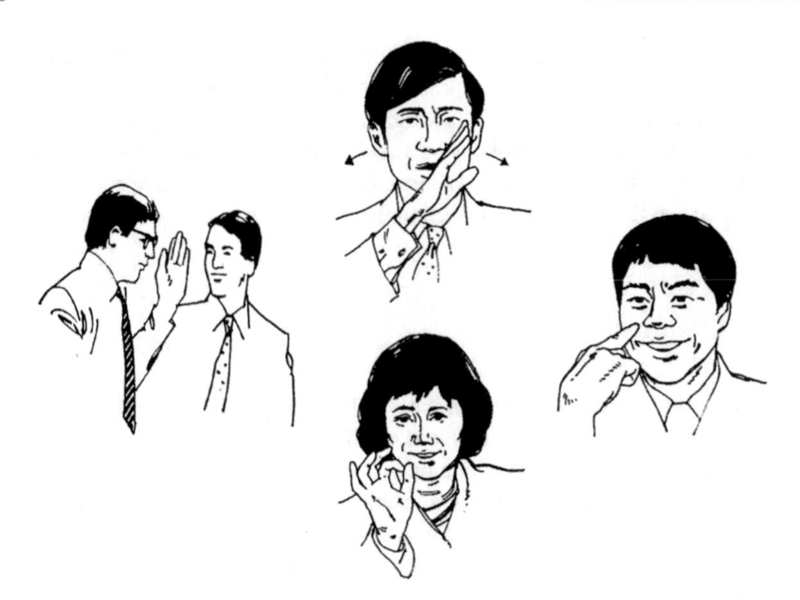

Fig. 15.2 Examples of Japanese communicative gestures (from [31]). Starting from top moving clockwise the gestures mean no (waving hand in front of face), I (pointing to nose), money, and apology for intruding personal space of other

Fig. 15.3 Examples of Italian communicative gestures (from [31]). Starting from top moving clockwise the gestures mean What is going on? something tastes very good, moderate threat, aggressive disinterest

One of the theories about the origins of human language is the gestural origin hypothesis [35]. It proposes that the use of gestures predates the evolution of verbal language. There is archeological, physiological, and behavioral evidence that support this theory. For example, paleo-archeological findings show differential growth in the brain and the vocal apparatuses [36]. Human babies exhibit gestural communication before they speak [37]. Bonobos and chimpanzees use gestures to communicate nonverbally without touching one another [38]. Apes and humans show a bias toward the usage of the right hand (left brain) when gesturing [39, 40]. In apes, the Brodmann area 44, a brain region that is activated during the production and perception of gestures, is enlarged in the left brain hemisphere [41].

These findings illustrate the high relevance of gestures for human–human communication. Gestures are deeply rooted in primate social evolution. In combination with facial expressions and vocal signals typical of apes and humans, they added a layer of flexibility to the behavioral repertoire that allows for great communicative complexity, which drove human social evolution.

15.1.5 Gestures in Human–Robot Interaction

The understanding of the importance of nonverbal communication, in combination with the technological progress of robot embodiments that allow the expressions of nonverbal signals, has lead in recent years to various approaches to implement and test communicative gestures in humanoid and non-humanoid robots. These implementations were done from different perspectives and were based on different research questions. In this section, we will discuss exemplary studies that aimed at developing gestures and other forms of nonverbal communication for different robotic environments.

Ono et al. [42] presented in their work a model of embodied communication, including both gestures and utterances. They tested their model with the Robovie platform, in an experimental setup in which the robot gestured to various degrees while explaining the route to a designated goal to a human interlocutor. They could show that (a) the more the robot gestured systematically, the more the human subjects' gestures increased in frequency, and (b) that the more the robot used gestures, the more the better the humans understood its utterance about how to reach the goal. Other research examined the role of gestures in the process of starting an interaction with a robot, maintaining it, and perceiving a connection to one another [43]. The results of these experiments showed that people direct their attention more frequently to robots and find their interactions with the robot more appropriate when gestures are present in the interaction. Riek et al. [44] tested the effect of different aspects of interactional gestures made by a robot on the ability of humans to cooperate with this robot. They found that humans were cooperating quicker when the robot made abrupt, front-oriented gestures.

Beck et al. [45] tested whether it is possible with for a robot to express emotions with body language in such a way that children are able to understand and interpret them. They used different body postures of the robot for typical emotional states like happiness, fear, anger, and pride. Their results underlined the importance of the position of specific body parts, i.e., the head position, during the expressed emotion in order to ensure the interpretability of the expression.

Another very interesting insight into how to use the body language and gestures during human–robot interaction comes from [46]. They use different gestures and gaze behaviors in order to test the persuasiveness of a storytelling robot. In their experiment, the participants listened to a robot telling a classical Greek fable. Their results showed that only a combination of appropriate social gaze and accompanying gestures increased the persuasiveness of the robot. The authors pointed out that in the condition the robot was not looking at the participants and only used gestures, the persuasiveness of the robot actually decreased because the participants did not feel like they were addressed.

This illustrates an important point for future HRI research. It is not sufficient to look only at different aspects of body language and then to model them separately on the robot, but it is at least as important to focus on their integration in order to achieve a holistic behavior expression during the interaction. Using video footage of professional actors, as was done in this study, is a good starting point for the modeling of these dynamics. Huang and Mutlu [47] used a robot narrator equipped with the ability to express different types of gestures. They designed deictic, beat, iconic, and metaphoric gestures following McNeill's terminology [32]. The results showed interesting effects for the different types of gestures. Deictic gestures, for example, improved the information recall rate of the participants, beat gestures contributed positively to the perceived effectiveness of the robots gestures, and iconic gestures increased the male participants' impression of the robot's competence and naturalness of the robot. An interesting aspect of their findings is that metaphoric gestures had a negative impact on the engagement of the participants with the robot. The authors state that a large number of arm movements involved in this type of gesture might have been a distraction for the participants.

These studies illustrate that researchers in HRI have recognized the importance of gestures for their field. Besides the insights this research gives into how humans use and understand gestures, and it also has a very practical and applied use. Specifically, the last five years have seen the deployment of a multitude of social robotic platforms in areas that range from shopping malls to schools and airports [48]. International projects like the Mummer project [49], for example, experiment with social signal processing, high-level action selection, and human-aware robot navigation by introducing the Pepper robot in a large public shopping for a long-term study. The result of this project was applications that enable the robot to talk to and to entertain customers with quizzes, and give guidance advice by describing and pointing out routes to specific goals in the shopping mall.

These examples illustrate that social robot need, for almost all of their future applications, to be able to interact with humans in human terms. Once the robots have left the laboratory and the factory, their communication capability needs to be appropriate for laymen users, i.e., they need to make themselves understood in an easy and intuitive way.

As pointed out on page 4, the frequency and type in the use of gestures are culturally dependent. If we imagine a social robot that is, for example, built in Europe, equipped with gesture libraries based on northern European social interaction dynamics and sold worldwide, it is easy to understand the issues that could arise. It is therefore important to stress that it is necessary to not only understand how to design gestures for social robots, but also to conduct comparative research and develop cultural sensitive gesture libraries. The result of an earlier study that was aimed at establishing a baseline for robot gestures during human–robot conversations [50] demonstrates this need. During the study, conversational pairs of humans were videotaped and their use of gestures was analyzed and compared. The research was conducted in Italy and in Japan, respectively. In this research, gestures were defined as nonlocomotory movements of the forearm, hand, wrist, or fingers with communicative value, following definitions from other behavioral research [38, 51], and communicative movements of the head like nodding up and down, shaking left to right, and swaying. The results showed expectedly quite severe differences not in the type, but also in the frequency and expressivity of the gestures used. Italians used their arms and hands considerably more during the conversations than the Japanese participants. While Italians used much more iconic and metaphoric gestures, the Japanese participants used small head movements to control and regulated the conversational dynamics.

Other studies found similar effects between participants from different cultural backgrounds.

Trovato et al. [52], for example, researched the importance of greeting gestures in human–robot interaction between Egyptian and Japanese participants. They could show that specifically during the robot's first interaction with a human it can be crucial to have a culturally sensitive gesture selection mechanism. They argue that once social robots will become mass-produced products, its cultural sensitivity in the behavior of the robot will determine its success rate. If users have the possibility to choose the robotic platform they are most comfortable with, then it stands to reason that they will choose one that exhibits cultural closeness. In another study, the same group presented a cultural sensitive greeting selection system [53]. Their system was able to learn new greeting behaviors based on their previous Japanese model. The research was conducted with German participants and the results showed that the model was able to evolve and to learn movements specific to German social interaction dynamics. The authors argue that this type of cultural sensitive customization will become more and more important and that robots should be able in the future to switch easily between different behavioral patterns depending on the cultural background of the human user.

In this first part of the chapter, we illustrated the importance of nonverbal communication and behavior coordination in human–human communication from a social, anthropological, and evolutionary perspective and showed how gestures, as one type of nonverbal-social signal, can be used during human–robot interactions. This is the framework in which we contextualize the second part of the chapter, which discusses an implementation of the theoretical concepts of behavior coordination and enaction in educational robotics.

15.2 Robots in Education

The previous part of this chapter was intended to give an overview of the role nonverbal communication and behavior coordination played in human social evolution, and to illustrate why the use of nonverbal communication signals for social robots that need to interact with humans in close physical and social proximity is important for the success of this technology. We looked at human–robot interaction research and saw an increasing awareness of the importance of social gestures for the field. In the following part, we will look at one field educational robotics and explore how social robots can be implemented in the teaching process and what role nonverbal communication and behavior coordination can play for the success of these robots. We will propose a new didactic framework, which represents an extension of the enactive approach to didactics [54] and ascribes to social robots a central role in the feedback process between teachers and students. It will become clear, why the use of robotic gestures in this framework is essential for the success of the enactive approach.

15.2.1 From Tools to Mediators

Since the development of Lego Mindstorms NXT [55], an increasing number of robots, have been deployed in schools, not only to teach programming, but also scientific subjects like physics or chemistry (e.g., [56, 57]). The integration of the Lego Mindstorms into school curricula followed a "constructionist" framework and the related "learning-by-making" methodology, as it was originally proposed by [58]. It has mainly been used in middle schools and high schools to teach students the basic principles of what robots are, how they work, and how software applications can be developed for them [59, 60]. This kind of uses of robot technology in schools enforced the kind project-based learning strategies [61], in which teachers usually engage their students into artifact or product building activities, and which we still see most frequently in technology-assisted STEM education.

However, the last ten years have seen more and more social robots being integrated into, for example, primary school language classes and in robot-assisted therapy settings for children with special needs. These robots are usually humanoid and serve in the function of social mediator.

As pointed out in section "Embodiment and Structural Coupling," in order for social interactions to be successful, behavior coordination is central. This is specifically true in educational contexts. Hence, mechanisms to provide appropriate feedback from robots in tutoring situations have moved into the focus of research on social robots in education (e.g., [62]). This feedback is usually based on different sensory inputs from human social signals, and on the processing of these social signals. Social signal processing with the goal of improving robot feedback has been at the center of various recent social robotic projects [49, 63].

In the specific case of long-term interactions between robots and children, the issue arises that the novelty effect of using robots wears off quickly and that the children subsequently become bored. In these circumstances, the robot does not only need to be reactive in a specific task, but additionally, it needs to provide appropriate emotional feedback. This kind of feedback needs to be based on memory models of the children's behavior over time. First successful attempts in this direction have been made to support vocabular learning in primary school students [64].

Different ways of classifying robots in educational contexts have been. For example, Mubin et al. [65] and Tanaka et al. [66] identify two different ways in which robots have been integrated into school curricula. As pointed out above one is as educational tools in themselves, e.g., to teach children the basic principles of programming, and one as educational agents. The latter category includes social robots like, for example, RoboVie [67], Tiro [68] and NAO [69]. A further classification of the roles of social robots in educational contexts has recently been given by Belpaeme et al. [70]. In their review, they found that this kind of robots mainly fulfills the roles of novices, tutors, or peers. When fulfilling the role of novice, a robot allows the students to act as tutor and to teach the robot a determined topic. This helps the children to rehearse specific aspects of the syllabus and to gain confidence in their knowledge [71, 72]. When the robot is fulfilling the role of tutor, its function is usually that of assistant for the teacher. Similar to robotic novices, robotic tutors have been used in language learning classes. Strategies used in robot-based tutoring scenarios include, for example, encouraging comments, scaffolding, intentional errors, and general provision of help [73]. The idea behind having robots assume a peer role for children is that this would be less intimidating. In these cases, the robot is presented as a more knowledgeable peer that guides the children along a learning trajectory [70], or as an equal peer that needs the support and help of the children [71].

Another very important field in which robots have been used to achieve educational goals is robot-assisted therapy (RAT) for children with special needs. Robots like KASPAR [74] fulfill the role of social mediator to facilitate social interaction among and between children with autism spectrum condition (ASC) (e.g., [75]). In this function, the robot teaches the children appropriate social behaviors via appropriate verbal and nonverbal feedback. RoboVie R3, on the other hand, has been used very successfully in the teaching of sign language to children with hearing disabilities. For this purpose, it was equipped with fully actuated five-fingered hands. In their study, from 2014, Köse et al. [76] describe comparative research between NAO and RoboVie R3. The mode of interaction between the robots and the participants was nonverbal, gesture-based turn-taking, and imitation games. Their results showed that the participants had no difficulty to learn from the robots, but that they found it easier to understand Robovie R3's performances due to it having five fingers, longer limbs, and being taller than NAO. These findings could be seen as evidence that for gesture-based communication, child-sized robots like RoboVie R3 and Pepper might be in an advantage given their better visibility and the apparent better interpretability of their movements. In follow up studies to their original research, Köse et al. [77] and Uluer et al. [78] replicated their original results using RoboVie R3 as an assistive social companion in sign language learning scenarios. They could additionally show that the interaction with the physical robot is more beneficial for the recognition rate of the gestures performed by the robots, when compared to a video representation.

As shown in Fig. 15.4, social robots are used in an area in which they are not considered as tools, i.e., subjects and part of the knowledge to be transmitted, but in the area where they are directly or indirectly transmitting knowledge. The function of the robot changes from object to educational agent involved in the generation of new knowledge. This moves the robot into the center of the teaching process. As we discussed on page 1 of this chapter, human culture has a cumulative nature and our social evolution is "ratcheted up" by active teaching [1]. This process is inherently human and the cultural techniques linked it to follow a trajectory that intuitively connects individuals and increases social cohesion in groups. They are necessarily based on verbal and nonverbal communication techniques and involve the entire human repertoire of social signaling. If we ascribe robots an active function in this process, it stands to reason that they need to be equipped at least to some extent with the capability to use body language and gestures.

Following this line of thought, it is noticeable that a lot of robots that are used as educational agents are either humanoid or semi-humanoid, such as NAO, Robovie R3 [79], or Maggie [80]. One of the reasons for this is that human features like a moveable head, moveable arms, and actuated hands are most suitable for the implementation of human nonverbal communication signals. However, this makes the development and implementation of this kind of fully embodied agents in education much more costly and difficult, than the use of robots similar to the ones that can be constructed from Lego Mindstorms. Herein lies the reason why, until now, the majority of robotic

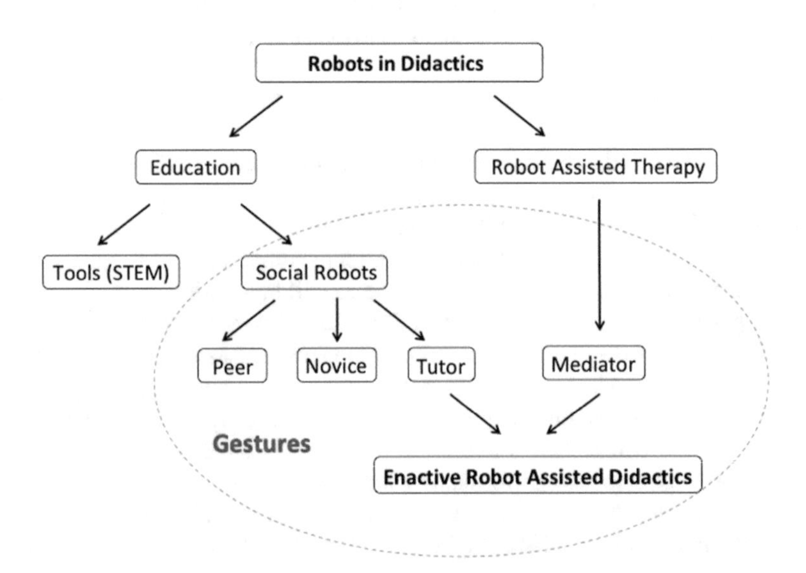

Fig. 15.4 Roles of robots in didactics. The red oval marks the space in which we propose robots should use gestures

technology was used as tools for STEM education in the past [81, 82]. However, with the readily availability of robots like NAO or Pepper, this is changing. These new types of robots lent themselves to be integrated into new existing theoretical approaches in the field of didactics. On such approach that gains momentum at the moment is enactive didactics. A detailed description of the enactive didactics approach can be found in Lehmann and Rossi [83].

15.2.2 Enactive Robot-Assisted Didactics

The enactive didactics approach focuses on the interactions between teacher and student during the knowledge creation process. The teacher is seen as the focal point that raises the awareness of an issue in the students. In the next step, the teacher and the students build an answer to the issue together. The trajectory along which this answer is constructed and sketched out by the teacher. She has the role of mediator between the world of the student and the new knowledge [84], and the task of activating a cognitive conflict [85] that bridges the student's knowledge, the new problems to address, and related new knowledge. After the new knowledge is established, it is crucial to validate it. In the enactive didactics approach, it is the function of the teacher to verify the epistemological correctness of the constructed knowledge, ensuring that it does not contradict the existing knowledge. In order to establish this validation, continuous feedback between the teacher and the students is necessary. The role of feedback is not only important for the student in this process, but also for the teacher, as each part of the teacher–learner dyad is seen as part of the structural coupling between the environment and, respectively, the teacher and the students (see Fig. 15.5a). Unfortunately, in reality, many interaction processes in education lack the space for interaction and feedback for various reasons. This absence of real feedback, however, produces self-referentiality, which is a characteristic of closed systems and diametrically opposed to the form of interaction between a subject and its environment as it is described in the enactive approach.

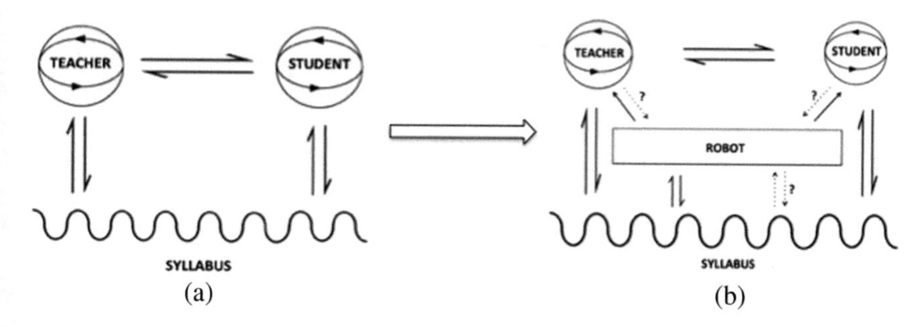

Fig. 15.5 Extension of the structural coupling characterizing the enactive didactics approach by integrating a robotic tutor (taken from Lehmann and Rossi [86])

As we proposed analytically elsewhere [86], integrating social robotics technology based on the enactive framework has the potential to remedy the problematic lack of feedback by reinforcing the reticular interactional structure described in the approach (see Fig. 15.5b). In other words, the integration of a robot in the function of social mediator will strengthen the communication between teacher, students, and syllabus (knowledge to be thought). Consequently, we describe this approach as enactive robot-assisted didactics (ERAD).

The central point of this idea is the strengthening of the communication between the human actors. In order for the robot to be successful, its attempts to initialize communication have to be intuitively understandable and most importantly nonintrusive or disruptive. The robot must be capable to catch the attention of the teacher or the students without disturbing the flow of the lecture and to intervene in a way that is perceived constructive and helpful.

In order to achieve this, we need to shift our attention to human–human nonverbal communication. As discussed before in this chapter, humans have an entire evolutionary history of using body posture, and more specifically head, arm, and hand movements to seek attention and transmit information to conspecifics. If robots ought to be successful in social mediator functions like the ones described here, they need to be enabled to tap into this behavioral repertoireand exploit the evolved human abilities to interpret the body movements of other. Since this ability to "read" our counterpart is limited, other humans,[1] this type of robots should be either humanoid or semi-humanoid (i.e., they should have a head, arms, and hands).

For ERAD, we propose a number of techniques that will enable the robot to collect data from the student and the teacher, but the central part is the communication abilities of the robot. Specifically, in noisy environments like the classroom, these abilities strongly depend on the robot's capability to use gestures. Since robots are already in the process of being integrated in such different cultural context like Japan and Western Europe, it will not be enough to equip robots only with one similar set of gestures. As pointed out by Trovato et al. [53], the only robots that are capable of adjusting their behaviors to a specific cultural background will be successful in an increasingly competitive market of robotic social mediators.

[1]Exceptions are species like dogs, with which we share a long evolutionary history and which have been bred selectively to understand human body language and to be understandable.

15.3 Conclusive Remarks

Since nonverbal communication signals and behavior coordination are from an evolutionary perspective such as important and integral part of human social interaction, it seems natural to use these concepts also in interactions with social robotic technology. It might even be necessary to rethink our approach to designing this type of interactive technology, following more a communication and coordination-driven perspective on the embodiments we construct. The research and theory discussed in this chapter underline the importance of cultural sensitive gesturing for social robots. In order for these robots to appear authentic and trustworthy, and to be intuitive to interact with, it will be necessary to equip them with a repertoire of nonverbal communication behaviors that are adequate for the cultural context they are used in. We argue that the way forward is a detailed analysis of the cultural specificities of each general population in order to generate the necessary behavioral libraries. Behavioral anthropologists have, for example, listed and described many cultural-specific gestures (e.g., [31]). The results of this research could be used and implemented in social robots. However, it is not enough to equip robots with specific executable, but their motion dynamics and frequencies in dependence of the reactions of their recipient need to be taken into consideration.

We chose the field of educational robotics for the illustration of how social robots could assume a central role in human interaction dynamics. The examples from educational robotics show the possibilities social robotic mediators and tutors have to ease and facilitate the approaching didactic shift caused by the rapid technologization of learning environments. Specifically, Asian countries like Japan, South Korea, and Singapore have embraced the use of robots in pre-schools and schools. Robots like TIRO and Robovie have been integrated in the school curricula and are supporting teachers in the classroom. The majority of the applications of these robots are linked to language learning and involve the robots linking new words and grammatical concepts to movements and gesturing and in this way multimodal anchoring the new knowledge in the memory of the children.

In order to put these applications on a sound theoretical didactic basis, we propose an extension of the current enactive didactics approach. We suggest to ascribe to social robots a central role in the feedback process between teacher and students in order to reinforce the reticular character of the structural coupling during the learning process. We argue that this central role requires from the robots embodied nonverbal communication competencies, whose character should be similar to this of humans to be easily understood and nondisruptive. This need for human similarity to human means that robots should be equipped with culturally sensitive social gesture libraries, which can be expressed best with a humanoid or semi-humanoid embodiment. A convergence in this point would also bear a further advantage. Even though there might be differences between the used robot embodiments, the general humanoid structure (i.e., head, torso, arms, and hands) would make the gestures not necessarily robot specific, but a general motion framework can be imagined, which could be used across platforms, similar to the Master Motor Map framework proposed by the KTU [87].

We plan to implement these ideas in a first step with the Pepper robot from Softbank Robotics. In order to develop and expand our enactive robot-assisted didactics approach, we are using Pepper with two main functionalities: (a) to give feedback about the structure of an ongoing lesson and (b) to enforce feedback between the teacher and students.

In scenario (a) Pepper helps, on one side, the teacher to maintain the predefined structure of a lecture and, on the other side, the students to understand the overall educational goal of the lesson. In order to do so, the robot gives an overview of what the content of the lecture will be at its beginning, and at the end of the lecture, it gives a summary of what has been discussed. Pepper uses gestures to illustrate the content of what it is saying. These gestures are specifically designed for the content of the lecture. During the lecture, the robot is used as an embodied timer. After a certain time, it will start to yawn. If the teacher does not react, it will move into a position that makes it appear tired. If the teacher still does not react, it will start to raise its arm, wave, and make the teacher verbally aware that it would be beneficial for the lecture to have a small break.

In scenario (b), we are using Pepper in combination with an audience response system (ARS). The ARSs are used for direct real-time feedback. Although their usefulness is undeniable, the feedback they provide, in form of simple statistics, is inherently unembodied and depends strongly on the willingness of the presenter to let the audience interfere with the presentation. We are using the robot in order to add an embodied component and enforce the integration of the feedback. For this concrete scenario, the lecture is structured into different sections. Each section is concerned with a specific topic. At the end of each section, the robot prompts the teacher to let the students fill in a short questionnaire about the content of the section in Google Forms with their mobile phones. After the data is collected, the robot then gives embodied feedback about the results. The prompting as well as the feedback is composed of verbalizations and informative gestures of increasing intensity.

These two examples illustrate the potential use of robots as embodied feedback devices and social mediators between students and teacher have. Many other scenarios are imaginable. The development toward a more and more embodied interaction with robots will generate intertwined human–robot ecologies, which will have potentially a profound impact on the social evolution of our species (e.g., [19]).

References

1. Tennie, C., Call, J., & Tomasello, M. (2009). *Ratcheting up the ratchet: On the evolution of cumulative culture. Philosophical Transactions of the Royal Society B, 364.*
2. Tomasello, M. (1999). *The cultural origins of human cognition*. Cambridge, MA: Harvard University Press.
3. Scheflen, A. E. (1972). *Body language and the social order; communication as behavioral control.*

4. Kacperck, L. (1997). Non-verbal communication: The importance of listening. *British Journal of Nursing, 6*(5), 275–279.
5. Van Vugt, M., & Kameda, T. (2012). Evolution and groups. *Group processes*, 297–332.
6. Burgoon, J. K., Guerrero, L. K., & Floyd, K. (2016). *Nonverbal communication*. Routledge.
7. Müller, C., Cienki, A., Fricke, E., Ladewig, S., McNeill, D. & Tessendorf, S., (Eds.). (2013). *Body-language-communication* (Vol. 1). Walter de Gruyter.
8. Argyle, M., & Ingham, R. (1972). Gaze, mutual gaze, and proximity. *Semiotica, 6*(1), 32–49.
9. Knapp, M. L., Hall, J. A., & Horgan, T. G. (2013). *Nonverbal communication in human interaction*. Cengage Learning.
10. Von Foerster, H. (2003). On self-organizing systems and their environments. In *Understanding Understanding* (pp. 1–19). New York, NY: Springer.
11. Maturana, H. R., & Varela, F. J. (1987). *The tree of knowledge: The biological roots of human understanding*. New Science Library/Shambhala Publications.
12. Varela, F. J., Thompson, E., & Rosch, E. (1991). *The embodied mind: Cognitive science and human experience*. MIT Press.
13. Oullier, O., De Guzman, G. C., Jantzen, K. J., Lagarde, J., & Scott Kelso, J. A. (2008). Social coordination dynamics: Measuring human bonding. *Social Neuroscience, 3*(2), 178–192.
14. Doughty, M. J. (2001). Consideration of three types of spontaneous eyeblink activity in normal humans: During reading and video display terminal use, in primary gaze, and while in conversation. *Optometry and Vision Science, 78*(10), 712–725.
15. Rizzolatti, G., & Craighero, L. (2004). The mirror-neuron system. *Annual Review of Neuroscience, 27,* 169–192.
16. Dumouchel, P., & Damiano, L. (2017). *Living with robots*. Harvard University Press.
17. Lehmann, H., Roncone, A., Pattacini, U., & Metta, G. (2016). Physiologically inspired blinking behavior for a humanoid robot. In *Proceedings of ICSR 2016*.
18. Riek, L. D., Rabinowitch, T. C., Chakrabarti, B., & Robinson, P. (2009). How anthropomorphism affects empathy toward robots. In *Proceedings of the 4th ACM/IEEE International Conference on Human Robot Interaction* (pp. 245–246).
19. Damiano, L., & Dumouchel, P. (2018). Anthropomorphism in human–robot co-evolution. *Frontiers in Psychology, 9,* 468.
20. Eyssel, F., & Kuchenbrandt, D. (2012). Social categorization of social robots: Anthropomorphism as a function of robot group membership. *British Journal of Social Psychology, 51*(4), 724–731.
21. Ekman, P., & Friesen, W. V. (1969). The repertoire of nonverbal behavior: Categories, origins, usage, and coding. *Semiotica, 1*(1), 49–98.
22. Hall, E. T. (1966). *The hidden dimension* (Vol. 609). Garden City, NY: Doubleday.
23. Lang, P. J., Bradley, M. M., & Cuthbert, B. N. (1990). Emotion, attention, and the startle reflex. *Psychological Review, 97*(3), 377.
24. Lynn, R. (2013). *Attention, arousal and the orientation reaction: International series of monographs in experimental psychology* (Vol. 3). Elsevier.
25. Argyle, M. (2013). *Bodily communication*. Routledge.
26. Emery, N. J. (2000). The eyes have it: The neuroethology, function and evolution of social gaze. *Neuroscience and Biobehavioral Reviews, 24*(6), 581–604.
27. Tomasello, M., Hare, B., Lehmann, H., & Call, J. (2007). Reliance on head versus eyes in the gaze following of great apes and human infants: The cooperative eye hypothesis. *Journal of Human Evolution, 52*(3), 314–320.
28. Carstensen, L. L., Löckenhoff, C. E., Ekman, P., Campos, J. J., Davidson, R. J., & de Waal, F. B. M. (2003). *Emotions inside out: 130 years after Darwin's 'The Expression of the Emotions in Man and Animals'*.
29. Ekman, P. (1972). *Nebraska symposium on motivation, 1971* (J. Cole, Ed.) (pp. 207–283). Lincoln, NE: Univ of Nebraska Press.
30. Kendon, A. (1988). How gestures can become like words. *Cross-cultural perspectives in nonverbal communication* (Vol. 1, pp. 131–141).
31. Morris, D. (1994). *Body talk: A world guide to gestures*. London: Jonathan Cape Ltd.

32. McNeill, D. (1992). *Hand and mind: What gestures reveal about thought*. University of Chicago press.
33. Morrel-Samuels, P., & Krauss, R. M. (1992). Word familiarity predicts temporal asynchrony of hand gestures and speech. *Journal of Experimental Psychology: Learning, Memory, and Cognition, 18*(3), 615.
34. Goldin-Meadow, S., Nusbaum, H., Kelly, S. D., & Wagner, S. (2001). Explaining math: Gesturing lightens the load. *Psychological Science, 12*(6), 516–522.
35. Corballis, M. C. (2002). *From hand to mouth: The origins of language*. NJ, US: Princeton.
36. Lieberman, P., Crelin, E. S., & Klatt, D. H. (1972). Phonetic ability and related anatomy of the newborn and adult human, Neanderthal man, and the chimpanzee. *American Anthropologist, 74*(3), 287–307.
37. Petitto, L. A., & Marentette, P. F. (1991). Babbling in the manual mode: Evidence for the ontogeny of language. *Science, 251*(5000), 1493–1496.
38. Pollick, A. S., & DeWaal, F. B. (2007). Ape gestures and language evolution. *Proceedings of the National Academy of Sciences, 104,* 8184–8189.
39. Annett, M. (1985). *Left, right, hand and brain: The right shift theory*. Psychology Press (UK).
40. Hopkins, W. D., & de Waal, F. B. (1995). Behavioral laterality in captive bonobos (Pan paniscus): Replication and extension. *International Journal of Primatology, 16*(2), 261–276.
41. Cantalupo, C., & Hopkins, W. D. (2001). Asymmetric Broca's area in great apes. *Nature, 414*(6863), 505.
42. Ono, T., Imai, M., & Ishiguro, H. (2001). A model of embodied communications with gestures between humans and robots. In *Proceedings of 23rd Annual Meeting of the Cognitive Science Society* (pp. 732–737). Citeseer.
43. Sidner, C. L., Lee, C., Kidd, C. D., Lesh, N., & Rich, C. (2005). Explorations in engagement for humans and robots. *Artificial Intelligence, 166,* 140–164.
44. Riek, L. D., Rabinowitch, T. C., Bremner, P., Pipe, A. G., Fraser, M., & Robinson, P. (2010). Cooperative gestures: Effective signaling for humanoid robots. In *2010 5th ACM/IEEE International Conference on Human-Robot Interaction (HRI)* (pp. 61–68). IEEE.
45. Beck, A., Cañamero, L., Damiano, L., Sommavilla, G., Tesser, F., & Cosi, P. (2011, November). Children interpretation of emotional body language displayed by a robot. In *International Conference on Social Robotics* (pp. 62–70). Berlin, Heidelberg: Springer.
46. Ham, J., Bokhorst, R., Cuijpers, R., van der Pol, D., & Cabibihan, J. J. (2011, November). Making robots persuasive: The influence of combining persuasive strategies (gazing and gestures) by a storytelling robot on its persuasive power. In *International Conference on Social Robotics* (pp. 71–83). Heidelberg: Springer, Berlin.
47. Huang, C. M., & Mutlu, B. (2013, June). Modeling and evaluating narrative gestures for humanlike robots. In *Robotics: Science and Systems* (pp. 57–64).
48. Tonkin, M., Vitale, J., Herse, S., Williams, M.A., Judge, W., & Wang, X. (2018, February). Design methodology for the UX of HRI: A field study of a commercial social robot at an airport. In *Proceedings of the 2018 ACM/IEEE International Conference on Human-Robot Interaction* (pp. 407–415).
49. Foster, M. E., Alami, R., Gestranius, O., Lemon, O., Niemelä, M., Odobez, J. M., & Pandey, A. K. (2016, November). The MuMMER project: Engaging human-robot interaction in real-world public spaces. In *International Conference on Social Robotics* (pp. 753–763). Cham: Springer.
50. Lehmann, H., Nagai, Y., & Metta, G. (2016). The question of cultural sensitive gesture libraries in HRI—An Italian—Japanese comparison. In *Proceedings of ICDL-Epi 2016*.
51. Tanner, J. E., & Byrne, R. W. (1996). Representation of action through iconic gesture in a captive lowland gorilla. *Current Anthropology, 37*(1), 162–173.
52. Trovato, G., Zecca, M., Sessa, S., Jamone, L., Ham, J., Hashimoto, K., et al. (2013). Cross-cultural study on human-robot greeting interaction: Acceptance and discomfort by Egyptians and Japanese. *Paladyn, Journal of Behavioral Robotics, 4*(2), 83–93.
53. Trovato, G., Zecca, M., Do, M., Terlemez, Ö., Kuramochi, M., Waibel, A., et al. (2015). A novel greeting selection system for a culture-adaptive humanoid robot. *International Journal of Advanced Robotic Systems*, *12*(4), 34.

54. Rossi, P. G. (2011). Didattica enattiva. *Complessità, teorie dell'azione, professionalità docente: Complessità, teorie dell'azione, professionalità docente*. FrancoAngeli.
55. Lau, K. W., Tan, H. K., Erwin, B. T., & Petrovic, P. (1999). Creative learning in school with LEGO (R) programmable robotics products. In *29th Annual IEEE Frontiers in Education Conference* (Vol. 2, pp. 12–26).
56. Balogh, R. (2010). Educational robotic platform based on arduino. In *Proceedings of Conference on Educational Robotics* (pp. 119–122).
57. Mukai, H., & McGregor, N. (2004). Robot control instruction for eighth graders. *IEEE, Control Systems, 24*(5), 20–23.
58. Papert, S., & Harel, I. (1991). Situating constructionism. *Constructionism, 36*(2), 1–11.
59. Hirst, A. J., Johnson, J., Petre, M., Price, B. A., & Richards, M. (2003). What is the best programming environment/language for teaching robotics using Lego Mindstorms? *Artificial Life and Robotics, 7*(3), 124–131.
60. Powers, K., Gross, P., Cooper, S., McNally, M., Goldman, K. J., Proulx, V., et al. (2006). Tools for teaching introductory programming: What works? *ACM SIGCSE Bulletin, 38*(1), 560–561.
61. Bell, S. (2010). Project-based learning for the 21st century: Skills for the future. *The Clearing House, 83*(2), 39–43.
62. Haas, M. D., Baxter, P., de Jong, C., Krahmer, E., & Vogt, P. (2017, March). Exploring different types of feedback in preschooler and robot interaction. In *Proceedings of the Companion of the 2017 ACM/IEEE International Conference on Human-Robot Interaction* (pp. 127–128).
63. Belpaeme, T., Kennedy, J., Baxter, P., Vogt, P., Krahmer, E. E., Kopp, S., et al. (2015). L2TOR-second language tutoring using social robots. In *Proceedings of the ICSR 2015 WONDER Workshop*.
64. Ahmad, M. I., Mubin, O., Shahid, S., & Orlando, J. (2019). Robot's adaptive emotional feedback sustains children's social engagement and promotes their vocabulary learning: A long-term child–robot interaction study. *Adaptive Behavior, 27*(4), 243–266.
65. Mubin, O., Stevens, C. J., Shahid, S., Al Mahmud, A., & Dong, J. (2013). A review of the applicability of robots in education. *Technology for Education and Learning, 1*, 1–7.
66. Tanaka, F., Isshiki, K., Takahashi, F., Uekusa, M., Sei, R., & Hayashi, K. (2015). Pepper learns together with children: Development of an educational application. In *IEEE-RAS 15th International Conference on Humanoid Robots* (pp. 270–275).
67. Ishiguro, H., Ono, T., Imai, M., Maeda, T., Kanda, T., & Nakatsu, R. (2001). Robovie: An interactive humanoid robot. *Industrial Robot: An International Journal, 28*(6), 498–504.
68. Han, J., & Kim, D. (2009) r-Learning services for elementary school students with a teaching assistant robot. In *4th ACM/IEEE International Conference on Human-Robot Interaction (HRI)* (pp. 255–256). IEEE.
69. Shamsuddin, S., Ismail, L. I., Yussof, H., Zahari, N. I., Bahari, S., Hashim, H., & Jaffar, A. (2011) Humanoid robot NAO: Review of control and motion exploration. In *2011 IEEE International Conference on Control System, Computing and Engineering (ICCSCE)* (pp. 511–516).
70. Belpaeme, T., Kennedy, J., Ramachandran, A., Scassellati, B., & Tanaka, F. (2018). Social robots for education: A review. *Science Robotics, 3*(21), eaat5954.
71. Tanaka, F., & Kimura, T. (2009) The use of robots in early education: A scenario based on ethical consideration. In *The 18th IEEE International Symposium on Robot and Human Interactive Communication. RO-MAN 2009* (pp. 558–560). IEEE.
72. Tanaka, F., & Matsuzoe, S. (2012). Children teach a care-receiving robot to promote their learning: Field experiments in a classroom for vocabulary learning. *Journal of Human-Robot Interaction, 1*(1), 78–95.
73. Leite, I., Castellano, G., Pereira, A., Martinho, C., & Paiva, A. (2012). Modelling empathic behaviour in a robotic game companion for children: An ethnographic study in real-world settings. In *Proceedings of 7th ACM/IEEE International Conference on HRI* (pp. 367–374). ACM.
74. Dautenhahn, K., Nehaniv, C. L., Walters, M. L., Robins, B., Kose-Bagci, H., Mirza, N. A., et al. (2009). KASPAR—A minimally expressive humanoid robot for human–robot interaction research. *Applied Bionics and Biomechanics, 6*(3–4), 369–397.

75. Iacono, I., Lehmann, H., Marti, P., Robins, B., &. Dautenhahn, K. (2011) Robots as social mediators for children with Autism—A preliminary analysis comparing two different robotic platforms. In *2011 IEEE International Conference on Development and Learning (ICDL)* (Vol. 2, pp. 1–6). IEEE.
76. Köse, H., Akalin, N., & Uluer, P. (2014). Socially interactive robotic platforms as sign language tutors. *International Journal of Humanoid Robotics, 11*(01), 1450003.
77. Köse, H., Uluer, P., Akalın, N., Yorgancı, R., Özkul, A., & Ince, G. (2015). The effect of embodiment in sign language tutoring with assistive humanoid robots. *International Journal of Social Robotics, 7*(4), 537–548.
78. Uluer, P., Akalın, N., & Köse, H. (2015). A new robotic platform for sign language tutoring. *International Journal of Social Robotics, 7*(5), 571–585.
79. Kanda, T., Hirano, T., Eaton, D., & Ishiguro, H. (2004). Interactive robots as social partners and peer tutors for children: A field trial. *Human-Computer Interaction, 19*(1–2), 61–84.
80. Gorostiza, J. F., Barber, R., Khamis, A. M., Malfaz, M., Pacheco, R., Rivas, R., & Salichs, M. A. (2006). Multimodal human-robot interaction framework for a personal robot. In *The 15th IEEE International Symposium on Robot and Human Interactive Communication. ROMAN 2006* (pp. 39–44). IEEE.
81. Benitti, F. B. V. (2012). Exploring the educational potential of robotics in schools: A systematic review. *Computers and Education, 58*, 978–988.
82. Benitti, F. B. V., & Spolaôr, N. (2017). How have robots supported stem teaching? In *Robotics in STEM Education* (pp. 103–129). Cham: Springer.
83. Lehmann, H., & Rossi, P. G. (2019a). Enactive Robot Assisted Didactics (ERAD): The role of the maker movement. In *Proceedings of the International Conference on Educational Robotics*.
84. Damiano, E. (2013) La mediazione didattica. *Per una teoria dell'insegnamento: Per una teoria dell'insegnamento*. FrancoAngeli.
85. Laurillard, D. (2012). Teaching as a design science. *Building pedagogical patterns for learning and technology.*
86. Lehmann, H. & Rossi, P. G. (2019b). Social robots in educational contexts: Developing an application in enactive didactics. *Journal of e-Learning and Knowledge Society 15*(2).
87. Terlemez, Ö., Ulbrich, S., Mandery, C., Do, M., Vahrenkamp, N., & Asfour, T. (2014). Master Motor Map (MMM)—Framework and toolkit for capturing, representing, and reproducing human motion on humanoid robots. In *2014 IEEE-RAS International Conference on Humanoid Robots* (pp. 894–901). IEEE.

Chapter 16
Priming and Timing in Human-Robot Interactions

Allison Langer and Shelly Levy-Tzedek

Abstract The way a person moves can have an impact on how other individuals move. This is termed "movement priming," and it can have important implications, i.e., for rehabilitation. Very little attention has so far been given to priming of human movement by robots: Does the movement of robots affect how people around them move? What are the implications of such priming, if it exists? Here, we briefly review the topic of human-human priming and then the evidence for robot-human priming. We dedicate a section to the timing of the robotic movement, as it both primes the movement of users (people move slower in the presence of a slow-moving robot, for example) and is also an important determinant in user satisfaction from the interaction with the robot. In fact, user satisfaction is affected not only by the timing of the robot's movements, but also by the timing of the robot's speech, and even by the timing of the errors it makes (e.g., at the beginning vs. at the end of the interaction with the user). We conclude with potential explanations for why robots prime the movements of humans, and why timing plays such an important role in human-robot interaction.

16.1 Introduction

Human movement is a complex orchestration of finely timed muscle activation patterns. It is affected by a variety of factors—age, fatigue, motivation, disease state, medication, etc. (e.g., [41–44, 82]) Kashi S., Feingold Polak R., Lerner B., Rokach L., Levy-Tzedek S. A machine-learning model for automatic detection ofmovement compensations in stroke patients. IEEE Transactions on Emerging Topics in Computing (2020, in press). DOI 10.1109/TETC.2020.2988945. One of the factors that

A. Langer · S. Levy-Tzedek (✉)
Recanati School for Community Health Professions, Department of Physical Therapy, Ben-Gurion University of the Negev, Beer-Sheva, Israel
e-mail: shelly@bgu.ac.il

S. Levy-Tzedek
Zlotowski Center for Neuroscience, Ben-Gurion University of the Negev, Beer-Sheva, Israel

Freiburg Institute for Advanced Studies (FRIAS), University of Freiburg, Freiburg, Germany

N. Noceti et al. (eds.), *Modelling Human Motion*,
https://doi.org/10.1007/978-3-030-46732-6_16

affects human movement is the movement of other humans around them, a phenomenon that is termed "movement priming." With the increasing presence of robots in various context of our lives, it is important to understand whether and how robots prime the movement of humans around them. Here, we review works documenting the presence of priming in human-human interactions, as well as works that show various ways in which the actions of a robot affect the human user. We focus on *motor priming*: the effect that the movement of the robot has on the movement of the person it interacts with. Recognizing the extent of priming and designing for priming is of prime importance to any researcher and engineer who works on human-robot interactions (HRI), in a variety of contexts. Priming can have important—and even detrimental—consequences when people work alongside robots in industry, or in the medical field. To give one example, there are robotic "nurses" being developed, which are designed to hand surgical tools to a surgeon [6]. The implication of the robot priming the human movement is that these robotic nurses should be designed to move with a velocity profile similar to that of a human, and not faster. Since the surgeon working with such a robotic nurse will be primed by the robotic nurse's movements, if these are too fast, or too sharp, the safety of the patient under operation may be compromised.

We first introduce the broad concept of *priming* and give examples from various types of human-human priming. We then expand on human-human *movement* priming, on its use in clinical settings for rehabilitation, and on robot-human movement priming. We conclude by reviewing works that focus on the role of timing in HRI—both how the timing of the robot's movement affects the movement of the person, and how the timing of other robot actions, such as speech, and even the timing of errors made by the robot, affects the person's response and satisfaction from the interaction.

16.2 What Is Priming?

Priming can be described as behavioral change generated by preceding stimuli [49]. Priming is a nonconscious process associated with learning, where exposure to a stimulus alters the response of another stimulus [49]. In classic sequential priming studies, participants are presented with a series of trials that each contain two stimuli: a prime and a target stimulus. The congruency between the prime and target is varied across trials with some prime-target trials congruent in meaning and others incongruent [34]. Priming is present when, in congruent trials, a target is processed more quickly when the prime is shown first. For the past several decades, researchers have sought to understand the effects and mechanisms of various types of priming, including semantic, stereotype, affective, visual, and more recently, motor priming.

16.3 Where Do We Find Priming?

Semantic priming occurs when the response to a target word is facilitated if it is preceded by a related word [54]. For example, semantic priming explains how the processing of words or pictures is more accurate or faster if semantically related information (e.g., bread) is presented prior to the target word (butter), compared to unrelated information (e.g., tire) [51]. Meyer and colleagues conducted what are considered to be the seminal experiments in semantic priming [52, 68, 69], which demonstrated the robustness of the phenomenon across naming, semantic categorization, and other lexical tasks [54]. For example, a related tendency called syntactic priming occurs when speakers re-use recently used or heard linguistic options whenever possible [75]. Decades of research have revealed evidence of both short- and long-term semantic priming effects, and researchers are currently searching for the mechanisms underlying the difference in durations of these effects [80].

Stereotype priming has been widely studied using the previously mentioned sequential priming paradigm. Social psychologists used stereotype categorizations (e.g., male or female) and items stereotypically associated with these groups (e.g., jobs, physical traits) as primes and targets [2, 14, 34]. Recently, researchers have studied ways in which priming can be used to overcome the effects of stereotype threat, which holds that fear of confirming a negative stereotype about the group to which they belong prevents people from reaching their full potential [73]. The impact of stereotype priming has been investigated in a variety of cases, including gender [48], race [81], elderly individuals' memory [23], and even HRI [57].

Visual priming has been described as follows: visual objects are perceived more quickly when they had been previously seen, regardless of whether one remembers having seen them before [20]. An early example of visual priming is Zajonc's now well-known "mere exposure" effect, where subliminal presentation of an otherwise neutral stimulus biased subsequent liking judgments [84]. Kunst-Wilson and Zajonc [38] later demonstrated how visual stimuli with emotional significance can be processed without being consciously perceived, which led to further research on emotion (affective) priming [24, 36, 55, 77].

Visuomotor priming occurs when perceptual processes, such as vision, affect congruent motor actions [11]. In Craighero et al. [11], participants were asked to grasp one of two objects based on a visual cue on a computer screen. The authors found that in a condition in which the "to-be-grasped" object was shown, participants reacted faster when initiating a grasping movement compared to when shown an irrelevant object, or no object at all. Response accuracy and latency are also enhanced by stimulus-response congruency with regard to a specific dimension (e.g., location, direction, and intention) of the observed components [29]. For example, when participants executed finger movements in response to observing either a finger tapping (compatible stimulus) or lifting action (incompatible stimulus), there was a pronounced reduction in reaction time for compatible trials [3] (see Fig. 16.1). Similarly, responses to human hand movement stimuli (e.g., a video image of a hand opening) are faster and more accurate when they involve execution of the same movement,

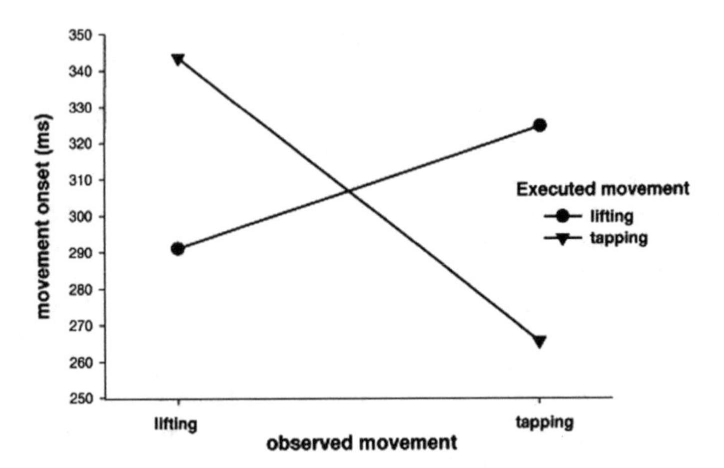

Fig. 16.1 Movement onset (response time) as a function of observed movement (tapping vs. lifting) and executed movement (tapping vs. lifting). Reprinted from "Movement observation affects movement execution in a simple response task," by Marcel Brass, Harold Bekkering, Wolfgang Prinz, Acta Psychologica, 106, p. 20. Copyright (2001), with permission from Elsevier

such as hand opening, than when they involve execution of an alternative movement, such as a hand closing [10].

This demonstrated that, similar to visual and semantic priming, motor priming is modulated by the compatibility of the stimulus and its target response, otherwise known as motor resonance. Motor resonance, the neural basis of which is the mirror neuron system, refers to the automatic activation, during action perception, of the perceiver's motor system [65]. Liuzza et al. [46] described motor resonance as the overlap of characteristics between the perceived action and the perceiver's actions [46]. For example, Calvo-Merino et al. [5] demonstrated that dancers' mirror neuron systems showed greater activity when dancers viewed moves from their own motor repertoire, compared to opposite-gender moves that they frequently saw but did not perform.

Visuomotor priming and motor resonance follow a conceptual framework called "ideomotor theory," which was developed by James et al. more than a century ago [30]. This theory has since been researched and developed in depth (for a review see [71]). The neural basis for this coupling of perception of action and execution of action was first studied in macaques, whose premotor cortex is activated both when a monkey performs a specific action and when it passively observes the experimenter perform that same action [64]. These mirror neurons are thought to contribute to our understanding of the goals and intentions of others by internal simulation of their actions [28].

Following Rizzolatti et al. [65]'s finding, researchers looked into what movement characteristics affect visuomotor priming. For example, Liuzza et al. [46] used a visuomotor priming paradigm to show that motor resonance in children is strengthened when observing a child's hand in action, rather than an adult's hand. One of the questions in visuomotor priming, beyond the effects of gender and age on priming effects, is whether it is important that the movement be similar to how humans or animals move—does movement need to resemble "biological motion" in order to produce visuomotor priming effects in humans? Edwards et al. [15] found that even movements which do not follow the "biological motion" profile can prime actions in others. An extensive review by Sciutti et al. [70] shows that while some studies did not find any motor resonance or priming when actions were performed by non-biological agents, more recent studies show that robotic agents can evoke similar mirror neuron activity as humans do. We will review previous work and discuss the implications of the findings later in this chapter.

16.3.1 Motor Priming in Clinical Settings

Motor priming is a relatively new topic of investigation in the fields of motor control and rehabilitation. When used as part of a therapeutic intervention, motor priming can lead to behavioral and neural changes [49], and can be used to improve function [60]. Madhavan and Stoykov [49] distinguish between motor priming and neurorehabilitative training by proposing that priming is performed first and is used to ready the brain to better respond to the neurorehabilitative training that follows. Specifically, priming interventions may prepare the sensorimotor system for subsequent motor practice, thereby enhancing its effects [60]. In stroke rehabilitation, motor priming has been shown to have beneficial effects on recovery. Stinear et al. [74] found that bilateral motor priming increased the rate, though not the magnitude, of recovery in the subacute phase of post-stroke rehabilitation. Motor priming is also a viable therapeutic tool to control involuntary movements in individuals with spinal-cord injury [17]. Compared to other approaches used in neurorehabilitation, such as noninvasive brain stimulation or pharmacological interventions [17], movement priming is safe and cost effective [49], making it a feasible choice for many individuals. For a review on the clinical applications and neural mechanisms of motor priming, see Madhavan and Stoykov [49].

16.3.2 Motor Priming and HRI

Early neuroimaging and behavioral studies that investigated robotic movement priming found that only human movement, but not robotic movement, gave rise to visuomotor priming [61]. For example, Castiello et al. [7] found that observation of a human grasping objects affected the subsequent performance of grasping movements, but observation of a robotic hand performing the same tasks did not influence subsequent movement execution. In Tai et al. [76], participants made arm movements while observing either a robot or another human making the same or different arm movements. Their results demonstrated that when humans, but not a robotic arm, made different arm movements, there was a significant interference effect on executed movements [76]. Similarly, Kilner et al. [35] showed that performance of sinusoidal arm movements in a vertical or horizontal plane was subject to interference from simultaneous observation of another human performing incompatible arm movements, i.e., movement in an orthogonal direction. However, when these incompatible movements were performed by a full-size robot—with a head, trunk, arms, and legs—rather than by a human, execution of the sinusoidal movements was unimpaired (see Fig. 16.2).

However, more recent studies that have looked into movement priming between humans and robots have shown repeated evidence for movement priming by robotic agents [16, 31, 56, 59, 62]. Results from Oberman et al. [56] suggest that robot actions, even those without objects, may activate the human mirror neuron system. Pierno et al. [59] found that children with autism exhibited faster movement duration when

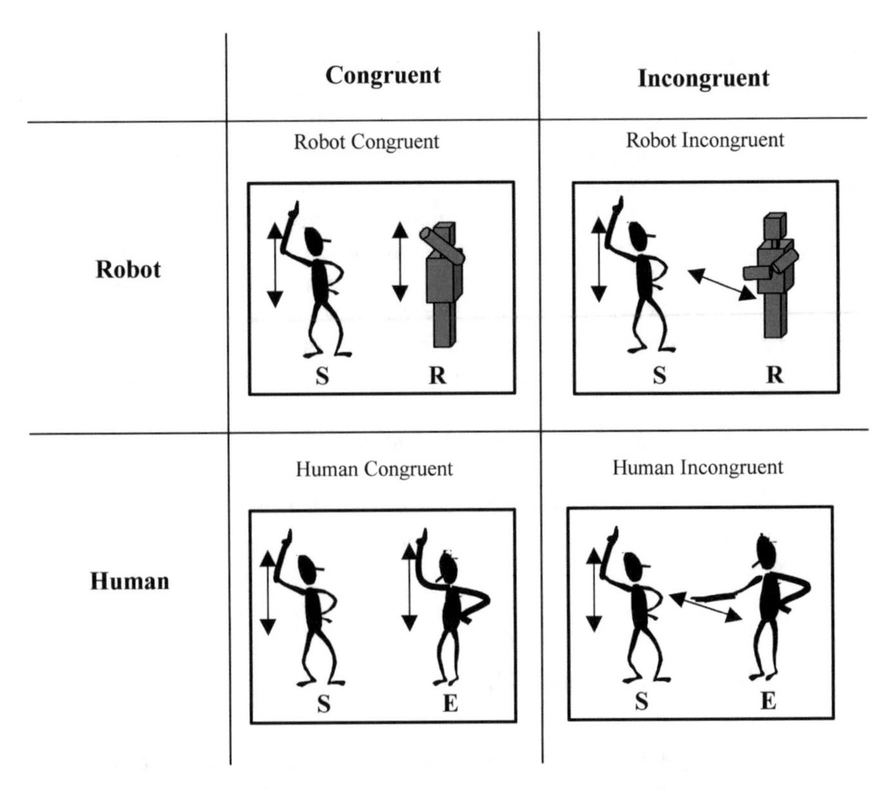

Fig. 16.2 Experimental design investigating two hypotheses: (1) interference should occur when an observed movement is qualitatively different from a simultaneously executed movement, and (2) interference effects are not simply a result of increased attentional demands or increased task complexity and that they are specific to observing biological incongruent movements. Reprinted from "An Interference Effect of Observed Biological Movement on Action" by J.M Kilner, Y Paulignan, S.J Blakemore, Current Biology, Vol. 13, p. 4., Copyright (2003), with permission from Elsevier

primed by a robotic but not by a human arm movement. More recently, Eizicovits et al. [16] demonstrated movement priming by a robotic arm; participants moved significantly slower when interacting with a slow robotic arm, compared to when they interacted with a fast-responding non-embodied system. In yet another experiment, when participants played the "mirror game" with a robotic arm, the movements of the robotic arm primed the subsequent movements performed by the participants [31]. Vannucci et al. [78] demonstrated motor priming through a joint task where participants and a humanoid robot worked together to fill a box with Lego bricks, and participants' movement speed varied according to the experimentally manipulated speeds of the robot.

These seemingly contradictory findings on the presence or absence of robot movement priming may be explained by differences in stimulus presentation. For example, Tai et al. [76] found that when participants watched an experimenter and an experimenter-controlled robot performing grasping actions, only the experimenter's actions activated the participants' mirror neuron system, as indicated by regional brain activation measured by positron emission tomography [48]. However, previous studies of the mirror neuron system in macaques found that the mirror neuron system does not respond when an action is performed indirectly (e.g., by using a tool) [19]. In Tai et al. [76], since study participants could see the experimenter explicitly controlling the robot with a button press, this could have rendered the robot being perceived as a tool [56], thus explaining why its actions did not prime the participants' actions. In a subsequent study that attempted to make a robot appear completely autonomous, Oberman et al. [56] found that robot actions activated the mirror neuron system.

The embodiment—or physical presence—of the robot, as well as how its movements are observed—through static or dynamic images—may also play a role in the degree to which a robot primes the movement of humans. Though Press et al. [61] found that watching a human perform an action resulted in a shorter reaction time than when seeing a robot perform the same action, the authors used still images of a robot in the observation phase. When Kashi and Levy-Tzedek [31] used an embodied, physically present robot that performed biological movements in a mirror-game joint task, they find a motor priming effect on the subsequent movements of the participants.

Robotic movement priming may be advantageous if it can be harnessed for rehabilitation by inducing the user to perform desirable movements [31]. Hsieh et al [27] conducted a clinical trial investigating priming effects where participants in the experimental group performed bilateral repetitive and symmetric movements using a robotic device before completing functional tasks. The results from the trial indicate that adding the technique of bilateral priming using the robotic device may facilitate better rehabilitation outcomes than a task-oriented approach alone. Given the novelty of the use of motor priming with robots in neurorehabilitation, very few studies have examined priming effects by robots in this context. However, with an increasing trend to integrate robots into rehabilitation [32], motor priming may be a promising future field of investigation.

16.4 Movement Timing in HRI

As noted above, timing is one of the movement characteristics that is primed in robot-human interactions (e.g., [16]). However, the importance of timing, when designing and studying interactions between humans and robots, extends beyond the effect of priming alone, and manifests itself also in how people respond to the robot and how motivated they are to continue interacting with it. For that reason, we dedicate the following section to an in-depth review of robot timing in HRI and its various implications for human-robot collaboration, including conveyance of intention and fluency of interaction.

16.4.1 Timing in Collaborative Tasks

For robots to be integrated in everyday life, to assist in daily tasks, or serve as teammates in collaborative work scenarios, it is important that they achieve a type of interaction fluency that comes naturally between humans [8]. Timing plays a central role here: the temporal synchronization of functional actions is necessary for sharing resources and affects how humans perceive robotic teammates [26]. Hoffman and Breazeal [25] argue that, in addition to being efficient, robotic teammates must be fluent in their coordinated actions, as measured by the time between human and robot actions, time spent moving together, and time the human spends waiting for the robot.

Robot-to-human handovers, where robots hand objects to human users, provide an illustrative example of the importance of achieving natural timing in collaborative human-robot tasks. Robots can perform handovers in a variety of contexts, including reaching for objects for the elderly, handing surgical instruments in an operating room, or handling tools in a factory. A high level of coordination is required of both the giver and the receiver's movements in any handover, and roboticists have made it a long-term goal to reach the same level of fluency in handovers between humans and robots as exists between humans themselves [1, 4, 39]. Several quantitative measures of fluency in a handover task with a robot correlate with a human's *sense* of fluency in the task, including minimal wait time resulting in efficient task execution [4]. Researchers have found that users' ability to unambiguously understand the movement goal of the robot increased fluency and eliminated failed attempts [4]. When users understand the goals of the robot, they can start their own action sooner, and these anticipatory actions have been shown to contribute greatly to fluency of interactions [21, 25].

Other nonverbal cues, such as gaze cues, have been used in handovers to decrease task-completion time. Moon et al. [53] found that participants reached for the offered object significantly earlier when a robot provided a shared-attention gaze cue during a handover. Admoni et al. [1] considered how altering the timing of the handover and gaze cue combination can be an effective strategy to communicate other information, such as where to place an object. The authors introduced a deliberate delay during the handover, where the robot holds an object longer than expected, to draw the user's attention to the robot's head, in order to convey, through eye gaze, where to place the object.

Interaction fluency is also a crucial component in turn-taking interactions between humans and robots, which are multimodal and reciprocal in nature [8]. When humans engage in turn-taking—when resources and or physical space must be shared—they are able to seamlessly use speech, gaze, gesture, and other modes of communication to move in coordinated time with a partner. The challenge for roboticists has been to match this seamlessness in human-robot teamwork. Chao and Thomaz [8] developed a system for an autonomous humanoid robot to collaborate with humans with speech and physical action and evaluated it using Towers of Hanoi, a turn-taking task that requires the human and robot to share the same resources and work space. When the robot "interrupted" its automatic actions in response to a human's hand in the workspace or in response to human speech, the researchers observed increased task efficiency and users felt a higher sense of interaction fluency. Future research on the timing dynamics in human-robot collaborative tasks will continue to reveal ways to improve interaction balance, leading to more efficient, and more naturalistic, robotic teammates.

16.4.2 Timing in Robotic Motion

The timing of robotic motion can be used to purposely express intention when interacting with a human user [86]. Zhou et al. [86] demonstrated a situation where different timing of the same motion appears to convey different information about the robot:

> Imagine seeing a robot arm carry a cup smoothly across the table [...]. Now, imagine seeing a different arm pausing and restarting, slowing down and then speeding back up [...]. The path might be the same, but the difference in timing might make us think very differently about the robots and about what they are doing. We might think that the second robot is less capable, or maybe that its task is more difficult. Perhaps it doesn't have as much payload, perhaps the cup is heavier, or perhaps it does not know what to do.

By manipulating factors related to timing, such as speed, changes of speed (in particular ways), and pausing (at particular times), these authors found effects on users' perceptions of the robot's disposition, naturalness, competence, capability, and carried object weight.

16.4.3 Timing of Interactions with Social Robots

Beyond its importance for effective teamwork, timing plays a fundamental role in the regulation of human-robot interaction and communication [37]. Early research on the timing of social robots' interaction characteristics drew on how humans naturally interacted with other humans. For example, in designing a robot guide for a museum, Yamazaki et al. [83] used the timing of the verbal and nonverbal actions of a human guide when interacting with visitors. The researchers found that visitors were likely to respond with natural gestures and speech in response to the robot when the robot itself performed head and gaze actions at time points that were meaningful to the interaction, rather than at random time points. This study stressed the importance of properly coordinated conversation dynamics in order for robots to elicit natural responses from humans and set the stage for future work on integrating robot guides into social spaces.

Many studies have since investigated user preferences for timing of robotic speech that have implications for HRI design. Shiwa et al. [72] found that: (1) people prefer one-second delayed responses rather than immediate responses, (2) using conversational fillers was an effective strategy to moderate negative impressions of the robot after an episode where the robot took long to respond, and (3) users' previous experiences with robots affected their timing preferences. Researchers have also sought to understand how the timing of robot speech errors affect the overall interaction. Based on their work, Gompei and Umemuro [22] suggest that the robot should not make speech errors in the early stage of engagement with human users, while some speech errors after the users become accustomed to the robot might be effective in improving users' perception of the familiarity of the robot. However, Lucas et al. [47] demonstrated conversational errors that occur later in a social robot's dialog hinder users from taking the robot's advice.

Context and user characteristics may also affect preferences for robot timing in HRIs. One practical application for social robots has been giving route directions to visitors in public spaces. Okuno et al. [58] found that people interacting with a route-directing robot preferred a speech pattern that included pauses, even if they may have been unnaturally long, in order to have time to understand the directions. Thus, though shorter reaction times may be regarded as more preferable for efficiency, there are certain contexts where a slower response may be warranted and even desirable. These studies demonstrate that, in HRI, timing of the robot's various functions—not only motor ones—is important and affects the user's response.

Preferences for robot response time may also be mediated by age. In a study examining user preferences in using either a robotic or a nonembodied computer-controlled system designed for upper limb rehabilitation, Eizicovits et al. [16] found differences between how the older (age 73.3 ± 6.2 years) and younger (age 25.6 $\pm$ 7 years) participants related to the response time of the system. The participants were asked to play a game of 3D tic-tac-toe with the opponent, which was either a robot, or a nonembodied computer system. During the game with the robot, the players—the human and the robot—took turns picking and placing colored cups on a 3D grid. When the opponent was nonembodied, only the human placed the cups on the grid, and the nonembodied opponent indicated its "move" by instantaneously turning on a colored LED light in the chosen grid location. Some of the participants in the young group expressed impatience with the time it took the robot to make its moves, while the participants in the older group, who themselves often perform slower movements [44], did not express dissatisfaction with the slower reaction time of the robotic system, compared to the computer-controlled one. Importantly, the robot's slow response seems to have affected the participants' willingness to keep playing with it; when asked to choose against which opponent they would like to play two more game sessions, both young and old participants preferred the robot, but this willingness decreased when asked against which one they would like to play ten more game sessions. Indeed, when asked what their preferences would be if the timing of both the robot and the nonembodied system would be equal, the young group overwhelmingly preferred the robot (>80%, see Fig. 16.3), demonstrating how timing of the robot can play a pivotal role in user preferences.

Recently, more attention has been given to understanding how robot errors, and the timing of these errors, influence user trust [40]. Robinette et al. [66] examined

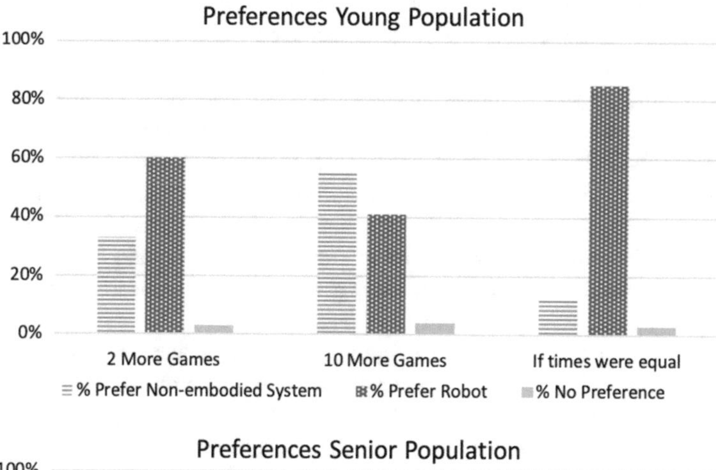

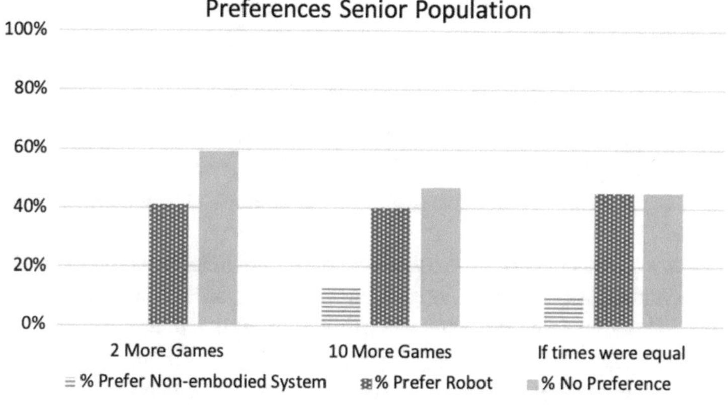

Fig. 16.3 Participant preferences for playing games with nonembodied system or robot. Reproduced from Restorative Neurology and Neuroscience, Vol. 36 no. 2, Eizicovits, Danny, Edan, Yael, Tabak, Iris, Levy-Tzedek, Shelly, Robotic gaming prototype for upper limb exercise: Effects of age and embodiment on user preferences and movement, Pages 261-274., Copyright (2018) with permission from IOS Press

three different robot trust-repair techniques that proved to be helpful in repairing trust after it was lost in an HRI, but the timing of the robot's delivery was a critical factor in determining how successfully trust was repaired. Furthermore, studies that measured trust continuously during the interaction with the robot found a primacy-recency bias: participants' trust ratings decreased more severely in response to the robot's reliability drops at the beginning and end of a task [13].

16.5 Conclusion

We conclude by briefly reviewing the main findings we covered in this chapter, and suggesting possible explanations for why the movements of robots prime the movements of humans, and why timing is such a crucial aspect of human-robot interaction. While the concept of priming has been widely investigated, studies of motor priming in HRI are limited and have produced contradictory findings. While the actions of users in some studies were primed by the robots' actions, others were not. As roboticists continue to engineer embodied, human-like robotic agents which can produce biologically-inspired movements, it is increasingly important to recognize and design for motor priming in HRI. It is worthwhile to elucidate the conditions that lead to movement priming, as well as use it to improve the interaction. How timing affects not only priming, but also HRI in general, should remain a focus in future research on human-robot motor priming, given its importance in shaping several aspects of humans' perceptions of robotic agents.

People tend to anthropomorphize robots, even when they are aware that they are virtual agents [12, 85]. This anthropomorphizing is amplified with robots [12], compared to computers [63] or virtual characters [50], presumably because of their embodiment [33] and physical movement [67]. With their ability to move in biologically realistic ways, it is not surprising that robotic actions can prime the actions of users around them. This priming effect can be harnessed for rehabilitative purposes and could potentially serve as an ecological quantitative measure of the natural, unconscious effects of the observation of robotic actions [70]. Future studies should continue to investigate the contexts and ways in which robots are priming the movements of their users to better understand the benefits—and potential risks—of this phenomenon.

Timing of actions appears to be one of the important determinants of both human-human and human-robot interactions. It appears that humans have a very accurate representation of time (e.g., [45]) and value time to the extent that people rate "wasting my time" as a major cause and reason for anger [9, 79]. It is thus rather sensible that when robots now assume roles in society—as they become teammates, care takers, etc. [18]—they are held to similar standards of time efficiency. Timing of robot actions is thus important on several levels: It conveys information (such as intended goals, weight of objects), it primes the movement of the human user, and it affects user motivation and trust [13] to further interact with the robot in the future.

References

1. Admoni, H., Dragan, A., Srinivasa, S. S. & Scassellati, B. (2014). Deliberate delays during robot-to-human handovers improve compliance with gaze communication. In 2014 *9th ACM/IEEE International Conference on Human-Robot Interaction (HRI), IEEE,* pp. 49–56.
2. Banaji, M. R., & Hardin, C. D. (1996). Automatic stereotyping. *Psychological Science, 7,* 136–141.
3. Brass, M., Bekkering, H., & Prinz, W. (2001). Movement observation affects movement execution in a simple response task. *Acta Psychologica, 106,* 3–22.
4. Cakmak, M., Srinivasa, S. S., Lee, M. K., Kiesler, S. & Forlizzi, J.(2011). Using spatial and temporal contrast for fluent robot-human hand-overs. In 2011 *6th ACM/IEEE International Conference on Human-Robot Interaction (HRI), IEEE.* pp. 489–496.
5. Calvo-Merino, B., Grèzes, J., Glaser, D. E., Passingham, R. E., & Haggard, P. (2006). Seeing or doing? influence of visual and motor familiarity in action observation. *Current Biology, 16,* 1905–1910.
6. Carpintero, E., Pérez, C., Morales, R., García, N., Candela, A. & Azorín, J. (2010). Development of a robotic scrub nurse for the operating theatre. In 2010 *3rd IEEE RAS & EMBS International Conference on Biomedical Robotics and Biomechatronics.* 26–29 Sept. pp. 504-509.
7. Castiello, U., Lusher, D., Mari, M., Edwards, M. & W. Humphreys, G. (2002). Observing a human or a robotic hand grasping an object: Differential motor priming effects.
8. Chao, C., & Thomaz, A. L. (2011). Timing in multimodal turn-taking interactions: Control and analysis using timed petri nets. *Journal of Human-Robot Interaction, 1,* 1–16.
9. Cherry, M. & Flanagan, O. (2017). The Moral Psychology of Anger.
10. Craighero, L., Bello, A., Fadiga, L., & Rizzolatti, G. (2002). Hand action preparation influences the responses to hand pictures. *Neuropsychologia, 40,* 492–502.
11. Craighero, L., Fadiga, L., Rizzolatti, G., & Umilta, C. (1998). Visuomotor priming. *Visual Cognition, 5,* 109–125.
12. Darling, K. (2015). 'Who's Johnny?'anthropomorphic framing in human-robot interaction, integration, and policy. *Anthropomorphic Framing in Human-Robot Interaction, Integration, and Policy (March 23, 2015). ROBOT ETHICS, 2.*
13. Desai, M., Kaniarasu, P., Medvedev, M., Steinfeld, A. & Yanco, H. (2013). Impact of robot failures and feedback on real-time trust. In *Proceedings of the 8th ACM/IEEE International Conference on Human-Robot Interaction, IEEE Press,* pp. 251–258.
14. Dovidio, J. F., Evans, N., & Tyler, R. B. (1986). Racial stereotypes: The contents of their cognitive representations. *Journal of Experimental Social Psychology, 22,* 22–37.
15. Edwards, M. G., Humphreys, G. W., & Castiello, U. (2003). Motor facilitation following action observation: A behavioural study in prehensile action. *Brain and Cognition, 53,* 495–502.
16. Eizicovits, D., Edan, Y., Tabak, I., & Levy-Tzedek, S. (2018). Robotic gaming prototype for upper limb exercise: Effects of age and embodiment on user preferences and movement. *Restorative neurology and neuroscience, 36,* 261–274.
17. Estes, S. P., Iddings, J. A., & Field-Fote, E. C. (2017). Priming neural circuits to modulate spinal reflex excitability. *Frontiers in neurology, 8,* 17–17.
18. Feingold-Polak, R., Elishay, A., Shahar, Y., Stein, M., Edan, Y. & Levy-Tzedek, S. (2018). Differences between young and old users when interacting with a humanoid robot: A qualitative usability study. *Paladyn, Journal of Behavioral Robotics.*
19. Gallese, V., Fadiga, L., Fogassi, L., & Rizzolatti, G. (1996). Action recognition in the premotor cortex. *Brain, 119,* 593–609.
20. Gauthier, I. (2000). Visual priming: the ups and downs of familiarity. *Current Biology, 10,* R753–R756.
21. Gielniak, M. J. & Thomaz, A. L. (2011). Generating anticipation in robot motion. 2011 RO-MAN, IEEE. pp. 449–454.
22. Gompei, T. & Umemuro, H. (2015). A robot's slip of the tongue: Effect of speech error on the familiarity of a humanoid robot. In 2015 *24th IEEE International Symposium on Robot and Human Interactive Communication (RO-MAN), IEEE.* pp. 331–336.

23. Hagood, E. W., & Gruenewald, T. L. (2018). Positive versus negative priming of older adults' generative value: Do negative messages impair memory? *Aging & Mental Health, 22,* 257–260.
24. Hermans, D., Spruyt, A., de Houwer, J., & Eelen, P. (2003). Affective priming with subliminally presented pictures. *Canadian Journal of Experimental Psychology/Revue Canadienne De Psychologie Expérimentale, 57,* 97.
25. Hoffman, G. & Breazeal, C. (2007). Effects of anticipatory action on human-robot teamwork efficiency, fluency, and perception of team. In *Proceedings of the ACM/IEEE international conference on Human-Robot interaction, ACM*. pp. 1–8.
26. Hoffman, G., Cakmak, M. & Chao, C. (2014). Timing in human-robot interaction. In *Proceedings of the 2014 ACM/IEEE International Conference on Human-Robot Interaction, ACM*, pp. 509–510.
27. Hsieh, Y.-W., Wu, C.-Y., Wang, W.-E., Lin, K.-C., Chang, K.-C., Chen, C.-C., et al. (2017). Bilateral robotic priming before task-oriented approach in subacute stroke rehabilitation: A pilot randomized controlled trial. *Clinical rehabilitation, 31,* 225–233.
28. Iacoboni, M., Molnar-Szakacs, I., Gallese, V., Buccino, G., Mazziotta, J. C., & Rizzolatti, G. (2005). Grasping the intentions of others with one's own mirror neuron system. *PLoS Biology, 3,* e79.
29. Itaguchi, Y., & Kaneko, F. (2018). Motor priming by movement observation with contralateral concurrent action execution. *Human Movement Science, 57,* 94–102.
30. James, W., Burkhardt, F., Bowers, F., & Skrupskelis, I. K. (1890). *The principles of psychology*. London: Macmillan.
31. Kashi, S., & Levy-Tzedek, S. (2018). Smooth leader or sharp follower? playing the mirror game with a robot. *Restorative Neurology and Neuroscience, 36,* 147–159.
32. Kellmeyer, P., Mueller, O., Feingold-Polak, R. & Levy-Tzedek, S. (2018). Social robots in rehabilitation: A question of trust. *Science Robot., 3*, pp. eaat1587.
33. Kidd, C. D. & Breazeal, C. (2005). Human-Robot interaction experiments: Lessons learned. In Proceeding of AISB. pp. 141–142.
34. Kidder, C. K., White, K. R., Hinojos, M. R., Sandoval, M., & CRITES JR, S. L. (2018). Sequential stereotype priming: A meta-analysis. *Personality and Social Psychology Review, 22,* 199–227.
35. Kilner, J. M., Paulignan, Y., & Blakemore, S.-J. (2003). An interference effect of observed biological movement on action. *Current Biology, 13,* 522–525.
36. Klauer, K. C., & Musch, J. (2003). Affective priming: Findings and theories. *The psychology of evaluation: Affective processes in cognition and emotion, 7,* 49.
37. Kose-Bagci, H., Broz, F., Shen, Q., Dautenhahn, K. & Nehaniv, C. L. (2010). As time goes by: Representing and reasoning about timing in human-robot interaction studies. In 2010 *AAAI Spring Symposium Series*.
38. Kunst-Wilson, W. R., & Zajonc, R. B. (1980). Affective discrimination of stimuli that cannot be recognized. *Science, 207,* 557–558.
39. Kupcsik, A., Hsu, D. & Lee, W. S. (2016). Learning dynamic robot-to-human object handover from human feedback. *arXiv preprint* arXiv:1603.06390.
40. Langer, A., Feingold-Polak, R., Mueller, O., Kellmeyer, P. & Levy-Tzedek, S. (2019). Trust in socially assistive robots: Considerations for use in rehabilitation. *Neuroscience & Biobehavioral Reviews*.
41. Levy-Tzedek, S. (2017a). Changes in predictive task switching with age and with cognitive load. *Frontiers in Aging Neuroscience, 9*.
42. Levy-Tzedek, S. (2017). Motor errors lead to enhanced performance in older adults. *Scientific Reports, 7,* 3270.
43. Levy-Tzedek, S., Krebs, H. I, Arle, J., Shils, J. & Poizner, H. (2011a). *Rhythmic movement in Parkinson's disease: Effects of visual feedback and medication state*.
44. Levy-Tzedek, S., Maidenbaum, S., Amedi, A., & Lackner, J. (2016). Aging and sensory substitution in a virtual navigation task. *PLoS ONE, 11,* e0151593.
45. Levy-Tzedek, S., Ben-Tov, M., & Karniel, A. (2011). Rhythmic movements are larger and faster but with the same frequency on removal of visual feedback. *Journal of Neurophysiology, 106,* 2120–2126.

46. Liuzza, M. T., Setti, A., & Borghi, A. M. (2012). Kids observing other kids' hands: Visuomotor priming in children. *Consciousness and Cognition, 21,* 383–392.
47. Lucas, G. M., Boberg, J., Traum, D., Artstein, R., Gratch, J., Gainer, A., et al. (2018). Getting to know each other: The role of social dialogue in recovery from errors in social robots. In *Proceedings of the 2018 ACM/IEEE International Conference on Human-Robot Interaction, ACM*, pp. 344–351.
48. Lungwitz, V., Sedlmeier, P., & Schwarz, M. (2018). Can gender priming eliminate the effects of stereotype threat? The case of simple dynamic systems. *Acta Psychologica, 188,* 65–73.
49. Madhavan, S. & Stoykov, M. E. (2017). Editorial: Motor priming for motor recovery: Neural mechanisms and clinical perspectives. *Frontiers in Neurology, 8.*
50. Mcdonnell, R., Jörg, S., Mchugh, J., Newell, F. & O'sullivan, C. (2008). Evaluating the emotional content of human motions on real and virtual characters. In *Proceedings of the 5th Symposium on Applied Perception in Graphics and Visualization, ACM*. pp. 67–74.
51. Mcnamara, T. P. (2005). In *Semantic Priming: Perspectives From Memory and Word Recognition*, Psychology Press.
52. Meyer, D. E., & Schvaneveldt, R. W. (1971). Facilitation in recognizing pairs of words: Evidence of a dependence between retrieval operations. *Journal of Experimental Psychology, 90,* 227–234.
53. Moon, A., Troniak, D. M., Gleeson, B., Pan, M. K., Zheng, M., Blumer, B. A., et al. (2014). Meet me where i'm gazing: how shared attention gaze affects human-robot handover timing. In *Proceedings of the 2014 ACM/IEEE International Conference on Human-Robot Interaction, ACM.* pp. 334–341.
54. Murphy, K. (2012). Examining semantic priming in a delayed naming task. *International Journal of Psychological Studies, 4,* 198.
55. Murphy, S. T., & Zajonc, R. B. (1993). Affect, cognition, and awareness: Affective priming with optimal and suboptimal stimulus exposures. *Journal of Personality and Social Psychology, 64,* 723.
56. Oberman, L. M., McCleery, J. P., Ramachandran, V. S., & Pineda, J. A. (2007). EEG evidence for mirror neuron activity during the observation of human and robot actions: Toward an analysis of the human qualities of interactive robots. *Neurocomputing, 70,* 2194–2203.
57. Ogunyale, T., Bryant, D. A. & Howard, A. (2018). Does removing stereotype priming remove bias? a pilot human-robot interaction study. *arXiv preprint* arXiv:1807.00948.
58. Okuno, Y., Kanda, T., Imai, M., Ishiguro, H. & HAGITA, N. (2009). Providing route directions: Design of robot's utterance, gesture, and timing. In *Proceedings of the 4th ACM/IEEE International Conference on Human Robot Interaction, ACM*, pp. 53–60.
59. Pierno, A. C., Mari, M., Lusher, D., & Castiello, U. (2008). Robotic movement elicits visuomotor priming in children with autism. *Neuropsychologia, 46,* 448–454.
60. Pomeroy, V., Aglioti, S. M., Mark, V. W., McFarland, D., Stinear, C., Wolf, S. L., et al. (2011). Neurological principles and rehabilitation of action disorders: Rehabilitation interventions. *Neurorehabilitation and neural repair, 25,* 33S–43S.
61. Press, C., Bird, G., Flach, R., & Heyes, C. (2005). Robotic movement elicits automatic imitation. *Cognitive Brain Research, 25,* 632–640.
62. Rea, F., Vignolo, A., Sciutti, A. & Noceti, N. (2019). *Human Motion Understanding for Selecting Action Timing in Collaborative Human-Robot Interaction.*
63. Reeves, B. & Nass, C. I. (1996). *The media equation: How people treat computers, television, and new media like real people and places*, Cambridge University Press.
64. Rizzolatti, G., & Craighero, L. (2004). The mirror-neuron system. *Annual Review of Neuroscience, 27,* 169–192.
65. Rizzolatti, G., Fadiga, L., Fogassi, L., & Gallese, V. (1999). Resonance behaviors and mirror neurons. *Archives Italiennes de Biologie, 137,* 85–100.
66. Robinette, P., Howard, A. M. & Wagner, A. R. (2015). Timing is key for robot trust repair. In *International Conference on Social Robotics*. (pp. 574–583). Springer.
67. Scheutz, M. (2011). *13 The Inherent Dangers of Unidirectional Emotional Bonds Between Humans and Social Robots* (p. 205). Robot ethics: The ethical and social implications of robotics.

68. Schvaneveldt, R. W. & Meyer, D. E. (1973). Retrieval and comparison processes in semantic memory. *Attention and performance IV*, pp. 395–409.
69. Schvaneveldt, R. W., Meyer, D. E., & Becker, C. A. (1976). Lexical ambiguity, semantic context, and visual word recognition. *Journal of Experimental Psychology: Human Perception and Performance, 2,* 243.
70. Sciutti, A., Bisio, A., Nori, F., Metta, G., Fadiga, L., Pozzo, T., et al. (2012). Measuring human-robot interaction through motor resonance. *International Journal of Social Robotics, 4,* 223–234.
71. Shin, Y. K., Proctor, R. W., & Capaldi, E. J. (2010). A review of contemporary ideomotor theory. *Psychological Bulletin, 136,* 943.
72. Shiwa, T., Kanda, T., Imai, M., Ishiguro, H., & Hagita, N. (2009). How quickly should a communication robot respond? delaying strategies and habituation effects. *International Journal of Social Robotics, 1,* 141–155.
73. Steele, C. M., & Aronson, J. (1995). Stereotype threat and the intellectual test performance of African Americans. *Journal of Personality and Social Psychology, 69,* 797–811.
74. Stinear, C. M., Petoe, M. A., Anwar, S., Barber, P. A., & Byblow, W. D. (2014). Bilateral Priming Accelerates Recovery of Upper Limb Function After Stroke. *Stroke, 45,* 205–210.
75. Szmrecsanyi, B. 2005. *Language users as creatures of habit: A corpus-based analysis of persistence in spoken English.*
76. Tai, Y. F., Scherfler, C., Brooks, D. J., Sawamoto, N., & Castiello, U. (2004). The human premotor cortex is' mirror'only for biological actions. *Current Biology, 14,* 117–120.
77. Tamietto, M., & de Gelder, B. (2010). Neural bases of the non-conscious perception of emotional signals. *Nature Reviews Neuroscience, 11,* 697.
78. Vannucci, F., Sciutti, A., Lehman, H., Sandini, G., Nagai, Y. & Rea, F. (2019). Cultural differences in speed adaptation in human-robot interaction tasks. *Paladyn, Journal of Behavioral Robotics.*
79. Wajcman, J. (2015). *Pressed for time: The acceleration of life in digital capitalism*, University of Chicago Press.
80. Was, C., Woltz, D., & Hirsch, D. (2019). Memory processes underlying long-term semantic priming. *Memory & cognition, 47,* 313–325.
81. White, K. R. G., Danek, R. H., Herring, D. R., Taylor, J. H., & Crites, S. L. (2018). Taking priming to task: Variations in stereotype priming effects across participant task. *Social Psychology, 49,* 29–46.
82. Yaffe, J. A., Zlotnik, Y., Ifergane, G. & Levy-Tzedek, S. (2020). Implicit task switching in Parkinson's disease is preserved when on medication. *PLoS ONE, 15(1), e0227555.*
83. Yamazaki, A., Yamazaki, K., Kuno, Y., Burdelski, M., Kawashima, M. & Kuzuoka, H. (2008). Precision timing in human-robot interaction: Coordination of head movement and utterance. In *Proceedings of the SIGCHI Conference on Human Factors in Computing Systems, ACM*, pp. 131–140.
84. Zajonc, R. B. (1968). Attitudinal effects of mere exposure. *Journal of Personality and Social Psychology, 9,* 1.
85. Zawieska, K., Duffy, B. R., & Strońska, A. (2012). Understanding anthropomorphisation in social robotics. *Pomiary Automatyka Robotyka, 16,* 78–82.
86. Zhou, A., Hadfield-Menell, D., Nagabandi, A. & Dragan, A. D. (2017). Expressive robot motion timing. In *Proceedings of the 2017 ACM/IEEE International Conference on Human-Robot Interaction, ACM*, pp. 22–31.

Index

N. Noceti et al. (eds.), *Modelling Human Motion*,
https://doi.org/10.1007/978-3-030-46732-6

Printed in the United States
by Baker & Taylor Publisher Services